WILLIAM F. MAAG LIBRARY
YOUNGSTOWN STATE UNIVERSITY

Minority Carriers in III–V Semiconductors: Physics and Applications

SEMICONDUCTORS
AND SEMIMETALS
Volume 39

Semiconductors and Semimetals

A Treatise

Edited by

R. K. Willardson
CONSULTING PHYSICIST
SPOKANE, WASHINGTON

Albert C. Beer
CONSULTING PHYSICIST
COLUMBUS, OHIO

Eicke R. Weber
DEPARTMENT OF MATERIALS SCIENCE AND MINERAL ENGINEERING
UNIVERSITY OF CALIFORNIA AT BERKELEY

Minority Carriers In III–V Semiconductors: Physics and Applications

SEMICONDUCTORS AND SEMIMETALS

Volume 39

Volume Editors

RICHARD K. AHRENKIEL

NATIONAL RENEWABLE ENERGY LABORATORY
GOLDEN, COLORADO

MARK S. LUNDSTROM

SCHOOL OF ELECTRICAL ENGINEERING
PURDUE UNIVERSITY
WEST LAFAYETTE, INDIANA

ACADEMIC PRESS, INC.
Harcourt Brace & Company, Publishers

Boston   San Diego   New York
London   Sydney   Tokyo   Toronto

This book is printed on acid-free paper. ∞

Copyright © 1993 by Academic Press, Inc.
ALL RIGHTS RESERVED.
NO PART OF THIS PUBLICATION MAY BE REPRODUCED OR
TRANSMITTED IN ANY FORM OR BY ANY MEANS, ELECTRONIC
OR MECHANICAL, INCLUDING PHOTOCOPY, RECORDING, OR
ANY INFORMATION STORAGE AND RETRIEVAL SYSTEM, WITHOUT
PERMISSION IN WRITING FROM THE PUBLISHER.

ACADEMIC PRESS, INC.
1250 Sixth Avenue, San Diego, CA 92101-4311

United Kingdom Edition published by
ACADEMIC PRESS LIMITED
24–28 Oval Road, London NW1 7DX

LIBRARY OF CONGRESS CATALOGING-IN-PUBLICATION DATA

Semiconductors and semimetals.—Vol.1—New York: Academic Press, 1966–

   v.: ill.; 24 cm.

Irregular.
Each vol. has also a distinctive title.
Edited by R. K. Willardson, Albert C. Beer, and Eicke R. Weber
ISSN 0080-8784 = Semiconductors and semimetals

   1. Semiconductors—Collected works. 2. Semimetals—Collected works.
I. Willardson, Robert K. II. Beer, Albert C. III. Weber, Eicke R.
QC610.9.S48   621.3815′2—dc19   85-642319
                                           AACR 2   MARC-S

Library of Congress   [8709]
ISBN 0-12-752139-9 (v. 39)

PRINTED IN THE UNITED STATES OF AMERICA
93 94 95 96 BB 9 8 7 6 5 4 3 2 1

Contents

List of Contributors		viii
Preface		ix

Chapter 1 Radiative Transitions in GaAs and Other III–V Compounds

Niloy K. Dutta

I.	Introduction	1
II.	Band Structure in Semiconductors	3
III.	Absorption and Emission Rates	5
IV.	Optical Gain	14
V.	InGaAsP Materials	18
VI.	Radiative Lifetime	22
VII.	Quantum-Well Structures	24
VIII.	Application to Lasers	32
	References	36

Chapter 2 Minority-Carrier Lifetime in III–V Semiconductors

Richard K. Ahrenkiel

I.	Introduction	40
II.	Recombination Mechanisms	42
III.	Lifetime Measurement Techniques	65
IV.	Time-Resolved Photoluminescence in Device Structures	71
V.	High-Injection Effects in Double Heterostructures	108
VI.	GaAs Minority-Carrier Lifetime	119
VII.	$Al_xGa_{1-x}As$ Lifetimes	141
VIII.	Summary	144
	Acknowledgments	144
	References	145

Chapter 3 High Field Minority Electron Transport in p-GaAs

Tomofumi Furuta

I.	Introduction	151
II.	Drift Velocity	153

III.	Energy Transfer Process	162
IV.	Ultrafast Energy Relaxation Process	170
V.	Monte Carlo Simulation Results	177
VI.	Summary	189
	Acknowledgment	190
	References	190

Chapter 4 Minority-Carrier Transport in III–V Semiconductors

Mark S. Lundstrom

I.	Introduction	194
II.	Minority-Carrier Transport in Compositionally Nonuniform Semiconductors	195
III.	Heavy Doping Effects and Minority-Carrier Transport	201
IV.	Coupled Photon/Minority-Carrier Transport	205
V.	Effects of Heavy Doping on Device-Related Materials Parameters	213
VI.	Minority-Carrier Transport in III–V Devices	228
VII.	Summary	254
	Acknowledgment	254
	References	255

Chapter 5 Effects of Heavy Doping and High Excitation on the Band Structure of Gallium Arsenide

Richard A. Abram

I.	Introduction	259
II.	Many-Body Effects in Bulk GaAs	261
III.	Many-Body Effects in GaAs Quantum Wells	279
IV.	Effects of the Impurity Centres	288
V.	Optical Experiments	301
VI	Electrical Experiments	309
	Acknowledgments	312
	References	312

Chapter 6 An Introduction to Nonequilibrium Many-Body Analyses of Optical and Electronic Processes in III–V Semiconductors

David Yevick and Witold Bardyszewski

I.	Introduction	318
II.	$k \cdot p$ Band Structure	319
III.	Second Quantization	322

IV.	Ensemble Properties	329
V.	Electron Self-Energy	343
VI.	Dielectric Function Models	350
VII.	Intraband Processes and Transport	358
VIII.	Interband Processes	364
IX.	Conclusions	380
	APPENDIX A	380
	APPENDIX B	385
	APPENDIX C	385
	References	386

INDEX 389
CONTENTS OF VOLUMES IN THIS SERIES 397

List of Contributors

Numbers in parentheses indicate pages on which the authors' contributions begin.

RICHARD A. ABRAM (259), *Department of Physics, University of Durham, Durham, DH1 3LE, United Kingdom*

RICHARD K. AHRENKIEL (39), *National Renewable Energy Laboratory, 1617 Cole Blvd., Golden, Colorado 80401*

WITOLD BARDYSZEWSKI (317), *Department of Theoretical Physics, Warsaw University, Hoza 69, 00 681 Warsaw, Poland*

NILOY K. DUTTA (1), *AT&T Bell Laboratories, 600 Mountain Avenue, Room 6E-414, Murray Hill, New Jersey 07974*

TOMOFUMI FURUTA (151), *NTT LSI Laboratories, 3-1 Morinosato Wakamiya, Atsugi-shi, Kanagawa 243-01, Japan*

MARK S. LUNDSTROM (193), *School of Electrical Engineering, Purdue University, West Lafayette, Indiana 47907*

DAVID YEVICK (317), *Department of Electrical Engineering, Queen's University, Kingston, Ontario K7L 3N6, Canada*

Preface

During the past decade, remarkable advances have occurred in III–V semiconductor science and devices. The physics of III–V semiconductors has reached a high level of development, and III–V technology has produced a wide variety of novel devices. Among the important III–V devices now finding applications are bipolar devices such as light-emitting diodes (LEDs), laser diodes, heterojunction bipolar transistors (HBTs), and photovoltaic (PV) cells. Many of these are now finding their way into commercial markets. For example, III–V light-emitting devices have already had a strong impact on consumer electronics. HBTs are finding applications in microwave, analog, and digital applications, and PV device research is the focus of a multimillion-dollar program sponsored by the U.S. Department of Energy. Photovoltaic devices offer an alternative to fossil fuels for the generation of electrical power and are the primary power sources for space applications and earth satellites.

Minority-carrier physics and devices are an important component of III–V technology. The generation, recombination, and transport of minority carriers play a central role in the operation of devices. A clear understanding of minority carrier physics and its relation to process technology is crucial to the continued development of III–V bipolar devices.

This volume focuses on subjects that are important for understanding minority-carrier properties of III–V semiconductors. Although the general features of the problem have been known in broad terms for some time, in the past few years there has been a rapid increase in our understanding of III–V minority carrier physics. Our objectives in this volume of Semiconductors and Semimetals is to review the present understanding of minority-carrier physics and, at the same, to identify important outstanding problems. Although the focus is on physics, the choice of topics is guided by the issues important for electronic and optoelectronic devices. The volume is divided into six chapters related to different aspects of minority-carrier physics and applications. Each chapter is written by one or more authors who are active and knowledgeable in the particular field. The topics range from semiconductor device theory to many-body analyses of optical processes.

The volume begins with Niloy Dutta's review of the fundamentals of radiative transitions. Dutta reviews the physics of optical absorption, spontaneous emission, and stimulated emission and shows how these concepts are applied to conventional and quantum-well lasers. The matrix elements for band-to-band transitions are derived, and the theory of band tail states is developed using the Kane and other models. The absorption, emission, and optical gain of GaAs and InGaAsP are discussed in detail, and a section is devoted to the theory of quantum wells and quantum-well lasers.

Richard Ahrenkiel discusses minority-carrier recombination in III–V semiconductors in Chapter II. The chapter focuses on the primary recombination mechanisms in III–V compounds with emphasis on the models of interface recombination. The analysis of time-resolved photoluminescence data occupies a significant portion of the chapter. The influence of photon recycling on minority-carrier lifetime in direct bandgap semiconductors is strongly emphasized. The chapter concludes with a review of the current status of the minority-carrier lifetime in doped and undoped GaAs and a brief review of the lifetime in the ternary alloy, AlGaAs.

In Chapter III, Tomofumi Furuta turns to the treatment of high-field, minority-carrier transport in GaAs. Results of time-of-flight measurements of minority electron transport in p-type GaAs are presented. The chapter includes a survey of recent experimental work on minority electron transport under high-field, hot-carrier conditions. The theory of relaxation processes and energy loss rate is developed. Electron–hole scattering is particularly important in direct-gap semiconductors such as GaAs and plays an important role in hot carrier relaxation and in high field transport. The chapter concludes with Monte Carlo simulations of hot electron transport, which includes the electron–hole interaction. These simulations are compared with experiment. The results show that the minority and majority velocity vs. electric field characteristics are quite different.

In Chapter IV, Mark Lundstrom turns to the issue of how minority-carrier physics influences the performance of bipolar devices. The topics include the impact of heavy doping and photon recycling on minority-carrier device performance. Calculations of the photon recycling effects on the internal quantum efficiency of p/n GaAs diodes are presented. The differing effects of heavy doping on bandgap shrinkage and the intrinsic carrier density in n- and p-type GaAs are considered. The latter is of special importance to the gain of III–V-based bipolar transistors. Measurement of the minority-carrier diffusivity in GaAs is also discussed here. The minority-carrier physics of bipolar transistors and solar cells is examined in some detail.

In Chapter V, Richard Abram examines the interesting optical and electronic properties of heavily doped and highly excited semiconductors. His chapter describes how many-body effects and carrier–impurity interactions alter the band structure of bulk GaAS and GaAs quantum wells. Use of the many-body theory to calculate bandgap narrowing in n- and p-type GaAs is developed here. The effects of dopant impurity centers on the band properties of GaAs are developed using several models. Calculations of bandgap shrinkage for n- and p-type GaAs are presented, and results of theoretical calculations are compared to optical and electrical measurements.

Finally, in Chapter VI, David Yevick and Witold Bardyszewski present a unified, theoretical treatment of the optical and transport properties of III–V semiconductors. They review a nonequilibrium, Green's function procedure that goes beyond the widely-used one-particle models. This model is applied to the calculation of minority-carrier transport, bandgap narrowing, radiative and Auger recombination, and the gain spectra of lasers.

As editors, we thank our colleagues, who did an excellent job of producing their draft chapters, revisions, and proof-readings with very little prodding from us. We also wish to thank our colleagues who reviewed the manuscript and otherwise contributed to this volume. These include H. C. Casey, Jr., W. Walukiewicz, C. M. Maziar, F. L. Lindholm, G. W. 't Hooft, P. W. Landsberg, W. Wettling, A. Ehrhardt, B. M. Keyes, and J. R. Lowney. Finally, the task of assembling this volume was greatly eased by the advice and encouragement of Ms. Jane Ellis, our editor at Academic Press.

R. K. Ahrenkiel
M. S. Lundstrom

To the memory of my sister Ruth Ahrenkiel Austin

*"My strength is as the strength of ten,
Because my heart is pure."*
Alfred, Lord Tennyson

R. K. A.

CHAPTER 1

Radiative Transitions in GaAs and Other III–V Compounds

Niloy K. Dutta

AT&T BELL LABORATORIES
MURRAY HILL, NEW JERSEY

I. INTRODUCTION		1
II. BAND STRUCTURE IN SEMICONDUCTORS		3
III. ABSORPTION AND EMISSION RATES		5
	1. Discrete Levels	5
	2. Semiconductors	6
	3. Spontaneous Recombination and Absorption	11
IV. OPTICAL GAIN		14
	4. Gain Spectra	15
	5. Gain-Current Relationship	17
V. InGaAsP MATERIALS		18
	6. Absorption and Emission Rates	18
	7. Optical Gain	19
VI. RADIATIVE LIFETIME		22
VII. QUANTUM-WELL STRUCTURES		24
	8. Energy Bands	25
	9. Density of States	27
	10. Absorption and Emission Rates	29
	11. Optical Gain	32
VIII. APPLICATION TO LASERS		32
	12. GaAs Lasers	33
	13. InGaAsP Lasers	34
	REFERENCES	36

I. Introduction

The electron–hole recombination mechanisms in direct-band gap semiconductors are described in this chapter. Recombinative mechanisms can in general be classified into two groups, radiative and nonradiative. *Radiative recombination* occurs when an electron in the conduction band recombines with a hole in the valence band and the excess energy is emitted in the form of a photon. Radiative recombination is thus the radiative transition of an electron in the conduction band to an empty state (hole) in the valence band. The optical processes associated with radiative transitions are (1) spontaneous emission, (2) absorption or gain, and (3) stimulated emission. Stimulated emission, in which the emitted photon has nearly the same energy and momentum as the incident photon, forms the basis for laser action.

In thermal equilibrium, a direct-band gap semiconductor (e.g., GaAs, InP, or GaSb) has few electrons in the conduction band ($\sim 10^6 \text{ cm}^{-3}$ for GaAs) and few holes (empty electron states) in the valence band. When a photon of energy greater than the bandgap ($h\nu > E_g$) passes through such a semiconductor, the photon has a high probability of being absorbed. The photon transfers its energy to an electron in the valence band and raises it to the conduction band. In principle, such a photon can produce an identical photon by stimulating the electron transition from the conduction to the valence band. However, external excitation is needed to increase the number of electrons in the conduction band enough that the probability of stimulated emission becomes higher than the probability of absorption. This situation corresponds to population inversion in a laser medium and is necessary for optical gain. The external excitation that generates a high density of electron–hole pairs in a semiconductor is usually provided by current injection. It can also be achieved by optical pumping (absorption of radiation higher in energy than the bandgap). Radiative recombination and its relationship to laser action have been discussed in the literature (Kressel and Butler, 1977; Casey and Panish, 1978; Thompson, 1980; Agrawal and Dutta, 1986).

Nonradiative recombination of an electron–hole pair, as the name implies, is characterized by the *absence* of an emitted photon in the recombination process. The nonradiative recombination processes that affect the performance of long-wavelength semiconductor lasers are Auger recombination, surface recombination, and recombination at defects. The Auger recombination mechanism involves four particle states (three electrons and one hole) and is believed to be important at high temperatures and for low-bandgap semiconductors.

The band structure of a III–V direct gap semiconductor is discussed in Section II. The absorption and emission rates of photons in GaAs for different models are described in Section III. The optical gain necessary for laser action is discussed in Section IV. Lasers fabricated using the InGaAsP material system emitting near 1.3 μm or 1.55 μm are used as sources for fiber optic transmission systems. The calculation and measurements of absorption and emission rates and optical gain for the InGaAsP material system are discussed in Section V. Measurements of radiative lifetime and the effect of photon recycling (self-absorption) in the interpretation of the data are discussed in Section VI. Radiative transitions, in quantum-well structures, are described in Section VII. The results of the calculation are then applied to the understanding of standard double heterostructure and quantum-well lasers.

II. Band Structure in Semiconductors

In a gaseous medium, the electronic transitions take place between two discrete levels. In a semiconductor, the recombining electrons and holes have a continuous band of energy eigenstates.

Figure 1 shows a simplified energy vs. wave vector diagram for a direct-bandgap semiconductor. An accurate description of the band structure requires sophisticated numerical techniques. A commonly used approximation of the exact band structure, in a direct-gap semiconductor, is the parabolic band model. In this model the energy-vs.-wave vector (E

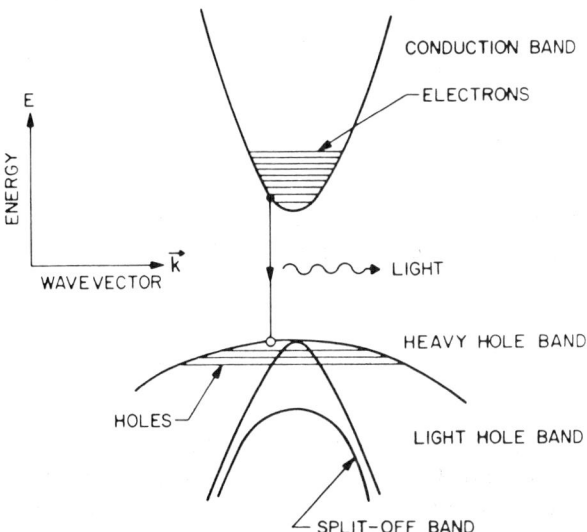

FIG. 1. Energy vs. wave vector diagram of a IV–V semiconductor. The conduction band and the three valence bands — heavy hole, light hole, split-off hole — are shown.

vs. **k**) relation is assumed to be parabolic, that is,

$$E_c = \frac{\hbar^2 k^2}{2m_c} \quad \text{for electrons,} \tag{1}$$

$$E_v = \frac{\hbar^2 k^2}{2m_v} \quad \text{for holes,} \tag{2}$$

where m_c and m_v are the effective masses of electron and holes, respectively, and k is the magnitude of the wave vector **k**.

In a direct-gap semiconductor, the minimum in the conduction band (E vs. **k**) curve and the maximum in the valence band curve occur at the same value of the wave vector **k** (**k** = 0 in Fig. 1). The photon carries negligible momentum compared with the carrier momentum $\hbar \mathbf{k}$. Also, radiative transitions occur between free electrons and free holes of essentially identical wave vectors. In the four band model shown in Fig. 1, the valence band is represented by three subbands. They are the heavy hole band, the light hole band, and the spin-split-off–hole band. The light and heavy hole bands are degenerate at **k** = 0. The split-off energy Δ is usually much larger than $k_B T$, where k_B is the Boltzmann constant and T is the temperature. Hence, the split-off band is full of electrons — i.e., no holes are present.

At a given temperature T, the available number of electrons and holes are distributed over a range of energies. The occupation probability f_c of an electron with energy E_c is given by the Fermi–Dirac statistics and is

$$f_c(E_c) = \frac{1}{\exp[(E_c - E_{fc})/k_B T] + 1}, \tag{3}$$

where E_{fc} is the quasi-Fermi level for the conduction band. Similarly, for the valence-band holes, the occupation probability of a hole with energy E_v is

$$f_v(E_v) = \frac{1}{\exp[(E_v - E_{fv})/k_B T] + 1}, \tag{4}$$

where E_{fv} is the quasi-Fermi level for the valence band. The notations used here assume that E_c and E_{fc} are measured from the conduction-band edge and are positive into the conduction band. On the other hand, E_v and E_{fv} are measured from the valence-band edge and are positive into the valence band. Note that f_v represents the occupation probability of the hole and not of the electron.

III. Absorption and Emission Rates

The spontaneous emission rate and the absorption rate for photons in a semiconductor are calculated using perturbation theory and summing over the available electron and hole states.

1. Discrete Levels

We first consider the case of an electronic transition between two discrete levels such as is found in a gaseous medium. The transition probability W of such a process is given by Fermi's golden rule:

$$W = \frac{2\pi}{\hbar} |H'_{if}|^2 \, \rho(E_f) \, \delta(E - E_i + E_f), \tag{5}$$

where H'_{if} is the matrix element $\langle i|H'|f\rangle$ of the perturbation Hamiltonian H' between the initial state $|i\rangle$ and final state $|f\rangle$, E_i and E_f are the energies of the initial and final state, respectively, and $\rho(E_f)$ is the density of electrons in the final state. For photons interacting with electrons of mass m_0, the square of the transition matrix element H'_{if} is given by

$$|H'_{if}|^2 = \frac{q^2}{m_0^2} \frac{2\hbar}{\varepsilon_0 \bar{\mu}^2 \omega} \frac{1}{4} |\langle i | \hat{\varepsilon} \cdot \vec{p} | f \rangle|^2, \tag{6}$$

where q is the charge of the electron, $\hbar$ is Planck's constant $h/2\pi$, ε_0 is the permittivity of free space, $\bar{\mu}$ is the effective refractive index of the medium, ω is the frequency of light, $\hat{\varepsilon}$ is the polarization vector, and $\vec{p}$ is the momentum of the electron. For spontaneous emission, the quantity $\rho(E_f)$ equals the number of states for photons of energy E per unit volume per unit energy. It is given by

$$\rho(E_f) = \frac{\bar{\mu}^3 \omega^2}{\pi^2 c^3 \hbar}, \tag{7}$$

where c is the velocity of light in free space. Using Eqs. (5)–(7), the spontaneous emission rate for photons of energy E between two discrete levels is given by

$$r_{sp}(E) = \frac{4\pi q^2 \bar{\mu} E}{m_0^2 \varepsilon_0 c^3 h^2} |M_{if}|^2 \, \delta(E_i - E_f - E). \tag{8}$$

For stimulated emission or absorption $\rho(E_f) = 1$, and the absorption or emission rate for photons of energy E is given by

$$W = \frac{\pi q^2}{m_0^2 \varepsilon_0 \bar{\mu}^2 \omega} |M_{if}|^2 \, \delta(E_i - E_f - E). \tag{9}$$

Since a photon travels a distance of $c/\bar{\mu}$ in 1 s, the number of photons absorbed per unit distance is the absorption coefficient $\alpha = \bar{\mu}\omega/c$. With the use of Eq. (9), the absorption coefficient becomes

$$\alpha(E) = \frac{q^2 h}{2\varepsilon_0 m_0^2 c \bar{\mu} E} |M_{if}|^2 \, \delta(E_i - E_f - E). \tag{10}$$

2. Semiconductors

We now calculate the spontaneous emission rate and absorption for direct-gap semiconductors such as GaAs. These quantities are obtained by integrating Eqs. (8) and (10) over the occupied electron and hole states of a semiconductor.

Consider the transition shown in Fig. 1 in the presence of an incident photon whose energy $E = h\nu = E_c + E_v + E_g$, where E_g is the bandgap. The photon can be absorbed, creating an electron of energy E_c and a hole of energy E_v. The absorption rate is given by

$$R_a = B(1 - f_c)(1 - f_v)\rho(E), \tag{11}$$

where B is the transition probability, $\rho(E)$ is the density of photons of energy E, and the factors $1 - f_c$ and $1 - f_v$ represent the probabilities that the electron and hole states of energy E_c and E_v are not occupied. The stimulated-emission rate of photons is given by

$$R_e = B f_c f_v \rho(E), \tag{12}$$

where the Fermi factors f_c and f_v are the occupation probabilities of the electron and hole states of energy E_c and E_v, respectively. The stimulated-emission process is accompanied by the recombination of an electron–hole pair. The condition for net stimulated emission or optical gain is

$$R_e > R_a. \tag{13}$$

Using Eqs. (11) and (12), this condition becomes

$$f_c + f_v > 1.$$

With the use of f_c and f_v given by Eqs. (3) and (4), this becomes

$$E_{fc} + E_{fv} > E_c + E_v. \tag{14}$$

If we add E_g to both sides and note that $E_c + E_v + E_g = h\nu$, it follows that the separation of the quasi-Fermi levels must exceed the photon energy in order for the stimulated-emission rate to exceed the absorption rate. This is the necessary condition for net stimulated emission or optical gain first derived by Bernard and Duraffourg (1961). The density of electrons (n) and holes (p) in the semiconductor is given by

$$n = \int \rho_c(E) f_c(E) \, dE \tag{15}$$

and

$$p = \sum_{i=1,n} \int \rho_{vi}(E) f_v(E) \, dE,$$

where $\rho_c(E)$ and $\rho_v(E)$ are the density of states in the conduction and valence bands, respectively. The second equation represents summation over both light and heavy hole bands. For the idealized case of parabolic bands (given by Eqs. (1) and (2)), the density of states is given by

$$\begin{aligned}\rho_c(E) &= (2)\frac{4\pi k^2}{(2\pi)^3}\frac{dk}{dE} \\ &= 4\pi(2m_c/h^2)^{3/2} E^{1/2},\end{aligned} \tag{16}$$

where $E = \hbar^2 k^2 / 2m_c$ and m_c is the conduction band effective mass. The factor 2 in Eq. (16) arises from the two electronic spin states.

a. Kane Model

Under high injected carrier densities or for high doping levels, the density of states is modified. This occurs because randomly distributed impurities create an additional continuum of states near the band-edge. A schematic of these band-tail states is shown in Fig. 2.

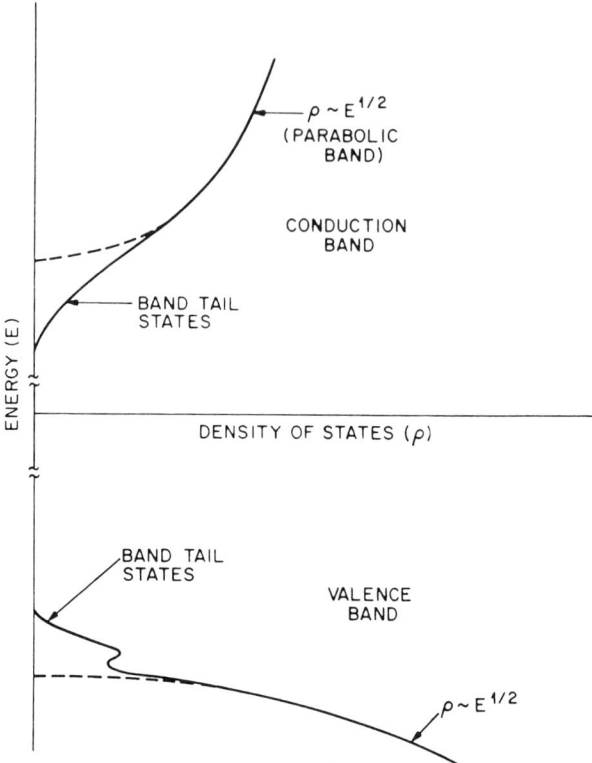

FIG. 2. Energy for the conduction and valence bands plotted against density of states. Dashed lines are for parabolic bands, and the solid lines show schematically the effect of band-tail states.

Several models for band-tail states exist in the literature. Principal among them are the Kane model (1963) and the Halperin–Lax (1966) model. Hwang (1970) has calculated the density of states for p-type GaAs with a hole concentration of 3×10^{18} cm^{-3} for the two models. The result is shown in Fig. 3. The material was heavily compensated with acceptor and donor concentrations N_A and N_D of 6×10^{18} cm^{-3} and 3×10^{18} cm^{-3}, respectively. Figure 3 shows that for the Kane model the band tails extend much more into the bandgap than for the Halperin–Lax model. The absorption measurements on heavily compensated samples indicate that the Halperin–Lax model is more realistic. However, this model is not valid for states very close to the band edge. To remedy the situation, Stern (1966) approximated the density of states by fitting a density-of-state equation of the Kane form (which is Gaussian) to the

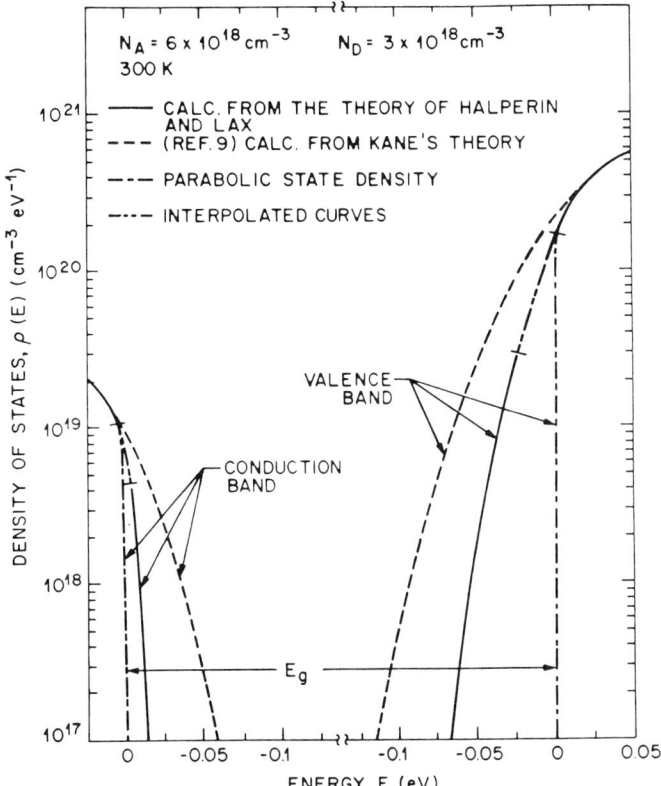

FIG. 3. Calculated density of states vs. energy in the conduction and valence bands for p-type GaAs with a net carrier concentration of 3×10^{18} cm^{-3}. (From Hwang, 1970.)

Halperin–Lax model of band-tail states. The fit is done at *one* energy value in the band tail, which lies within the validity of the Halperin–Lax model. The Kane form of band tail merges with the parabolic band model above the band edge, as seen in Fig. 3.

The importance of band-tail states to radiative transitions arises from two considerations. First, band-tail states can contribute significantly to the total spontaneous-emission rate. Second, band-tail transitions can have significant absorption, especially for heavily doped semiconductors at low temperatures.

b. Matrix Element

The matrix element for transitions between band-tail states differs from that involving free electron and hole states in one important aspect. The former does not satisfy momentum conservation, since the electron and holes states are not states of definite momentum. Therefore, the **k**-selection rule does not apply. By contrast, for transitions involving parabolic band states, the initial and final particle states obey the **k**-selection rule. Thus, the exact matrix elements for optical transitions in heavily compensated semiconductors must not invoke the **k**-selection rule for transitions involving band-tail states. At the same time, the matrix elements should extrapolate to that obeying the **k**-selection rule for above–band-edge transitions involving parabolic band states. The model for such a matrix element has been developed by Stern (1971a, 1971b, 1976) in a series of papers.

When the **k**-selection rule is obeyed, $|M_{if}|^2 = 0$ unless $k_c = k_v$. If we consider a volume V of the semiconductor, the matrix element $|M_{if}|^2$ is given by

$$|M_{if}|^2 = |M_b|^2 \frac{(2\pi)^3}{V} \delta(k_c - k_v). \tag{17}$$

The δ function accounts for the momentum conservation between the conduction-band and valence-band states. The quantity $|M_b|$ is an average matrix element for the Bloch states. Using the Kane model (Kane, 1957), $|M_b|^2$ in bulk semiconductors is given by

$$|M_b|^2 = \frac{m_0^2 E_g(E_g + \Delta)}{6m_c(E_g + 2\Delta/3)} = \xi m_0 E_g, \tag{18}$$

where m_0 is the free-electron mass, E_g is the band gap, and Δ is the spin–orbit splitting. For GaAs, using $E_g = 1.424$ eV, $\Delta = 0.33$ eV, $m_c = 0.067\, m_0$, we get $\xi = 2.6$. Yan et al. (1990) have recently shown that the matrix element $|M_b|^2$ is given by

$$|M_b|^2 = \frac{m_0}{6}\left(\frac{m_0}{m^*} - 1\right) \frac{E_g(E_g + \Delta)}{E_g + \frac{2}{3}\Delta}, \tag{19}$$

where $m^* = 0.053\, m_0$. Using the values just given, one gets $|M_b|^2 = 3.38\, m_0 E_g$, which is 1.27 times the value given by Eq. (18).

3. Spontaneous Recombination and Absorption

Using Eqs. (8) and (10) and summing over all the variable states, the spontaneous emission rate and absorption coefficient are given by

$$r_{sp}(E) = \frac{2\bar{\mu}q^2 E |M_b|^2}{\pi m_0^2 \varepsilon_0 h^2 c^3} \left(\frac{2m_r}{\hbar^2}\right)^{3/2} (E - E_g)^{1/2} f_c(E_c) f_v(E_v), \tag{20}$$

$$\alpha(E) = \frac{q^2 h |M_b|^2}{4\pi^2 \varepsilon_0 m_0^2 c \bar{\mu} E} \left(\frac{2m_r}{\hbar^2}\right)^{3/2} (E - E_g)^{1/2} [1 - f_c(E_c) - f_v(E_v)], \tag{21}$$

where

$$E_c = \frac{m_r}{m_c}(E - E_g),$$

$$E_v = \frac{m_r}{m_{hh}}(E - E_g),$$

$$m_r = \frac{m_c m_{hh}}{m_c + m_{hh}},$$

and m_{hh} is the effective mass of the heavy holes. The total spontaneous emission rate is given by

$$R = \int r_{sp}(E) \, dE. \tag{22}$$

The results of the calculation of the spontaneous-emission rate and the absorption coefficient are now described using the Gaussian Halperin–Lax band-tail model and Stern's matrix element.

The optical absorption (or gain) coefficient for transitions between the valence and conduction band is obtained by integrating Eq. (10) over all states, and is given by

$$\alpha(E) = \frac{q^2 h}{2\varepsilon_0 m_0^2 c \bar{\mu} E} \int_{-\infty}^{\infty} \rho_c(E') \rho_v(E'') |M_{if}|^2 [1 - f_v(E'') - f_c(E')] \, dE', \tag{23}$$

where $E'' = E' - E$, and ρ_c and ρ_v are the densities of states per unit volume per unit energy for the conduction and valence bands, respectively. The integral is evaluated numerically and summed for both light-hole and heavy-hole bands. The matrix element M_{if} can be expressed as a product of two terms, $M_{if} = M_b M_{env}$, where M_b is the previously defined contribution (Eq. (18)) from the band-edge Bloch functions. M_{env} is the matrix element of the envelope wave functions, representing the effect of band-tail states. The envelope wave function is a plane wave above the band edge and takes the form of the ground state of a hydrogen atom band-tail for impurity states. The calculation described here uses the envelope matrix element of Stern (1971a, 1971b). The square of the Bloch-function matrix element $|M_b|^2$ is given by Eq. (18) for the Kane model. Correction to the value of $|M_b|^2$ can arise from the contribution of other conduction bands. This has been described by Yan *et al.* (1990), Herrmann and Weisbuch (1977), and Chadi *et al.* (1976). These corrections would simply scale the gain values calculated using Eq. (23).

a. Application to GaAs

In the previous section, the formalism for calculating emission and absorption spectrum has been described. The results are now applied to GaAs. The emission spectrum can also be obtained from the measured absorption spectrum, $\alpha(E)$, using the detailed balance approach of van Roosbroeck and Shockley (1954), according to which the emission intensity $I(E)$ is given by

$$I(E) = \frac{8\pi n^2 E^2 \alpha(E)}{h^3 c[\exp(E/kT) - 1]}. \tag{24}$$

Figure 4 shows the calculated and measured α (from Casey and Panish, 1978) for *p*-GaAs for an acceptor concentration of 1.2×10^{18} cm^{-3}. The matrix element used by Casey and Panish (1978) is given by (18). The corrections to the matrix element by Yan *et al.* (1990) multiplies by results obtained using Eq. (18) by 1.27. Thus, the measured results differ from the calculation by a factor of 1.6/1.27 ($=1.26$) when the new matrix element is used.

Shown in Fig. 5 is the spontaneous emission spectrum calculated using the model and using Eq. (24) in conjunction with experimental $\alpha(E)$. The agreement is quite good. Also shown is the measured photoluminescence spectrum of a GaAs sample with net acceptor concentration of 1.2×10^{18} cm^{-3}.

Casey and Stern (1976) have carried out extensive experimental and theoretical studies of absorption and emission rates in *n*- and *p*-type

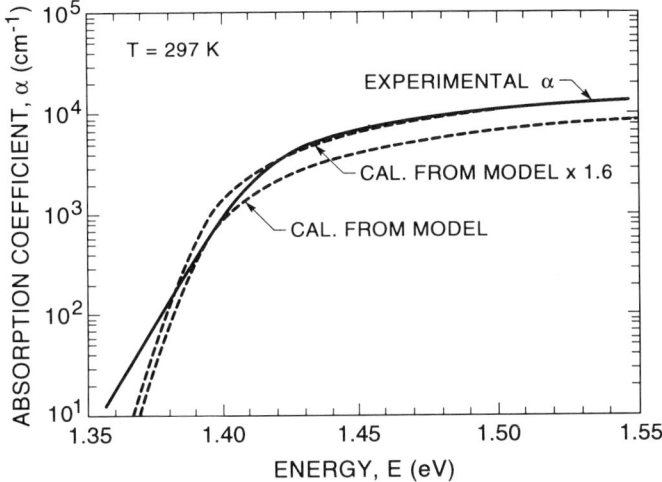

FIG. 4. Comparison of the measured and calculated absorption coefficient for p-type GaAs with carrier concentration of 1.2×10^{18} cm^{-3}. (From Casey and Stern, 1976.) The absorption data are from Casey *et al.* (1975).

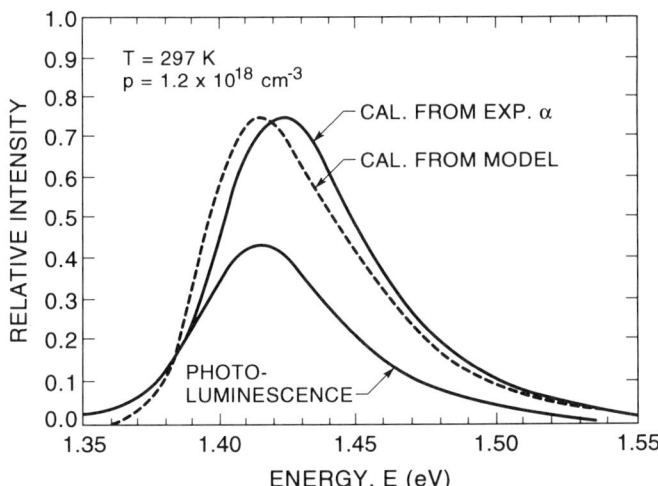

FIG. 5. Comparison of emission spectrum calculated using experimental $\alpha(E)$ and Eq. (24) and that calculated using the model. Also shown is the measured photoluminescence spectrum. (From Casey and Stern, 1976.)

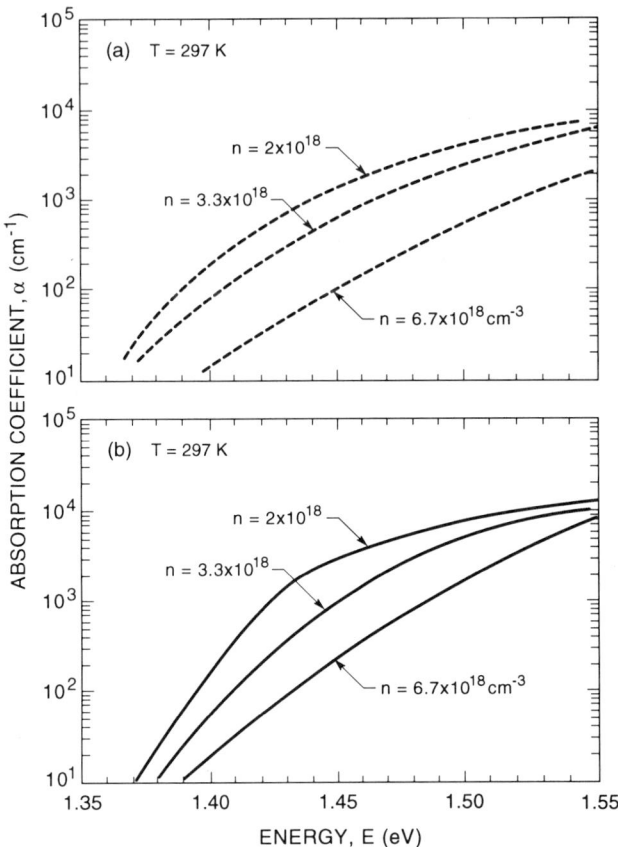

FIG. 6. (a) Calculated and (b) measured absorption coefficients at the indicated electron concentration. (From Casey and Stern, 1976.)

GaAs. The calculated and measured absorption coefficients for n-type GaAs for different doping levels are shown in Fig. 6. The general features of the measured and calculated absorption spectra, such as the Burstein shift (Burstein, 1954), are similar.

IV. Optical Gain

At sufficiently high injected carrier densities, Eq. (14), which represents the condition for net optical gain, is satisfied. The concept of optical gain is also associated with semiconductor lasers, which undergo stimulated emission for injection currents higher than a certain value (threshold current). As carriers are injected into a semiconductor, the spontaneous recombination rate increases. For the purpose of making a connection with the laser threshold current, a quantity called nominal current density J_n was defined by Stern (1976). J_n equals the current density needed to compensate the total spontaneous recombination rate per cm^2–μm at unity quantum efficiency.

$$J_n(\text{A/cm}^2 - \mu m) = q \int r_{sp}(E) \, dE, \tag{25}$$

where q is the electron charge. The total spontaneous recombination rate per unit volume is also expressed as

$$R = \int r_{sp}(E) \, dE = Bnp, \tag{26}$$

where n and p are the electron and hole concentrations, respectively, per cm^3, and B is the radiative recombination coefficient in cm^3 s^{-1}. In a double heterostructure laser, with active layer thickness of d(in μm), the actual current J (in A/cm^2) is related to J_n by $J = J_n d/\eta$, where η is the quantum efficiency.

4. GAIN SPECTRA

As the injected carrier density into the semiconductor is increased, the conduction and valence bands fill up to higher energies where the density of states is also higher. This causes the maximum gain to shift to higher energies with increasing carrier density and hence with increasing current density.

Figure 7 shows the calculated gain or absorption coefficient, as a function of photon energy, for several different excitation levels. The photon energy for maximum gain increases with increasing excitation. The calculation is carried out using a Gaussian Halperin–Lax model of the band tails and Stern's matrix element for p-type GaAs with $N_A - N_D = 4 \times 10^{17}$ cm^{-3} and $N_A + N_D = 10^{18}$ cm^{-3}. N_A and N_D are the acceptor and donor concentrations.

For nondegenerate semiconductors (at high temperature), the radiative recombination coefficient B in Eq. (26) is approximately independent of carrier density. However, at low temperatures, B is a strong function of carrier density. Figure 8 shows the dependence of B on the injected

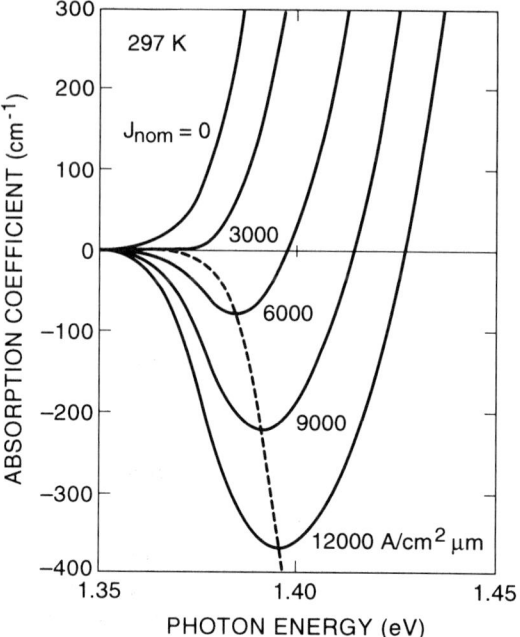

FIG. 7. Calculated gain (or absorption) as a function of photon energy for GaAs for different nominal current densities. (From Stern, 1976.)

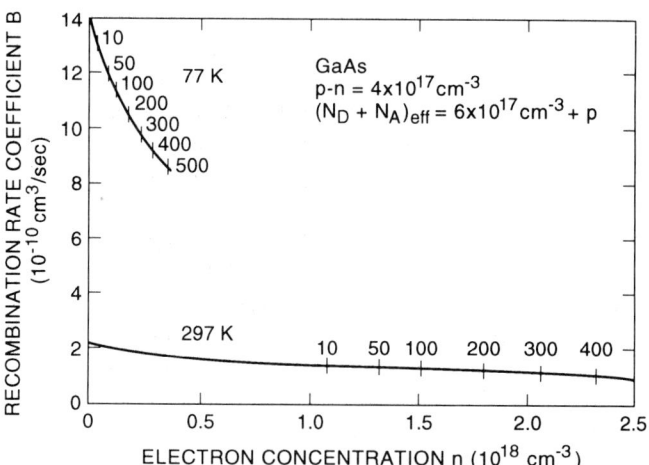

FIG. 8. Calculated radiative recombination coefficient B as a function of minority carrier (electron) concentration at 297 K and 77 K. The vertical bars mask the carrier concentrations at which the indicated values of gain are reached. (From Stern, 1976.)

electron density at 77 K and 297 K. At high temperature, B is a much weaker function of carrier density than at low temperature. The vertical bars in Fig. 8 mark the carrier densities at which the indicated values of the gain coefficient g are reached.

5. Gain–Current Relationship

The knowledge of the dependence of optical gain on injection current is important for understanding and improving the performance of devices such as lasers, optical amplifiers, and light-emitting diodes. The optical gain depends on the doping of the material. As the injection current into the material is increased, the carrier density increases and a net optical gain is reached for carrier densities that satisfies Eq. (14). The calculated gain for undoped GaAs as a function of nominal current density at various temperatures is shown in Fig. 9. The current density needed for transparency ($g = 0$ in Fig. 9) increases with increasing temperature, as expected from Eq. (14).

Stern (1976, 1973) has calculated the gain for different compensation levels. Osinski and Adams (1982) have developed a simplified band-tail model that is very useful for understanding the effect of doping and compensation levels on gain.

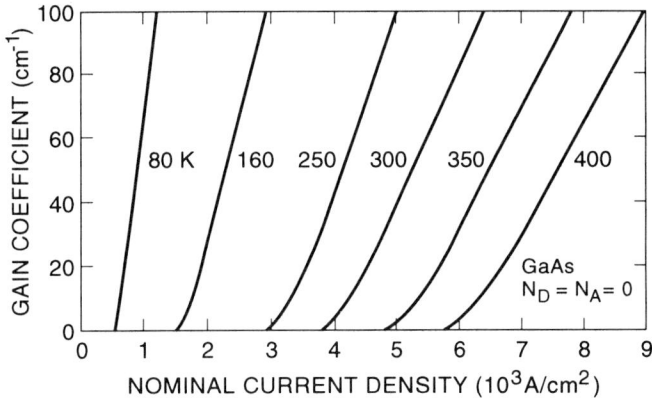

FIG. 9. Calculated maximum gain plotted as a function of nominal current density at various temperatures. (From Stern © 1973 IEEE.)

V. InGaAsP Materials

The quaternary alloy $In_{1-x}Ga_xAs_yP_{1-y}$ is of great interest for use in infrared devices. It can be grown lattice-matched on InP over a wide range of compositions. The resulting band gap (1.35 − 0.75 eV) covers a spectral range that contains the region of lowest loss and lowest dispersion in modern optical fibers (Payne and Gambling, 1975; Horiguchi and Osanai, 1976). This property makes $In_{1-x}Ga_xAs_yP_{1-y}$ very attractive as a semiconductor laser and detector material for future fiber-communication systems. An extensive understanding of the absorption, emission, and gain spectra of this quaternary system is therefore very important.

6. Absorption and Emission Rates

We calculate here the absorption, emission, and gain in $In_{0.72}Ga_{0.28}As_{0.6}P_{0.4}$ (a composition that lattice-matches InP) over a wide range of impurity concentration and doping. This particular composition is chosen because its bandgap ($E_g = 0.96$ eV, $\lambda = 1.3$ μm) is at a region of low loss and minimum dispersion in optical fibers (Horiguchi and Osanai, 1976). The parameters used in the calculation are listed in Table I. We first present the results for p-type InGaAsP. The calculation assumes acceptors are fully ionized, i.e., $N_A - N_D = p$, where N_A, N_D are the acceptor and donor concentrations, respectively. Figure 10 shows the absorption spectra of p-InGaAsP ($E_g = 1.3$ μm) for various majority carrier concentrations. With increasing carrier concentrations, a larger number of band-tail states are available to the holes, which manifests itself as absorption tail at energies below the nominal bandgap.

Figure 11 shows the calculated spontaneous emission spectra for

TABLE I

Band Structure Parameters of $In_{1-x}Ga_xAs_yP_{1-p}$. Lattice Matched to InP, $x = 0.4528y/(1 - 0.031y)$

Energy gap at zero doping	E_g (eV) = $1.35 - 0.72y + 0.12y^2$
Heavy-hole mass	$m_v/m_0 = (1-y)[0.79x + 0.45(1-x)] + y[0.45x + 0.4(1-x)]$
Light-hole mass	$m_h/m_0 = (1-y)[0.14x + 0.12(1-x)] + y[0.082x + 0.026(1-x)]$
dc dielectric constant	$\varepsilon = (1-y)[8.4x + 9.6(1-x)] + y[13.1x + 12.2(1-x)]$
Spin–orbit coupling	Δ (eV) = $0.11 + 0.31y - 0.09y^2$
Conduction-band mass	$m_\infty/m_0 = 0.080 - 0.039y$

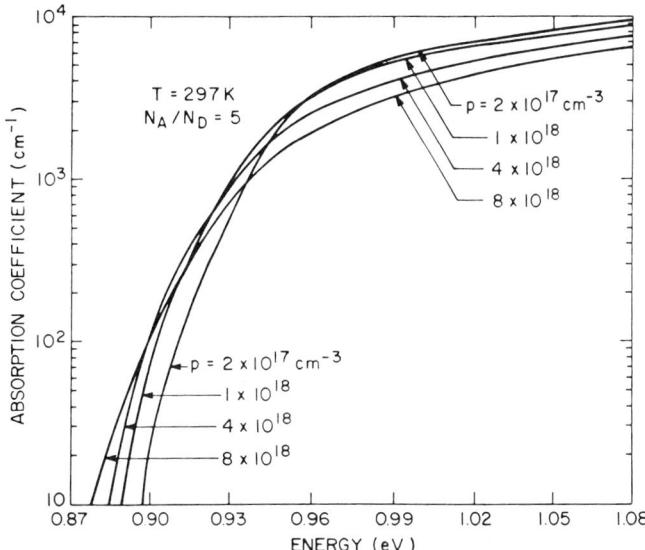

FIG. 10. Calculated absorption spectrum for p-type InGaAsP with various net acceptor concentrations. (From Dutta, 1980.)

various hole concentrations. Note that with increasing hole concentration, the peak of the spontaneous emission shifts to lower energies, while its height (maximum emission intensity) decreases. The width of the emission spectrum increases with increasing carrier concentration as a result of band-tail states. The effect of acceptor concentration on the width of the emission spectrum is shown in Fig. 12. The measured data points obtained from light-emitting diodes are also shown.

So far we have discussed the emission and absorption spectra for p-InGaAsP. The emission and absorption spectra of n-type semiconductors have essentially the same features as those for p-type semiconductors.

7. Optical Gain

A calculation of optical gain is important for analyzing the performance of double heterostructure lasers. Dutta (1981) carried out these calculations for InGaAsP materials. The active region of an InGaAsP laser is typically undoped. The calculated gain or absorption spectrum as a function of injected carrier density for undoped InGaAsP ($E_g = 0.96$ eV, $\lambda - 1.3$ μm) is shown in Fig. 13. The photon energy at which

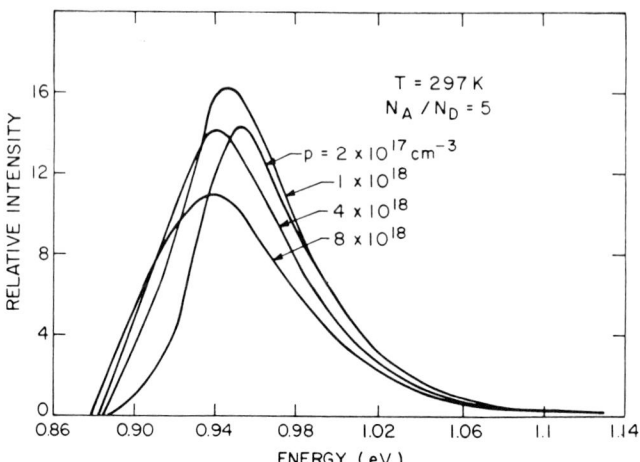

FIG. 11. Calculated spontaneous emission spectrum for p-type InGaAsP ($\lambda - 1.3\ \mu$m) with various net acceptor concentrations. (From Dutta, 1980.)

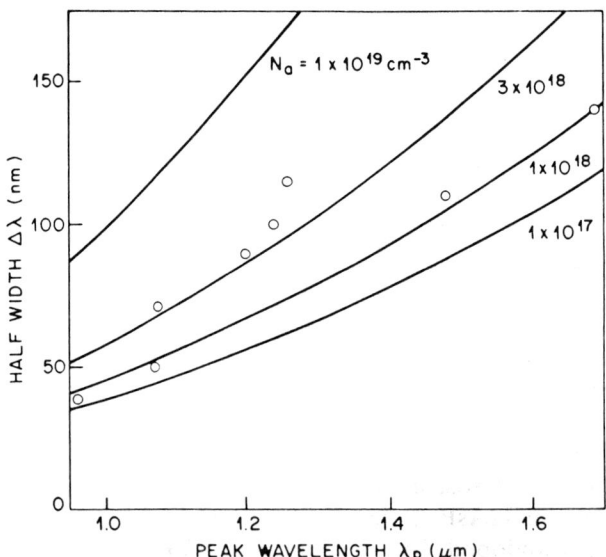

FIG. 12. Effect of acceptor concentration on the width of the spontaneous emission spectrum. (From Takagi, 1979.)

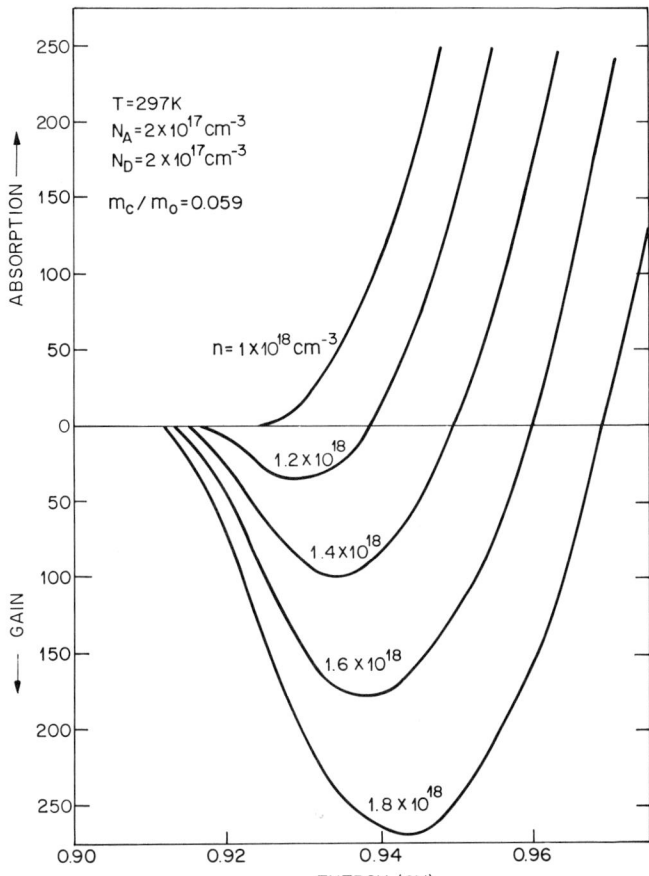

FIG. 13. Optical gain (or absorption) plotted as a function of photon energy for various injected carrier densities. (From Dutta, 1980.)

maximum gain occurs increases with increasing injection. This effect is similar to the shift in absorption edge to lower energies with increasing doping, as discussed earlier.

The calculated maximum gain as a function of nominal current density at various temperatures is shown in Fig. 14. The current needed for zero gain is known as the transparent current density. Note that the transparency current density increases with increasing temperature and the slope of the gain vs. current curve decreases with increasing temperature.

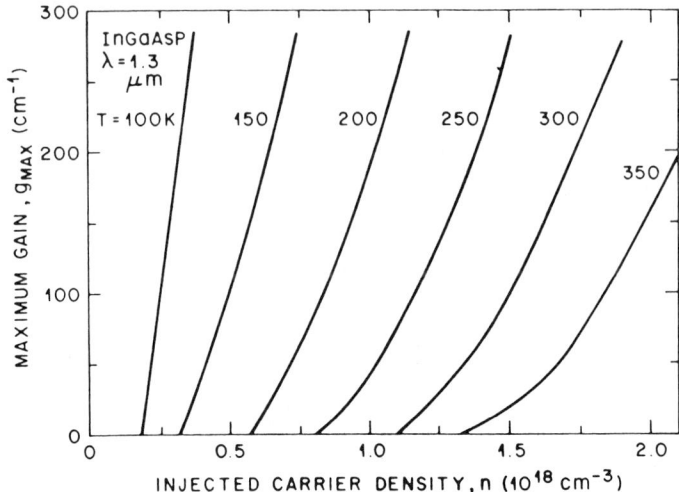

FIG. 14. Maximum gain plotted as a function of nominal current density for undoped InGaAsP ($\lambda - 1.3 \mu$m) at various temperatures. (From Dutta and Nelson, 1982.)

VI. Radiative Lifetime

When a small density of electron (δn) and hole (δp) pairs are created in a semiconductor by optical or electrical excitation, they recombine. The radiative lifetime is defined as

$$\tau_r = \delta n/R, \qquad (27)$$

where R is the total radiative recombination rate. For p-type material, R is approximately given by (for $\delta n, \delta p \ll p$)

$$R = Bp\delta n, \qquad (28)$$

where B is the radiative recombination coefficient and p is the net carrier concentration. Thus, $\tau_r = 1/Bp$, and the coefficient B, in general, depends weakly on the carrier concentration at 300 K. The radiative lifetime can be obtained from the decay of the luminescence spectrum for sufficiently thin ($<0.3 \mu$m) double heterostructures. A detailed analysis of radiative decay in double heterostructure is presented in Chapter 2 of this volume.

The minority carrier diffusion length is given by $L = (D\tau)^{1/2}$, where τ is the lifetime and D is the diffusivity. The quantity D is related to mobility μ by $D = \mu kT/q$. Thus, from a measurement of mobility and diffusion length it is possible to infer the value of the lifetime.

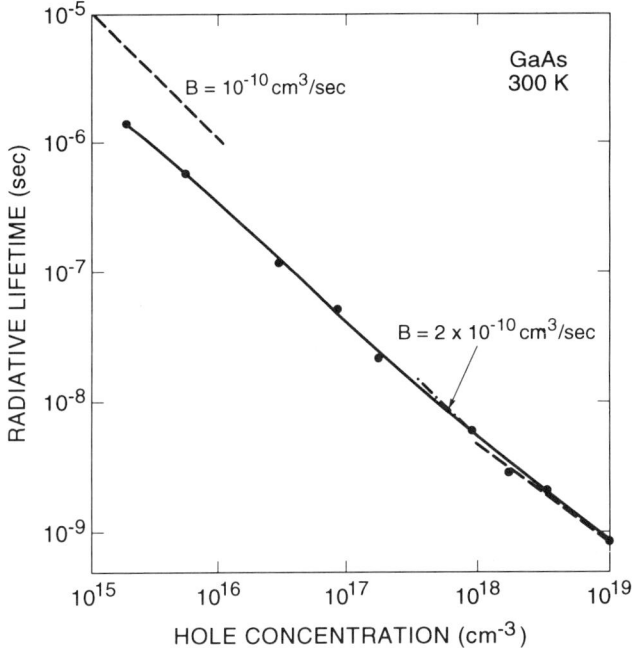

FIG. 15. Measured carrier lifetime in p-GaAs as a function of hole concentration. (From Nelson and Sobers, 1978.)

Figure 15 shows the radiative lifetime data of Nelson and Sobers (1978) for p-type GaAs. Also shown for comparison are calculated values by Stern (1976) and Casey and Stern (1976). The departure of the measured data from a straight line indicates the dependence of the radiative recombination coefficient B on carrier concentration.

The measured photoluminescence decay times are affected by self-absorption of emitted photons if the active region is thick. The self-absorption process can be explained as follows. Consider a photon generated in a GaAs/AlGaAs double heterostructure due to electron–hole recombination. If the photon reaches the air–semiconductor interface at an angle greater than the critical angle, it is reflected back into the GaAs material, where it can be absorbed, creating an electron–hole pair. This effect has been called "photon recycling" by numerous authors. This process, in effect, increases the lifetime of the primary electron–hole pair. Asbeck (1977) has calculated the effect of self-absorption on the measured photoluminescence lifetime. The dashed curve in Fig. 16 is the calculated dependence of the photoluminescence lifetime on the active region thickness of a GaAs/AlGaAs double heterostructure. Note that

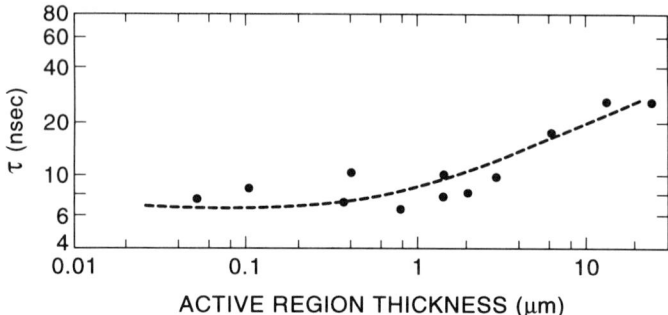

FIG. 16. Recombination lifetime versus active region thickness in a GaAs/AlGaAs double heterostructure (dashed curve). The dots are experimental data. (From Asbeck, 1977.)

the apparent increase in lifetime due to self-absorption is large for large active layer thicknesses, as expected from the preceding physical considerations. The experimental data points are also shown. Self-absorption (photon recycling) theory and experiment are presented in detail in Chapter 2.

VII. Quantum-Well Structures

A double heterostructure laser consists of an active layer, which emits the light, sandwiched between two higher gap cladding layers. When the thickness of the active layer becomes comparable to the deBroglie wavelength $(\lambda - h/p)$, quantum-mechanical effects are expected to occur. In double heterostructure lasers with a very thin active region (~ 100 Å), the carrier (electron or hole) motion normal to the active layer is restricted. As a result, the kinetic energies of the carriers moving in that direction are quantized into discrete energy levels like those of the well-known quantum-mechanical, one-dimensional potential well. This effect is observed in the absorption and emission (including laser action) characteristics and transport characteristics, including phenomena such as tunneling. Double heterostructure lasers, with a very thin active region or multiple thin active regions, are known as quantum-well lasers.

The optical properties of semiconductor quantum-well double heterostructures were initially studied by Dingle et al. (1974, 1976). Since then, extensive work on GaAlAs quantum-well lasers has been done by Holonyak et al. (1980a, 1980b), Tsang (1981, 1986), and Hersee et al. (1984). Much of the optical and transport properties of quantum wells using GaAlAs material system is reviewed by Capasso and Margaritondo (1987). A recent text (Weisbuch and Vanter, 1990) develops the quantum-mechanical properties of quantum wells.

Quantum-well (QW) lasers fabricated using the GaAlAs material system have improved performance characteristics, such as lower threshold current and higher quantum efficiency, compared to regular double heterostructure lasers. These improved performance characteristics have been demonstrated for the InGaAsP material system using the metalorganic chemical vapor deposition (MOCVD) epitaxial growth technique. Early work on InGaAsP MQW lasers was done using liquid phase epitaxy growth (Dutta et al., 1985a,b,c,d).

8. ENERGY BANDS

The schematic of a quantum-well structure, along with the band diagram, is shown in Fig. 17. L_Z and L_B are well and barrier thicknesses, respectively.

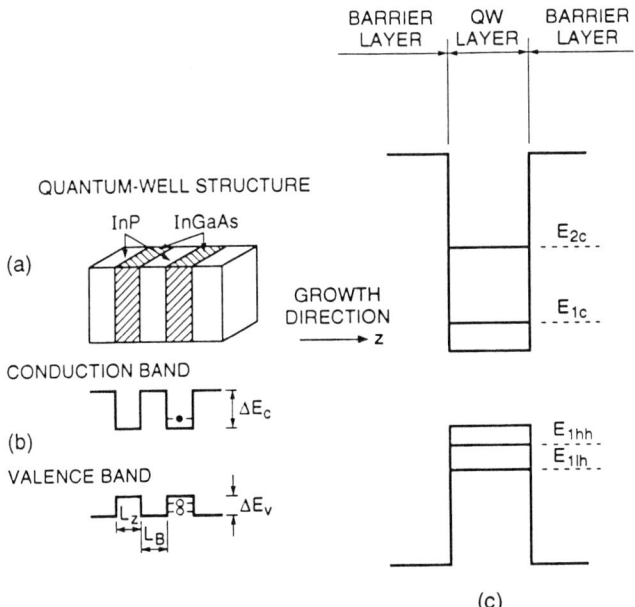

FIG. 17. InGaAs/InP quantum-well structure. (a) Layer structure; z direction is the growth direction. (b) Band structure for conduction and valence band. (c) Energy levels in the conduction and valence bands.

A carrier (electron or hole) in a double heterostructure is confined in a three-dimensional potential well. The energy levels of such carriers are obtained by separating the Hamiltonian into three parts, corresponding to the kinetic energyes in x-, y-, and z-directions, each of which form a continuum of states. When the thickness of the heterostructure (L_z) is comparable to the de Broglie wavelength, the kinetic energy corresponding to the particle motion along the z-direction is quantized. The energy levels can be obtained by separating the Hamiltonian into energies corresponding to x-, y-, and z-directions. For the x, y direction, the energy levels form a continuum of states given by

$$E = \frac{\hbar^2}{2m}(k_x^2 + k_y^2), \tag{29}$$

where m is the effective mass of the carrier and k_x and k_y are the wave vectors along the x and y directions, respectively. Thus, the electrons or holes in a quantum well may be viewed as forming a two-dimensional Fermi gas.

The energy levels in the z-direction are obtained by solving the Schrödinger equation for a one-dimensional potential well. It is given by

$$\begin{aligned} -\frac{\hbar^2}{2m}\frac{d^2\psi}{dz^2} &= E\psi \quad \text{in the well} \quad (0 \le z \le L_z), \\ -\frac{\hbar^2}{2m}\frac{d^2\psi}{dz^2} + V\psi &= E\psi \quad \text{outside the well} \quad (z \ge L_z; z \le 0), \end{aligned} \tag{30}$$

where ψ is the Schrödinger wavefunction and V is the depth of the potential well. For the limiting case of an infinite well, the energy levels and the wavefunction are

$$E_n = \frac{\hbar^2}{2m}\left(\frac{n\pi}{L_z}\right)^2 \quad \text{and} \quad \psi_n = A \sin\frac{n\pi z}{L_z} \quad (n = 1, 2, 3), \tag{31}$$

where A is a normalization constant. For very large L_z, Eq. (31) yields a continuum of states, and the system no longer exhibits quantum effects.

For a finite well, the energy levels and wavefunction can be obtained from Eq. (30) using the boundary conditions that ψ and $d\psi/dz$ are continuous at the interfaces $z = 0$ and $z = L$.

The potential well for electrons (ΔE_c) and holes (ΔE_v) in a double heterostructure depends on the materials involved. A knowledge of ΔE_c

and ΔE_v is necessary for accurate calculation of the energy levels. For an InGaAsP–InP double heterostructure, the values obtained by Forrest *et al.* (1984) are

$$\Delta E_c/\Delta E = 0.39 \pm 0.01, \qquad \Delta E_v/\Delta E = 0.61 \pm 0.01, \tag{32}$$

where ΔE is the bandgap difference between the confining layers and the active region.

The energy eigenvalues for a particle confined in the quantum well are

$$E(n, k_z, k_y) = E_n + \frac{\hbar^2}{2m_n^*}(k_z^2 + k_y^2), \tag{33}$$

where E_n is the nth confined-particle energy level for carrier motion normal to the well, and m_n^* is the effective mass. Figure 17 shows schematically the energy levels E_n of the electrons and holes confined in a quantum well. The confined-particle energy levels (E_n) are denoted by E_{1c}, E_{2c} for electrons, E_{1hh} for heavy holes, and E_{1lh} for light holes. These quantities can be calculated by solving the eigenvalue equations, Eqs. (30), for a given potential barrier (ΔE_c or ΔE_v), as described earlier.

Electron–hole recombination in a quantum well follows the selection rule $\Delta n = 0$, i.e., the electrons in states E_{1c} (E_{2c}, E_{3c}, etc.) can combine with the heavy holes E_{1hh} (E_{2hh}, E_{3hh}, etc.) and with light holes E_{1lh} (E_{2lh}, E_{3lh}, etc.).

The separation between the lowest conduction band level and the highest valence band level for an infinite well is given by

$$E \cong E_g + \frac{h^2}{8L_z^2}\left(\frac{1}{m_c} + \frac{1}{m_{hh}}\right). \tag{34}$$

Thus, the energy of the emitted photons can be varied by simply varying the well width L_z. Figure 18 shows the experimental results for InGaAs quantum-well lasers with various well thicknesses bounded by InP cladding layers (Temkin *et al.*, 1985). As well thickness is reduced, the laser emission shifts to higher energies.

9. Density of States

In a quantum-well structure, a series of energy levels and associated subbands are formed because of the quantization of electron energy in

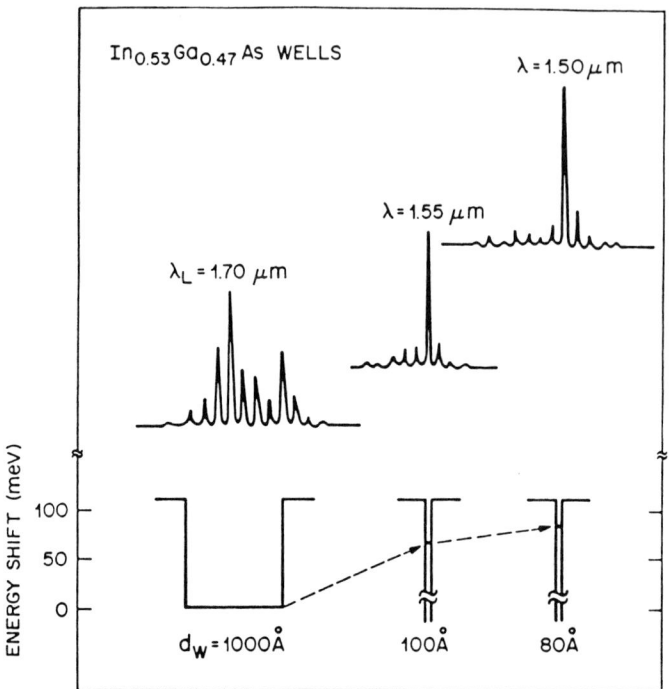

FIG. 18. Lasing wavelength for various well thicknesses of a InGaAs/InP multiquantum-well structure.

the direction normal to the well. The density of states of such electrons, in a single quantum well (SQW), is given by (Dutta, 1982; Arakawa and Yariv, 1985)

$$g_c(E) = \sum_{n=1}^{\infty} \frac{m_c}{L_z \pi \hbar^2} H[E - E_n], \tag{35}$$

where $H(x)$, m_c, $\hbar$, and E_n are the Heaviside function, the effective mass of electrons, Planck's constant (h) divided by 2π, and the energy level for the nth subband, respectively. A similar equation holds for the holes in the valence band. For the regular three-dimensional case, the density of states is given by Eq. (16). A comparison of Eqs. (35) and (16) shows that the density of states in a quantum well is independent of carrier energy and temperature. The modification of the density of states in a quantum well is sketched in Fig. 19. This modification can significantly alter the recombination rates in a QW relative to a regular DH.

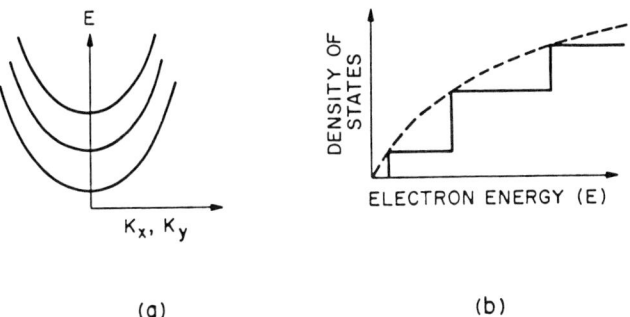

FIG. 19. (a) Energy versus wave vector for each subband. (b) Schematic representation of the density of states in a quantum well.

If one uses a multiquantum-well (MQW) structure instead of the single quantum well, the density of states is modified. When the barrier layers between the wells are thick, each well is independent, and the density of states for the entire MQW structure is simply N times the density of the states for a SQW (Eq. (35)), where N is the number of wells. However, if the barrier is thin or the barrier height is small, the energy levels in the adjacent wells are coupled, which splits each single well level into N different energy levels. In this case, the density of states is given by

$$g_c(E) = \sum_{n=1}^{\infty} \sum_{k=1}^{N} \frac{m_c}{L_2 \pi \hbar^2} H[E - E_{nk}], \qquad (36)$$

where E_{nk} ($k = 1, \ldots N$) are the energy levels that split from a single-well energy level. The difference between the maximum and minimum values of E_{nk} indicates the broadening of each QW level due to coupling. For the steplike density of states (Fig. 19) to be preserved in a MQW structure, the broadening due to coupling between the wells must be smaller than the broadening due to intraband relaxation. For the discussion of recombination rates, one assumes that the interwell coupling in MQW structure is weak enough so that the density of states for each well is described by Eq. (35).

10. Absorption and Emission Rates

The absorption coefficient $\alpha(E)$ and the spontaneous emission rate $r_{sp}(E)$ in a quantum-well structure can be calculated by integrating Eqs.

(8) and (10) over the available electron and hole states in the quantum wells. The modification of the density of states in a QW relative to a regular DH plays a significant role in this summation.

For simplicity, one neglects the effect of the band-tail states so that all transitions obey the k-selection rule. Consider an area A of the semiconductor in the x–y direction. Then the volume V is given by $V = AL_z$. The matrix element $|M|^2$ is given by

$$|M|^2 = |M_b|^2 \frac{(2\pi)^2}{A} \delta(k_c - k_v)\delta_{nn'}. \tag{37}$$

The last δ-function arises from the selection rule for the confined states in the z-direction. The quantity $|M_b|$ is an average matrix element for the Bloch states of the bands (in the x, y direction), as in the corresponding three-dimensional case. One assumes $|M_b|^2$ in Eq. (37) is given by (similar to Eq. (18))

$$|M_b|^2 = \xi m_0 E_q. \tag{38}$$

The uncertainty in the value ξ represents the accuracy of our results. Summing over all the states in the band, one gets

$$r_{sp}(E) = \frac{4\pi\mu q^2 E}{m_0^2 \varepsilon_0 h^2 c^3} |M_b|^2 \frac{(2\pi)^2}{A} \cdot \left(\frac{2A}{(2\pi)^2}\right)^2$$

$$\times \int f_c(E_c) f_v(E_v) d^2k_c d^2k_v \delta(k_c - k_v)\delta(E_i - E_f - E), \tag{39}$$

where f_c, f_v denote the Fermi factors for the electron at energy E_c and hhole at energy E_v. The factor 2 arises from two spin states. The integral in Eq. (39) can be evaluated with the following results:

$$r_{sp}(E) = \frac{32\pi^2 \mu q^2 E |M_b|^2 m_r}{m_0^2 \varepsilon_0 h^4 c^3} A f_c(E_c) f_v(E_v), \tag{40}$$

with

$$E_c = \frac{m_r}{m_c}(E - E_q),$$

$$E_v = \frac{m_r}{m_v}(E - E_q),$$

and

$$m_r = m_c m_v/(m_c + m_v),$$

Since $A = V/L_z$, the total spontaneous emission rate per unit volume is given by

$$R_{sp}(E) = \frac{32\pi^2 \mu q^2 m_r |M_b|^2}{m_0^2 \varepsilon_0 h^4 c^3 L_z} I, \qquad (41)$$

where

$$I = \int_{E_q}^{\infty} E f_c(E_c) f_v(E_v) \, dE.$$

The expression for the absorption $\alpha(E)$ is given by (after summing over the available states)

$$\alpha(E) = \frac{q^2 h}{2\varepsilon_0 m_0^2 c \mu E} |M_b|^2 \frac{(2\pi)^2}{A} \left(\frac{2A}{(2\pi)^2}\right)\left(\frac{1}{V}\right)$$

$$\times \int (1 - f_c - f_v) d^2k_c d^2k_v \delta[k_c - k_v] \delta[E_i - E_f - E]$$

$$= \frac{2q^2 m_r |M_b|^2}{\varepsilon_0 m_0^2 c h \mu E L_z} [1 - f_c(E_c) - f_v(E_v)], \qquad (42)$$

where E_c, E_v are given by Eq. (40). The radiative component of the current or the nominal current density J_n at threshold is given by

$$J_R = J_n = e R_{sp}(E). \qquad (43)$$

The preceding formulation assumes that the electron–electron interaction is strong enough so that an electron temperature T_e can be defined. Also, the electron and hole temperatures are assumed to the equal, although they can be different from the lattice temperature T_h. The transitions obey a selection rule $\Delta n = 0$, where n denotes the confined state quantum number in the z direction. Also, it has been assumed that only the lowest confined level is appreciably populated, so that the electron density is given by

$$n = g_c \int f_c(E_c) \, dE_c. \qquad (44)$$

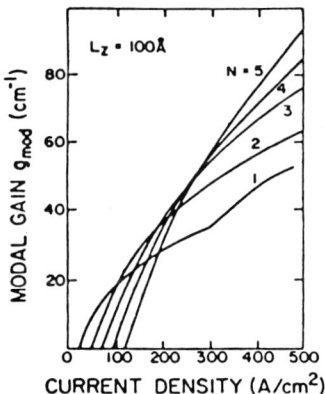

FIG. 20. The modal gain (Γ_g) as a function of injected current density for various numbers of quantum wells. (From Arakawa and Yariv © 1986 IEEE.)

11. Optical Gain

The optical gain and radiative recombination rate in a QW structure has been calculated using a single level approximation and assuming no broadening of each level as described in the previous section. In order to explain the saturation of gain at high currents, one needs to take into account the broadening of each level due to intraband relaxation. The calculated modal gain, as a function of current density, is shown in Fig. 20. N denotes the number of wells in the MQW structure. The calculation takes into account level broadening. The modal gain (G_m) is defined as $G_m = \Gamma_g$, where Γ is the confinement factor and g is the optical gain. The confinement factor for the calculation in Fig. 20 is $\Gamma = 0.03N$, which is a typical value for a 100 Å thick well.

VII. Application to Lasers

The detailed calculation and measurements of the absorption and emission characteristics of direct-gap III–V semiconductors have been historically motivated by the desire to have a better understanding of the performance of semiconductor lasers. The results of the previous sections is now applied to the calculation of threshold current of semiconductor lasers.

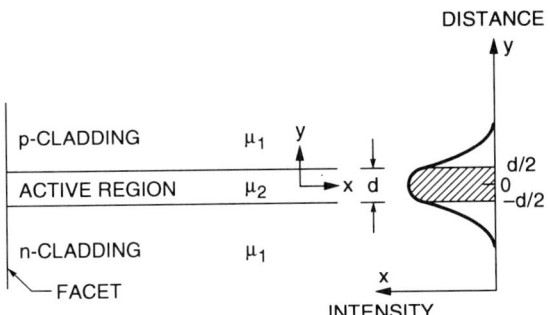

FIG. 21. Schematic of the waveguide of a semiconductor laser, which consists of a GaAs active region surrounded by *p*-type and *n*-type AlGaAs cladding layers. The intensity distribution of the fundamental mode is also shown. The hatched region represents the fraction of the mode within the active region.

12. GaAs Lasers

A typical GaAs double heterostructure laser consists of an active region (GaAs) that emits the light, sandwiched between *p*-type and *n*-type AlGaAs cladding layers. The AlGaAs material has both higher bandgap and lower refractive index than GaAs. This allows both optical confinement of light (resulting in the formation of a waveguide) and electrical confinement of carriers. The schematic of the layer structure is shown in Fig. 21. The intensity profile of the fundamental mode is also sketched. A fraction of the fundamental mode is confined within the active region. The condition for laser action is that optical gain equals the round-trip loss in the cavity, i.e.,

$$\Gamma g = \alpha + \frac{1}{L} \ln\left(\frac{1}{\sqrt{R_1 R_2}}\right), \qquad (45)$$

where Γ is the mode confinement, g is the optical gain in the active region, α is the sum of the absorption and scattering losses, L is the cavity length, and R_1, R_2 are mirror reflectivities. For a 250 μm long GaAs laser with a 0.2 μm thick active region, and with cleaved facets, the typical values are $\alpha = 10 \text{ cm}^{-1}$, $\Gamma = 0.65$, $R_1 = R_2 = R = 0.35$. The calculated g from Eq. (45) is 80 cm^{-1}.

Figures 9 and 14 shows that the optical gain g can be approximated by

$$g = b(J - J_0), \qquad (46)$$

where J_0 is the nominal current density at transparency and b is slope of

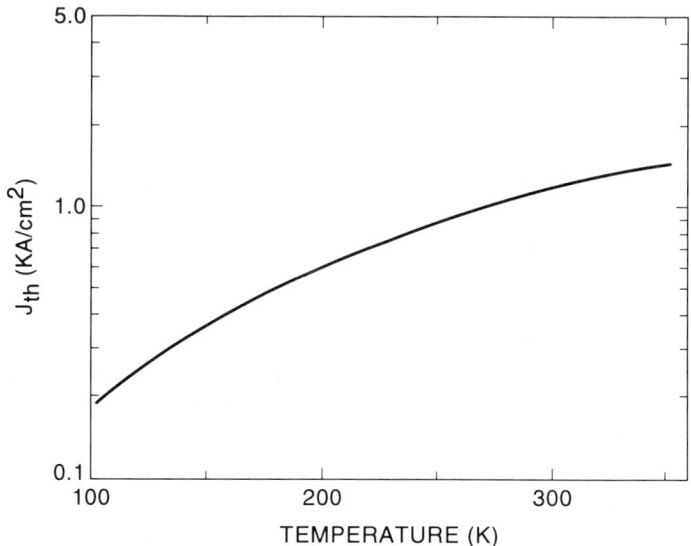

FIG. 22. Calculated threshold current as a function of temperature for a GaAs/AlGaAs double heterostructure laser.

the gain vs. current relation in Fig. 9. Using Eqs. (45) and (46), the threshold current density J_{th} of a laser with cleaved facets is given by

$$J_{th} = J_0 d + \frac{d}{b\Gamma}\left(\alpha + \frac{1}{L}\ln\left(\frac{1}{R}\right)\right), \qquad (47)$$

where R is the reflectivity of the cleaved facets and d is the active layer thickness. The calculated J_{th}, as a function of temperature using Eq. (47) and Fig. 9, is shown in Fig. 22 for a GaAs/AlGaAs laser with a 0.2 μm thick active layer. The threshold current increases with increasing temperature.

13. InGaAsP Lasers

As mentioned in the introduction, electrons and holes can also recombine nonradiatively. An example of a nonradiative recombination process is Auger recombination (Fig. 23). Three types of Auger recombination processes are shown in Fig. 23. The CCCH mechanism involves three electrons and a heavy hole and is dominant in n-type III–V materials. The CHHS process involves one electron, two heavy

1. RADIATIVE TRANSITIONS IN GaAs

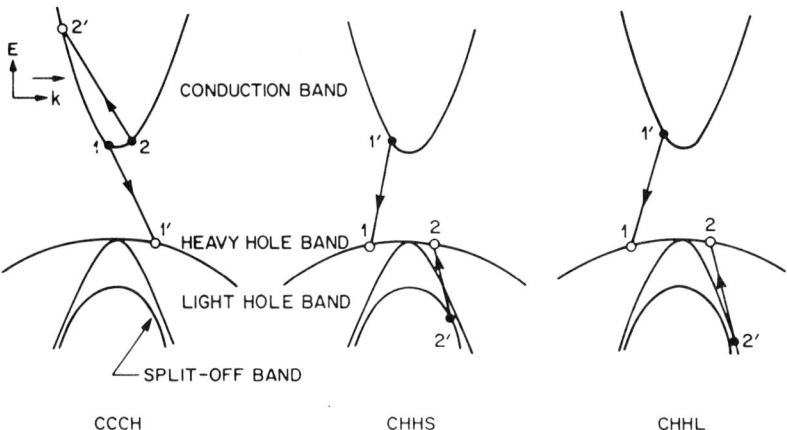

FIG. 23. Various band-to-band Auger recombination processes in a III–V semiconductor, shown schematically.

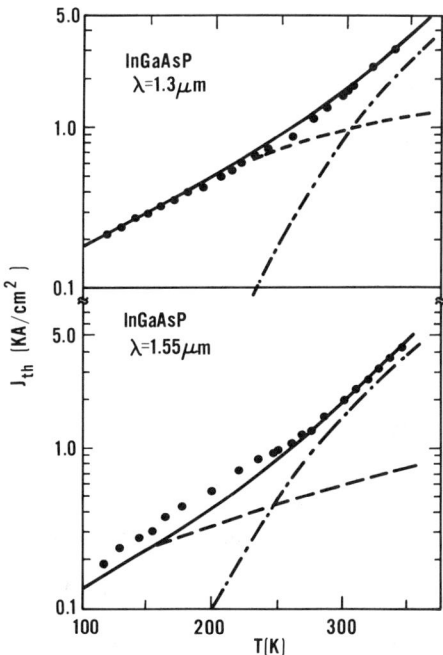

FIG. 24. Threshold current as a function of temperature for an InGaAsP/InP ($\lambda \sim$ 1.3 μm) double heterostructure laser.

holes, and a split-off bound hole and is dominant in p-type III–V material. Auger recombination is an important electron–hole recombination process for narrow gap semiconductors ($E_g > 1\ \mu$m), and it represents an important current loss mechanism for InGaAsP lasers, especially at high temperatures (Thompson and Henshall, 1980; Sugimura, 1981; Dutta and Nelson, 1981, Mozer *et al.*, 1983). The current lost to nonradiative recombination is given by

$$J_{nr} = \frac{q n_{th}}{\tau_{nr}}, \tag{48}$$

where n_{th} is the carrier density at threshold and τ_{nr} is the nonradiative lifetime. For Auger recombination, $\tau_{nr} = 1/Cn^2$, where C is the Auger coefficient and n is the carrier density. Thus, $J_{nr} = qCn_{th}^3$ represents an additional current loss mechanism for the InGaAsP laser at threshold. The threshold current of a InGaAsP/InP laser is the sum of the radiative component of the current (given by Eq. (47)) and the nonradiative component. The calculated and measured threshold current of a InGaAsP/InP double heterostructure laser, with a 0.2 μm thick active region, is shown in Fig. 24 as a function of temperature. The calculated threshold current has a radiative component (dashed line) and a nonradiative Auger component (dot-dashed line). The Auger component increases rapidly with increasing temperature because the Auger coefficient (C) and also the carrier density at threshold increase with increasing temperature.

References

Abramowitz, M., and Stegun, I. (1965). *Handbook of Mathematical Functions*, pp. 446–478. Dover, New York.
Agrawal, G. P., and Dutta, N. K. (1986). *Long Wavelength Semiconductor Laser.* van Nostrand Reinhold, New York.
Arakawa, Y., and Yariv, A. (1985). *IEEE JQE* **QE-21,** 1666.
Arakawa, Y., and Yariv, A. (1986). *IEEE JQE* **QE-22,** 1887.
Ashbeck, P. (1977). *J. Appl. Phys.* **48,** 821.
Bernard, M. G. A., and Duraffourg, G. (1961). *Phys. States Solidi* **1,** 699.
Burstein, E. (1954). *Phys. Rev.* **93,** 632.
Capasso, F., and Margaritondo, G., eds. (1987). *Heterojunction Band Discontinuation: Physics and Applications,* North Holland, Amsterdam.
Casey, H. C., Jr., and Stern, F. (1976). *J. Appl. Phys.* **47,** 631.
Casey and Panish (1976) p. 13 (twice).
Casey, H. C., Jr, and Panish, M. B. (1978). *Heterostructure Lasers, Parts A and B.* Academic Press, New York.

Casey, H. C., Jr., Sell, D. D., and Wecht, K. W. (1975). *J. Appl. Phys.* **46**, 250.
Chadi, D. J., Clark, A. H., and Burnham, R. D. (1976). *Phys. Rev.* **B13**, 4466.
Dingle, R., Wiegmann, W., and Henry, C. H. (1974). *Phys. Rev. Lett.* **33**, 827.
Dingle, R., and Henry, C. H. U.S. Patent 3,982,207, September 21, 1976.
Dutta, N. K. (1980). *J. Appl. Phys.* **51**, 6095.
Dutta, N. K. (1981). *J. Appl. Phys.* **52**, 55.
Dutta, N. K. (1982). *J. Appl. Phys.* **53**, 7211.
Dutta, N. K., and Nelson, R. J. (1981). *Appl. Phys. Lett.* **38**, 407.
Dutta, N. K., and Nelson, R. J. (1982). *J. Appl. Phys.* **53**, 74.
Dutta, N. K., Napholtz, S. G., Yen, R., Brown, R. L., Shen, T. M., Olsson, N. A., and Craft, D. C. (1985a). *Appl. Phys. Lett.* **46**, 19.
Dutta, N. K., Napholtz, S. G., Yen, R., Wessel, T., and Olsson, N. A. (1985b). *Appl. Phys. Lett.* **46**, 1036.
Dutta, N. K., Wessel, T., Olsson, N. A., Logan, R. A., Yen, R., and Anthony, P. J. (1985c). *Electron. Lett.* **21**, 571.
Dutta, N. K., Wessel, T., Olsson, N. A., Logan, R. A., and Yen, R. (1985d). *Appl. Phys. Lett.* **46**, 525.
Forrest, S. R., Schmidt, P. H., Wilson, R. B., and Kaplan, M. L. (1984). *Appl. Phys. Lett.* **45**, 1199.
Halperin, B. I., and Lax, M. (1966). *Phys. Rev.* **148**, 722.
Hermann, C., and Wiesbuch, C. (1977). *Phys. Rev.* **B15**, 823.
Hersee, S. D., DeCremoux, B., and Duchemin, J. P. (1984). *Appl. Phys. Lett.* **44**, 476.
Holonyak, N., Jr., Kolbas, R. M., Dupuis, R. D., and Dapkus, P. D. (1980a). *IEEE J. Quantum Electron.* **QE-16**, 170.
Holonyak, N., Jr., Kolbas, R. M., Laidig, W. D., Vojak, B. A., Hess, K., Dupuis, R. D., and Dapkus, P. D. (1980b). *J. Appl. Phys.* **51**, 1328.
Horiguchi, M., and Osanai, H. (1976). *Electron Lett.* **12**, 310.
Hwang, C. J. (1970). *Phys. Rev.* **B2**, 4126.
Kane, E. O. (1957). *J. Phys. Chem. Solids* **1**, 249.
Kane, E. O. (1963). *Phys. Rev.* **131**, 79.
Kressel, H., and Butler, J. K. (1977). *Semiconductor Lasers and Heterojunction LEDs*. Academic Press, New York.
Mozer, A., Romanek, K. M., Hildebrand, O., Schmid, W., and Pilkhum, M. H. (1983). *IEEE JQE* **QE-19**, 913.
Nelson, R. J., and Sobers, R. G. (1978). *J. Appl. Phys.* **49**, 6103.
Osinski, M., and Adams, M. J. (1982). *IEEE Proc. I* **129**, 229.
Payne, D. N., and Gambling, W. A. (1975). *Electron Lett.* **11**, 176.
Stern, F. (1966). *Phys. Rev.* **148**, 186.
Stern, F. (1971a). *Phys. Rev.* **B3**, 2636.
Stern, F. (1971b). *Phys. Rev.* **B3**, 3559.
Stern, F. (1973). *IEEE JQE* **QE-9**, 290.
Stern, F. (1976). *J. Appl. Phys.* **47**, 5382.
Sugimura, A. (1981). *IEEE JQE* **QE-17**, 627.
Sugimura, A. (1984). *IEEE J. Quantum Electron* **QE-20**, 336.
Takagi, T. (1979). *Jpn. J. Appl. Phys.* **18**, 2017.
Temkin, H., Panish, M. B., Petroff, P. M., Hamm, R. A., Vandenberg, J. M., and Sumski, S. (1985). *Appl. Phys. Lett.* **47**, 394.
Thompson, G. H. B. (1980). *Physics of Semiconductor Laser Devices*. John Wiley and Sons, New York.
Thompson, G. H. B., and Henshall, G. D. (1980). *Electron. Lett.* **16**, 42.

Tsang, W. T. (1981). *Appl. Phys. Lett.* **39**, 786.
Tsang, W. T. (1986). *IEEE J. Quantum Electron* **QE-20**, 1119.
van Roosbroeck, W., and Shockley, W. (1954). *Phys. Rev.* **94**, 1558.
Weisbuch and Vanter (1990) p. 20
Yan, R., Corzine, S. W., Coldren, L. A., and Suemone, I. (1990). *IEEE JQE* **26**, 213.
Yariv, A., Lindsey, C., and Sivan, V. (1985). *J. Appl. Phys.* **58**, 3669.

CHAPTER 2

Minority-Carrier Lifetime in III–V Semiconductors

Richard K. Ahrenkiel

NATIONAL RENEWABLE ENERGY LABORATORY
GOLDEN, COLORADO

I. INTRODUCTION	40
II. RECOMBINATION MECHANISMS	42
1. *Thermodynamic Equilibrium and Minority-Carrier Lifetime*	42
2. *Shockley–Read–Hall Recombination*	47
3. *Surface and Interface Recombination*	51
4. *Auger Recombination*	56
5. *Radiative Recombination and Photoluminescence*	57
6. *Radiative Lifetime*	60
III. LIFETIME MEASUREMENT TECHNIQUES	65
7. *General Techniques*	65
8. *Time-Correlated Single Photon Counting*	66
9. *Inductively Coupled Probe*	69
10. *Photoluminescence Spectroscopy*	69
IV. TIME-RESOLVED PHOTOLUMINESCENCE IN DEVICE STRUCTURES	71
11. *Minority-Carrier and Photoluminescence Lifetime*	71
12. *Confinement Structures*	72
13. *Transient Analysis of the Isotype Double Heterostructure*	74
14. *Self-Absorption and Photon Recycling*	84
15. *The PL Lifetime in Junction Devices*	97
16. *Thick Films and Bulk Crystals*	102
V. HIGH-INJECTION EFFECTS IN DOUBLE HETEROSTRUCTURES	108
17. *Radiative Lifetime*	108
18. *Intensity-Dependence of the Shockley–Read–Hall Lifetime*	112
VI. GaAs MINORITY-CARRIER LIFETIME	119
19. *GaAs Wafers*	119
20. *Undoped* GaAs *Diagnostic Structures*	121
21. *Lifetime of Electrons in* p-*Type, Epitaxial* GaAs	129
22. *Lifetime of Holes in Epitaxial,* n-*Type* GaAs	132
23. *Interface Recombination in* GaAs *Structures*	134
24. *The Photon Recycling Effect in* GaAs	137
VII. $Al_xGa_{1-x}As$ LIFETIMES	141
VIII. SUMMARY	144
ACKNOWLEDGMENTS	144
REFERENCES	145

I. Introduction

The properties of minority carriers are basic to understanding the physics of the III–V semiconductors. In order to design minority-carrier devices, accurate and reliable measurements of minority-carrier parameters are required. The performance of minority-carrier devices depends on these parameters. One example is the current gain of the heterojunction bipolar transistor (HBT), which depends on the minority-carrier diffusion length (L) in the base. The HBT high-frequency cutoff is inversely proportional to the base minority-carrier diffusivity. The efficiency of photovoltaic devices depends strongly on the minority-carrier lifetime of the semiconducting components. Improving minority-carrier parameters, in a cost-effective manner, is the focus of much of the current photovoltaic research and development. One can also include light-emitting diodes (LEDs) in the category of minority-carrier devices. The efficiencies of both lasers and LEDs are intimately tied to the ratio of nonradiative recombination rates to the radiative rates.

Minority-carrier transport, including mobility and diffusivity, is discussed in detail in Chapters 3 and 4 of this volume. The minority-carrier lifetime in semiconductors is perhaps the most widely differing of minority-carrier parameters. The lifetime can vary by many orders of magnitude in a semiconductor, depending on the purity and growth technique. For direct bandgap materials such as GaAs and InP, the dominant recombination mechanism, in high-quality material, is radiative recombination. Recombination in GaAs is contrasted with silicon, which has an indirect bandgap, and the radiative recombination is very weak. The silicon lifetime is defect-controlled except at very high injection levels. In undoped silicon, the largest reported low-injection lifetime is 20 ms (del Alamo, 1988). At high injection, a lifetime of 40 ms was reported (Yablonovitch and Gmitter, 1986). The lifetime in undoped silicon is controlled by recombination at localized sites that have energy levels lying near the center of the forbidden gap. These localized states originate from either impurities or mechanical defects such as dislocations. In addition, major sites of minority-carrier recombination are surfaces or interfaces that contain bonding defects. The energy levels of surface defects often lie in the forbidden gap, producing surface or interface states. The physics of the recombination at deep level defects is discussed in Section II.

GaAs has been more thoroughly studied than any other semiconductor except silicon and will be used as a case study in this chapter. Because GaAs is a direct bandgap semiconductor, the minority-carrier lifetime generally has been measured by the decay rate of the transient photoluminescence (PL). The relationship between PL decay and minority-carrier lifetime is discussed in Section IV.

Early lifetime measurements on GaAs wafer material indicated that nanosecond lifetimes were typical. Hwang (1971) reported that the room-temperature hole lifetime was about 20 ns and was independent of electron concentration up to doping levels, N_D, of about 2×10^{18} cm^{-3}. Between 2×10^{18} and 6.5×10^{18} cm^{-3}, the wafer lifetime decreased inversely with the donor density.

With the development of the epitaxial technologies, GaAs minority-carrier lifetimes increased rapidly. Nelson and Sobers (1976b) grew isotype double heterostructures (DH) by liquid phase epitaxy (LPE) with the structure $Al_xGa_{1-x}As/GaAs/Al_xGa_{1-x}As$. A lifetime of 1.3 µs was measured in an undoped DH structure. In ongoing work, the GaAs "active layer" doping was systematically varied from being undoped to an acceptor density N_A of 1×10^{19} cm^{-3}. Measurements indicated the lifetime decreased almost inversely with the hole concentration. These data were the first to indicate that the minority-carrier lifetime varies with the inverse of the majority-carrier density in high quality material.

Nelson and Sobers (1978a) calculated the effects of the self-absorption and re-radiation of photoluminescence, or the so-called "photon recycling" effect. Photon recycling greatly complicates (Garbuzov, 1986) the measurement of minority-carrier lifetime in direct-bandgap semiconductors. Asbeck (1977) calculated the photon recycling effect in DH structures as a function of layer thickness. These calculations indicated that the effective radiative lifetime increases by factors of 5 to 10 for 10-µm-thick DH devices. Photon recycling calculations are developed in Section IV.

The primary emphasis of this chapter is the development of mathematical tools for analyzing time-resolved photoluminescence data. These tools are developed in Sections II, IV, and V. Section II contains an abbreviated discussion of the primary recombination mechanisms that are important in III–V minority-carrier physics. Section II also includes a discussion of the important aspects of light emission. Section IV focuses on the PL decay analysis of various device structures, and Section V focuses on high-injection effects.

Measurement techniques are the subject of Section III. The current, preferred method of lifetime measurement is time-resolved photoluminescence. However, this method has not proven to be applicable to weak light emitters and the smaller-bandgap materials. The latter problem is instrument-related and may well be solved by new technology in the near future.

Section VI highlights the experimental data on lifetime in GaAs DH structures. When interface recombination is taken into account, the low-injection lifetime is controlled by intrinsic, radiative recombination in high-quality material. State-of-the-art epitaxy has produced room-temperature lifetimes of 0.5 µs to about 5.0 µs in thin, undoped $Al_xGa_{1-x}As/GaAs$ DH structures. Lifetimes of 10 to 15 µs were recently found in GaInP/GaAs undoped DH structures.

Section VII reviews current data related to minority-carrier lifetime in $Al_xGa_{1-x}As$. The lifetime in $Al_xGa_{1-x}As$ is limited by impurity recombination. Other important III–V semiconducting materials are not discussed in this chapter. Some reasons for this omission are brevity and limited data bases. Lifetime data for InP are quite limited and have been discussed in several recent reviews (Ahrenkiel, 1988, 1991). Lifetime data for InGaAs are also very limited and were discussed in a recent review (Ahrenkiel, 1988). It is to be hoped that a number of other important ternaries and quaternaries, which will not be discussed here, will be discussed in future reviews.

II. Recombination Mechanisms

1. THERMODYNAMIC EQUILIBRIUM AND MINORITY-CARRIER LIFETIME

In a nondegenerate semiconductor, the equilibrium density of electron–hole pairs is constant at a given temperature. The law of mass action (Spenke, 1958) arises from thermodynamics and relates the product of electrons and holes to the intrinsic density, which is specific to a material. The law of mass action is written as

$$np = n_i^2(T). \qquad (1)$$

Here n is the free-electron density, p is the free-hole density, and $n_i(T)$ is the intrinsic density per cm^3 at the absolute temperature T in degrees Kelvin. It is shown in semiconductor textbooks that n_i is related to the effective conduction-band (N_c) and valence-band (N_v) densities of states by the relationship

$$n_i^2(T) = N_c N_v \exp(-E_g/kT). \qquad (2)$$

The intrinsic density is independent of the doping level, provided that the semiconductor is nondegenerate. At doping levels that are large enough to produce degeneracy, n_i becomes a function of n and p. Degeneracy also produces a decrease in E_g called bandgap shrinkage. These effects are discussed in detail by Landsberg (1991). The GaAs intrinsic density, at degenerate doping levels, has been recently discussed in the literature (Lundstrom *et al.*, 1990).

Excess carriers can be electrically or optically injected into the semiconductor, creating a nonequilibrium concentration:

$$np > n_i^2(T). \tag{3}$$

Thermodynamics drives the system toward the equilibrium described by Eq. (1). In the absence of further injection, equilibrium is restored via various recombination mechanisms that will be discussed in this chapter. In an intrinsic material or under *high-injection* conditions, $n \sim p$, and the recombination process is called bimolecular decay. Bimolecular recombination rates will be derived in this section. If the medium is extrinsically doped with donors or acceptors, the minority-carrier carriers are holes or electrons, respectively. The majority-carrier density produced by shallow donors (N_D) or acceptors (N_A) is approximately equal to the doping concentration. Therefore,

$$n_0 = N_D, \qquad p_0 = \frac{n_i^2}{N_D} \qquad \text{for } n\text{-type semiconductors}$$

and

$$p_0 = N_A, \qquad n_0 = \frac{n_i^2}{N_A} \qquad \text{for } p\text{-type semiconductors.} \tag{4}$$

Here, n_0 is the free electron density, and p_0 is the free hole-density that is derived from the law of mass motion. If both donors and acceptors are present, compensation occurs:

$$n_0 = (N_D - N_A) \qquad \text{or} \qquad p_0 = (N_A - N_D) \tag{5}$$

Assume that at $t = 0$, $\Delta n, \Delta p$ electron, hole pairs per unit volume are produced by an injection pulse, where

$$\Delta n(x, t) \equiv n(x, t) - n_0,$$
$$\Delta p(x, t) \equiv p(x, t) - p_0. \tag{6}$$

Here, $n(x, t)$ and $p(x, t)$ are the nonequilibrium electron and hole concentrations, respectively.

The solutions for $\Delta n(x, t)$ are found by solving the time-dependent continuity equation (Sze, 1981). For *field-free* or quasi-neutral regions of the semiconductor, the continuity equation reduces to the time-dependent *diffusion equation*. The diffusion equation is a partial differential equation relating the net diffusion of minority carriers into a volume and the net recombination of the carriers within the volume. After the nearly instantaneous generation of electron–hole pairs at $t = 0$, the diffusion equation for $\Delta n(r, t)$ is given by

$$\frac{\partial \Delta n(r, t)}{\partial t} = D \nabla \Delta n(r, t) - \frac{\Delta n(r, t)}{\tau}. \tag{7}$$

Here, τ is the recombination lifetime, which may also be a function of other parameters including Δn. Also, D is the minority electron diffusivity. Complete solutions of the diffusion equation are discussed for specific device structures in Section IV. For a general solution of this equation, one first assumes that the time and spatial variables can be separated. The electron concentration (Δn) can be written

$$\Delta n(r, t) = f(t)g(r). \tag{8}$$

Substituting Eq. (8) into the diffusion Eq. (7), it is easy to show that

$$\frac{1}{f(t)}\frac{df(t)}{dt} + \frac{1}{\tau} = D\frac{\nabla^2 g(r)}{g(r)} = C. \tag{9}$$

The left-hand side of Eq. (9) is a function of time only, and the right-hand side is a function of spatial variables only. Thus, each side of the equation must equal a constant that is designated by the value C. One can write an ordinary differential equation as a function of time for the left-hand side. This part will examine solutions of the time-dependent components of $\Delta n(t)$ assuming that $C = 0$:

$$\frac{d\Delta n}{dt} = -\frac{\Delta n}{\tau}. \tag{10}$$

For n-type material (with $n = N_D$), one may write the np product at $t = 0$ as

$$pn = (N_D + \Delta n)(p_0 + \Delta p) > n_i^2. \tag{11}$$

2. MINORITY-CARRIER LIFETIME IN III–V SEMICONDUCTORS

Low injection is defined for the situation

$$\Delta p, \Delta n \ll N_D, \quad (12)$$

whereas *high injection* is defined as

$$\Delta p, \Delta n \gg N_D. \quad (13)$$

The left-hand side of Eq. (10) is a rate equation for the excess carriers. One can write the recombination rate as

$$\frac{dp}{dt} = \frac{dn}{dt} = -R(n, p) \quad \text{where} \quad R(n, p) \equiv r(n, p)[np - n_i^2]$$
$$(\text{cm}^{-3}\,\text{s}^{-1}). \quad (14)$$

Here, $R(n, p)$ is the recombination rate of electron-hole pairs and equals the loss rate of the electron–hole pairs from the volume. Also, $r(n, p)$ is a volume recombination rate function specific to a particular recombination mechanism and has units of cm^3/s. In general, r is a function of n and p, but the first case to be analyzed here is an idealized case of r being a constant and independent of the carrier concentrations. The solution of (14) is particularly simple for this situation. The functional form of r for the other common recombination mechanisms will be discussed later in this section. The constant-r case is approximately valid for radiative recombination, as is shown later in this section.

For the injection of Δn, Δp electron–hole pairs into n-type material, the recombination rate of minority holes is given by Eq. (14):

$$\frac{dp}{dt} = \frac{d\Delta p}{dt} = -r[(N_D + \Delta n)(p_0 + \Delta p) - n_i^2], \quad (15)$$

where $r = $ constant. Here, $N_D p_0$ cancels the n_i^2 term. Then

$$\frac{d\Delta p}{dt} = -r(N_D \Delta p + \Delta p^2). \quad (16)$$

The term $\Delta n\, p_0$ is dropped because p_0 is usually very small compared to both N_D and Δp. In *low-injection*, the Δp^2 term can be neglected with the result

$$\frac{d\Delta p}{dt} = -rN_D \Delta p. \quad (17)$$

The excess minority-carrier density then decays with a time constant τ.

$$\Delta p(t) = \Delta p_0 e^{-rN_D t} = \Delta p_0 e^{-t/\tau}. \tag{18}$$

The injected hole density at $t = 0$ is Δp_0. The decay time τ is defined as the minority-carrier lifetime. For the case of r being constant:

$$\tau = \frac{1}{rN_D}. \tag{19}$$

Later analysis will show that this result is approximately valid for radiative recombination. In that case, r is a parameter of the semiconductor known as the B-coefficient. The low-injection recombination rate $r(n, p)$ at deep defect levels will be shown to be approximately $N_t \sigma_p v_{th}/N_D$, where N_t is the defect density, σ_p is the capture cross-section for holes, and v_{th} is the thermal velocity of the hole. In this case, the total recombination rate $R(n, p)$ is $N_t \sigma_p v_{th} \Delta p$ and is independent of the doping level.

Combining Eqs. (17) and (19), the lifetime is given by

$$\tau = -\frac{\Delta p(t)}{\left(\dfrac{d\Delta p(t)}{dt}\right)}. \tag{20}$$

When $\Delta p(t)$ data are acquired, the lifetime is obtained by plotting $\log[\Delta p(t)]$ versus time and measuring the slope of the plot.

The quantity $r(n, p)$ will be discussed for radiative and nonradiative recombination mechanisms and may be a simple constant or a more complex function of material parameters. The concept of a single minority-carrier lifetime obviously applies only to low injection conditions for extrinsic or doped semiconductors. For certain situations such as high-injection, the recombination rate becomes nonlinear. A single lifetime model does not apply here, but I will define an instantaneous lifetime according to Eq. (20).

Minority-carrier recombination, at any injection level, is described by a general solution to Eq. (16). The right-hand expression must be integrated to find $\Delta p(t)$:

$$\Delta p(t) = \frac{\Delta p_0 \exp(-t/\tau)}{1 + \dfrac{\Delta p_0}{N_D}(1 - \exp(-t/\tau))}. \tag{21}$$

2. MINORITY-CARRIER LIFETIME IN III–V SEMICONDUCTORS

At high injection levels ($\Delta p_0 \gg N_D$), the initial decay ($t \ll \tau$) is nonexponential and is approximated by the expression

$$\Delta p(t) = \frac{\Delta p_0}{1 + rN_D t} = \frac{\Delta p_0}{1 + \frac{t}{\tau}}. \tag{22}$$

Here τ is the previously defined low-injection minority-carrier lifetime. For long times ($t \gg \tau$), Eq. (21) again becomes exponential and $\Delta p(t)$ varies as $\exp(-t/\tau)$. The instantaneous injection level is defined as the ratio of $\Delta p(t)/N_D$:

$$I(t) \equiv \frac{\Delta p(t)}{N_D}. \tag{23}$$

Equation (21) can then be rewritten in terms of $I(t)$:

$$I(t) = \frac{I_0 \exp(-t/\tau)}{1 + I_0(1 - \exp(-t/\tau))}. \tag{24}$$

All of the preceding equations were derived for excess holes in n-type semiconductors. The equivalent expressions for electrons in p-type semiconductors are written by simply substituting Δn for Δp and N_A for N_D. To make most of the subsequent equations symmetric and applicable to either n-type or p-type semiconductors, I will use the following notation:

$$\rho \equiv \Delta n \text{ or } \Delta p \quad \text{and} \quad N \equiv N_A \text{ or } N_D. \tag{25}$$

In this notation, ρ is the electron density in a p-type semiconductor and the hole density in an n-type semiconductor. Likewise, N is the free hole density in p-type semiconductors and the free electron density in n-type semiconductors.

2. SHOCKLEY–READ–HALL RECOMBINATION

The recombination of electron–hole pairs at defect levels in the bandgap was first discussed by Shockley and Read (1952) and by Hall (1952). Later work by Sah *et al.* (1957) added to the understanding of the role of such defects. This common mechanism in minority-carrier kinetics is frequently called the Shockley–Read–Hall (SRH) recombination mechanism. The physics involves minority-carrier capture at defects that have quantum levels in the bandgap of the semiconductor. The SRH recombination rate is derived in most semiconductor textbooks (Grove, 1967; Sze, 1981) and is shown to be

$$R_{\text{SRH}} = \frac{\sigma_p \sigma_n v_{\text{th}} N_t [pn - n_i^2]}{\sigma_n \left[n + n_i e^{\frac{E_t - E_i}{kT}} \right] + \sigma_p \left[p + n_i e^{\frac{E_i - E_t}{kT}} \right]} \quad (26a)$$

and has units of $cm^{-3} s^{-1}$. Here, N_t is the volume density of deep levels; σ_p and σ_n are the hole and electron capture cross-sections, respectively; and E_t is the energy level of the trap. The electron and hole concentrations are n and p, respectively, and v_{th} is the thermal velocity of the electron or hole. Here, then, $r(n, p)$ is written as

$$r_{\text{SRH}} = \frac{\sigma_p \sigma_n v_{\text{th}} N_t}{\sigma_n \left[n + n_i e^{\frac{E_t - E_i}{kT}} \right] + \sigma_p \left[p + n_i e^{\frac{E_i - E_t}{kT}} \right]}. \quad (26b)$$

If electron–hole pairs of volume density $\rho = \Delta n, \Delta p$ are injected, then the instantaneous values of p and n can be written as

$$p = p_0 + \rho, \qquad n = n_0 + \rho. \quad (27)$$

For a doped semiconductor, one has either $n_0 \gg p_0$ (n-type) or $p_0 \gg n_0$ (p-type). First, taking the case of an *n-type* material, one can derive the recombination rate for holes at a single energy level at E_t in the forbidden gap. This case is completely symmetrical to that of electron recombination in p-type material. In this expression, $n_0 = N_D$ and $p_0 \sim 0$. Therefore,

$$R_{\text{SRH}} = \frac{\sigma_p \sigma_n v_{\text{th}} N_t [\rho N_D + \rho^2]}{\sigma_n \left[N_D + \rho + n_i e^{\frac{E_t - E_i}{kT}} \right] + \sigma_p \left[\rho + n_i e^{\frac{E_i - E_t}{kT}} \right]}. \quad (28)$$

A logarithmic plot of R_{SRH} versus $E_t - E_i$ shows a steep peak at $E_t = E_i$. The maximum recombination rate occurs at defects levels that lie at or near midgap. When the defect energy level is near either band, thermal emission to the band quenches the recombination process. For the preceding case, when E_t lies near E_v, the denominator term n_i

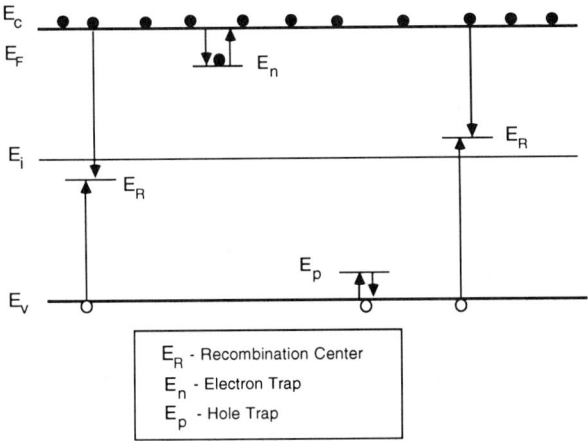

FIG. 1. Schematic diagram of impurity-related energy levels within the forbidden gap of a semiconductor. Levels are labeled as to whether the defect is likely to be a trap or a recombination center according to the Shockley–Read–Hall model.

$\exp(E_i - E_t)/kT$ becomes very large. This term describes the emission of captured holes to the valence band. Because of hole emission, the hole occupancy is small and the recombination rate is decreased. The defect level, in this case, is called a trap.

Figure 1 shows the forbidden gap of a semiconductor that has several types of impurity levels. Those near the midgap position E_i are labeled recombination centers. Also shown are levels that are designated as electron traps and hole traps. These lie near the conduction band and the valence band, respectively.

The defect level described by Eq. (28) may be either a trap or a recombination center, depending on E_t. The defect level is a trap when the absolute value $|E_i - E_t|$ is large ($\sim E_g/2$) and the energy level E_t lies near either band. A captured hole or electron is thermally emitted back to the valence band or to the conduction band, respectively. In either type of trap, an emission term in the denominator of Eq. (28) dominates, and R_{SRH} decreases many orders of magnitude relative to E_t being midgap. The demarcation between trapping and recombination centers is described in detail elsewhere (Rose, 1978). If the energy level E_t lies near midgap E_i, R_{SRH} becomes very large, and recombination is the most probable event. Thus, midgap centers ($E_t \sim E_i$) are very effective recombination sites.

One can divide the numerator and denominator of Eq. (28) by the product $\sigma_p \sigma_n v_{th} N_t$ and use the following definitions of SRH lifetime;

$$\tau_n = \frac{1}{\sigma_n N_t v_{th}} \quad \text{and} \quad \tau_p = \frac{1}{\sigma_p N_t v_{th}} \tag{29}$$

Equation (28) is rewritten

$$R_{SRH} = \frac{[\rho N_D + \rho^2]}{\tau_p \left[N_D + \rho + n_i e^{\frac{E_t - E_i}{kT}} \right] + \tau_n \left[\rho + n_i e^{\frac{E_i - E_t}{kT}} \right]}. \tag{30}$$

Generally, N_D is much greater than the terms containing n_i in Eq. (30). Therefore,

$$R_{SRH} = \frac{[\rho N_D + \rho^2]}{\tau_p [N_D + \rho] + \tau_n \rho}. \tag{31}$$

For midgap centers, one can use the symmetric notation with the majority-carrier doping equal to N. Therefore,

$$R_{SRH} = \frac{[\rho N + \rho^2]}{\tau_{min}[N + \rho] + \tau_{maj} \rho}. \tag{32}$$

Here, τ_{min} and τ_{max} are the minority- and majority-carrier SRH lifetimes, respectively. The minority-carrier injection level has not been specified in Eq. (32). For the low-injection case of $\rho \ll N$, Eq. (32) simplifies to

$$R_{SRH} = \frac{\rho}{\tau_{min}}. \tag{33}$$

The rate equation can then be written as

$$\frac{d\rho}{dt} = -R_{SRH} = \frac{-\rho}{\tau_{min}}. \tag{34}$$

The solution for $\rho(t)$ is exponential and is written

$$\rho(t) = \rho_0 \exp(-t/\tau) = \rho_0 \exp(-\sigma N_t v_{th} t). \tag{35}$$

Here, σ is the capture cross-section for the minority carrier.

The SRH lifetime is usually approximated by either τ_p or τ_n in n-type or p-type semiconductors, respectively. The recombination rate at the SRH level is controlled by the capture of the minority carrier by the defect. Thus, the product of the capture cross-section σ and the defect density N_t appears in the recombination rate. In low injection, the majority-carrier density N is much greater than the minority-carrier density p. After minority-carrier capture, the center is reset by the subsequent capture of a majority carrier. Because of the high concentration of majority carriers, capture occurs at a very high rate. Thus, in low injection, the SRH lifetime is determined solely by the minority-carrier capture rate. In high injection or with large cross-section ratios, this assumption cannot be made, as will be discussed in Section V.

When more than one deep-level defect type is present, the total SRH low-injection lifetime is described by a single minority-carrier lifetime τ, where:

$$\frac{1}{\tau} = \sum_i \frac{1}{\tau_i}. \tag{36}$$

Here, τ_i is the Shockley–Read–Hall lifetime for each specific type of defect. Equation (36) results from the notion that recombination probabilities are additive.

The SRH defects here are assumed to be uniformly distributed in the volume of the semiconductor. They may be localized at specific locations in the material. In particular, recombination levels are usually generated at surfaces and interfaces because of the high concentration of impurities and bonding defects. These defects are the origin of the parameter called the surface or interface recombination velocity.

3. SURFACE AND INTERFACE RECOMBINATION

The interface between the semiconductor and the outside world is an inherent source of deep-level defects. Early theory by Tamm (1932) and by Shockley (1932) showed that the interruption of periodicity of the crystal lattice produces a continuum of localized states. Consequently, these states have been called Tamm or Shockley states. More commonly however, they are called surface states. Tamm used a Kronig–Penney model of a one-dimensional crystal to show the origin of states in the forbidden gap. These states are caused by dangling bonds at the crystal surface that result from the interruption of periodicity. The potential density of surface states is that of the two-dimensional density of surface atoms, or about 10^{15} cm^{-2}. Early experimental work on silicon found that the crystal surface had a very high recombination rate, and the concept of surface recombination velocity (S) was introduced. The states induced in the energy gap of a semiconductor by the surface are called surface states.

Besides the inherent defects from the break in periodicity, the surface is usually a getter for impurities. These impurities range from atmospheric gases to metals. The surface impurities may also be a source of surface states and may coexist with dangling bond states.

The interface in solid/solid systems is at least equally important to that of a free surface. The terms surface and interface are often used somewhat interchangeably to describe the effects of the semiconductor boundary. The semiconductor interfaces greatly influence minority-carrier transport. Pioneering work by Bardeen (1947) indicated that the potential barrier at semiconductor interfaces was controlled more by charges in interface states than by the contact potential difference of the materials. The most investigated interface, in current semiconductor technology, is that between silicon (Si) and thermally grown silicon dioxide (SiO_2). These states at the Si/SiO_2 interface are, of course, called interface states and have been the subject of intense experimental and theoretical studies. The reduction of Si/SiO_2 interface states has been the focus of many years of research and development.

Early work indicated that the formation of SiO_2 reduced the interface state density by many orders of magnitude. The controlled growth (Nicollian and Brews, 1982) of SiO_2 on silicon, combined with various thermal annealing processes, reduces the Si/SiO_2 interface state density by many orders of magnitude. The removal of dangling bonds and their associated interface states is called passivation. Silicon dioxide is basic to the phenomenal development of the silicon technology, as it reduces S by more than six orders of magnitude (Nicollian and Brews, 1982). SiO_2 has many other uses, including a gate insulator in field effect transistors (FETs). In metal-oxide semiconductor field effect transistors (MOSFETs), the surface potential is intentionally modulated to control the current flow between source and drain.

Early experimental work (Casey and Buehler, 1977) on the free surface of GaAs indicated that S exceeded 10^7 cm/s. In recent PL decay experiments by Ehrhardt and co-workers (1991), on GaAs wafers, an S of about 3×10^7 cm/s was measured. Early measurements of minority-carrier lifetime in GaAs indicated that bulk lifetimes were dominated by the surface effects. The Illinois group of N. Holonyak and co-workers (Zwicker *et al.*, 1971a, 1971b) was among the first to look at PL lifetimes in epitaxial GaAs. This group also attempted to reduce the dominating effects of the large recombination velocity of the bare surface. One of the first confinement structure devices was reported by Dapkus and co-workers (1970). They fabricated $n^+/n/n^+$ homostructures in which the n-layer is undoped GaAs. Confinement was produced by the upward band bending in the undoped layer, thereby confining holes and producing a "surface free" active layer. This device is limited in that the confinement barrier becomes very small for other than an undoped active layer. Later work by the Illinois group (Keune et al., 1971) used epitaxially grown $Al_xGa_{1-x}As$ as a wide-bandgap confinement structure. Epitaxial $Al_xGa_{1-x}As$ has become the standard window/confinement semiconductor material and is essential to GaAs device technology. The operation of many GaAs devices depends upon the epitaxial growth of the $Al_xGa_{1-x}As/GaAs$ interface. Studies of this interface and the $Al_xGa_{1-x}As/GaAs$ recombination velocity are described in Section VI. This technology is important for all devices that are dependent on minority carrier transport. The surface or interface recombination velocity is important in the operation of most compound semiconductor devices.

Surface recombination effects have been analyzed in detail in a book by Many et al. (1971). Although surface states are usually represented by a continuum of states in the forbidden gap, a model using a single level at energy E_t is useful in describing the recombination effects (Many et al., 1971). In analyzing the surface or interface recombination effects, one can assume that these single-level SRH defects lie in a two-dimensional plane bounding the semiconductor. The limitations of the single-level approximation are compared with those of a more exact continuum model in a recent paper (Landsberg and Browne, 1988).

In this derivation, I have assumed that all recombination centers lie in two thin sheets of unit (1 cm^2) cross-sectional area and thickness Δt. The fractional active volume containing defect centers is $\Delta t/d$, where d is the total thickness of the active region. The recombination rate for the volume containing defects can be written from Eq. (26).

$$R = \sigma v_{th} N_t \frac{2\Delta t}{d} \frac{\rho}{1 + \frac{2n_i}{N} \cosh \frac{E_t - E_i}{kT}}. \tag{37}$$

As before, the minority-carrier density is ρ, and the majority-carrier density is N. The defect-containing volume becomes a surface upon shrinking Δt to 0. In this limit, one writes

$$\lim \left| \frac{d\rho_s}{dt} = \sigma v_{th} N_t \frac{2\Delta t}{d} \frac{\rho_s}{1 + \frac{2n_i}{N} \cosh \frac{E_t - E_i}{kT}} \right|_{\Delta t \to 0}. \qquad (38)$$

The excess volume density ρ in Eq. (37) became ρ_s, the excess density (cm^{-3}) at the surface. One can define a planar density N_{st} as the limiting value of the product $N_t \Delta t$ as Δt goes to zero. N_{st} is called the interface (or surface) state density. Equation (37) is a volume recombination rate, as it is an average over the entire volume. However, the rate is zero except for the surface contribution that is labeled R_{vs}.

$$R_{vs} = \frac{2}{d} \frac{\sigma v_{th} N_{st} \rho_s}{1 + \frac{2n_i}{N} \cosh \frac{E_t - E_i}{kT}}. \qquad (39)$$

The units of R_{vs} in Eq. (39) are cm^{-3} s^{-1}. The recombination rate R_{vs} may be used to define the surface recombination velocity S:

$$R_{vs} \equiv \frac{2}{d} R_s = \frac{2S}{d} \rho_s. \qquad (40)$$

Here, R_s is the surface recombination rate in cm^{-2} s^{-1}. The surface recombination rate R_s is clearly $\rho_s S$, where S has units of cm/s. Substituting R_{vs} from Eq. (39) into Eq. (40) and solving for S, one gets

$$S = \frac{\sigma v_{th} N_{st}}{1 + \frac{2n_i}{N} \cosh \frac{E_t - E_i}{kT}}. \qquad (41)$$

For most real surfaces or interfaces, S involves a summation over a number of near-midgap states. The summation is replaced in this simple model by a single state at energy E_t. Assuming that there are N_k discrete energy levels and each level has a density N_{sk}(cm^{-2}), Eq. (41) becomes

$$S = \sum_{j=1}^{N_k} \frac{\sigma_j v_{th} N_{sj}(E_j)}{1 + \frac{2n_i}{N} \cosh \frac{E_j - E_i}{kT}}. \qquad (42)$$

Here, E_j is the energy and σ_j is the capture cross-section of the jth level. Shallow states or traps do not contribute to S and may be neglected in the summation. For a single midgap level with $E_t \sim E_i$, the recombination velocity S becomes $\sigma v_{th} N_{st}$.

Because of surface band bending, the surface charge density ρ_s is not usually equal to the bulk or "flat band" density ρ. The surface minority-carrier density ρ_s may be found by solving the steady-state continuity equation for the specific surface potential. Defining an downward bending surface potential as V_s, solutions for $\rho_s(V_s)$, the equilibrium minority-carrier surface density, were developed by Dhariwal and Mehrotra (1988). When the surface band-bending is repulsive for minority-carriers, these workers find

$$\rho_s = \rho \exp\left(\frac{-qV_s}{kT}\right). \tag{43}$$

Here, ρ is the minority-carrier density in the bulk quasi-neutral region, and ρ_s is the surface minority-carrier density per cubic centimeter. Thus, one can write S, in terms of the surface potential, as

$$S = \frac{\sigma v_{th} N_{st} \exp\left(-\frac{qV_s}{kT}\right)}{1 + \frac{2n_i}{N} \cosh \frac{E_t - E_{is}}{kT}}. \tag{44}$$

Here, E_{is} is the energy of the interface state at the surface. If the energy level or levels are located near midband ($E_t \sim E_{is}$), Eq. (44) can be simplified:

$$S = \sigma v_{th} N_{st} \exp\left(-\frac{qV_s}{kT}\right). \tag{45}$$

For a given surface density N_{st}, the surface potential V_s will be dependent upon the doping density N and the injection level ρ/N. Therefore, S will be dependent upon the doping density and injection level. Experimentally, S is usually found to be doping-dependent, as data in Section VI show. In addition, S is dependent on the injected carrier density ρ because of other mechanisms. This mechanism is the dependence of surface density ρ_s on the injection level as described here and by Dhariwal and Mehrotra (1988). The second mechanism is the injection dependence of the SRH recombination rate, which is discussed in Section V.

4. Auger Recombination

Auger recombination is a nonradiative recombination process in which the energy of the electron–hole pair is transformed into the kinetic energy of a free particle. A large variety of Auger processes involve not only the free electron and hole, but also phonons and localized states (traps). Four-carrier Auger recombination processes in III–V semiconductors are described schematically in Chapter 1 (Fig. 23) of this volume. A compilation of the various Auger processes has been described in an early review (Landsberg and Robbins, 1978). The minority-carrier lifetime in undoped crystalline silicon is dominated by Auger recombination at injection levels exceeding about 1×10^{17} cm^{-3} (Willander and Grivickas, 1988). The recombination mechanisms of undoped silicon, over a range of injection levels, are described by a combination of SRH, band-to-band Auger, and trap-Auger effects (Landsberg, 1987).

Auger recombination effects are most frequently seen at high injection levels in LEDs and solid state lasers. The recombination rate varies with n^2p in n-type or np^2 in p-type nondegenerate semiconductors. Thus, the Auger process is usually seen at relatively high carrier concentrations. The complicating effects of degeneracy on Auger recombination rates are developed by Landsberg (1991). The three-carrier Auger recombination lifetime, τ_a, for n- and p-type semiconductors varies as (Pankove, 1971)

$$\frac{1}{\tau_a} = Anp + Cn^2 \quad (n\text{-type}), \qquad \frac{1}{\tau_a} = Anp + Cp^2 \quad (p\text{-type}). \quad (46)$$

Here, the coefficient C is a cross section that is proportional to the transfer efficiency of energy to the majority carrier. The A coefficient represents the transfer of energy to the minority carrier. Henry and co-workers (1984) found evidence that the nonradiative component of the minority-carrier lifetime in InGaAs was dominated by Auger recombination. The Auger cross-section is significantly larger in p-type than in n-type InGaAs layers. Larger Auger effects are seen in the quaternary semiconductor InGaAsP (Henry et al., 1982); their experimental results are consistent with the Auger cross-section calculations of Sugimura (1983).

Evidence of the Auger effect was observed at very high injection levels in GaAs LEDs and lasers (Su and Olshansky, 1980). Other calculations have considered the details of the Auger process in semiconductors of the zincblende structure (Gel'mont et al., 1983). A heavy hole transition to the spin–orbit split-off band appears to be the most probable Auger process. The Auger coefficient for GaAs was calculated by Haug (1983). There is very little experimental data relevant to the Auger effect in GaAs. However, the mechanism is probably significant in GaAs devices at high-injection levels (Lundstrom et al., 1990).

5. RADIATIVE RECOMBINATION AND PHOTOLUMINESCENCE

Radiative recombination is the process by which a conduction band electron and valence band hole recombine to produce a photon of energy $hv \sim E_g$. The band-to-band luminescence process is referred to by a variety of terms in the literature, depending on the excitation mechanism. When the combining electron–hole pairs are produced by a forward-biased electrical current in a p–n junction, the light emission is called electroluminescence. Electroluminescence from GaAs p–n junctions was observed many years ago (Moss, 1957; Pankove and Massoulie, 1962; Loebner and Poor, 1959; Braunstein, 1955) and provides the mechanism for light-emitting and laser diodes. Photoluminescence is the common term for the optical generation of recombination luminescence. Photoluminescence has been observed for almost every semiconductor, but the intensity is much greater for direct than for indirect bandgap materials. Photoluminescence in GaAs crystals was also first reported about 30 years ago (Nathan and Burns, 1963). Room-temperature photoluminescence spectra usually show an intense band peaking near E_g. This peak comes from the direct recombination of electron and holes. Other peaks that are observed at energies less than E_g are related to transitions from impurity or exciton states (Knox, 1963). Excitons in GaAs, which thermally disassociate at room temperature, become observable at cryogenic temperatures.

A derivation of the matrix elements describing band-to-band electronic transitions is presented in Chapter 1 of this volume. It was shown that the photon emission rate R_L for band-to-band transitions is given by

$$R_L = r_{\text{rad}} np = Bnp \quad (\text{cm}^{-3}\,\text{s}^{-1}). \tag{47}$$

Here, p is the hole density, n is the electron density, and B is a term that comes from summing the dipole matrix elements connecting the valence and conduction bands. B has units of cm^3/s and is the r of Eq. (14) that is specific to radiative recombination. Obviously, B is much larger for direct than for indirect bandgap semiconductors. Calculations by Dumke (1957) provided the first calculations of B for both direct and indirect bandgap semiconductors. The books by Pankove (1971) and Landsberg (1991) list the B-coefficients for a number of common semiconductors. For example, B is about 10^{-10} cm^3/s for GaAs, compared to 10^{-15} cm^3/s for silicon. An early calculation by Hall (1960) produced the B-coefficient for direct semiconductors in terms of the band effective masses, bandgap, dielectric constant, and temperature as

$$B = 0.58 \times 10^{-12} \sqrt{\varepsilon} \left(\frac{1}{m_p + m_n}\right)^{1.5} \left(1 + \frac{1}{m_p} + \frac{1}{m_n}\right)\left(\frac{300}{T}\right)^{1.5} E_g^2 \quad \text{(cm}^3\text{/s)}, \tag{48}$$

where ε is the dielectric constant, and m_p and m_n are the hole and electron effective masses, respectively, in units of the free electron mass. Also, T is the absolute temperature, and E_g is bandgap in electron volts. Applying this equation to GaAs, and using effective masses $m_n = 0.08$ and $m_p = 0.5$, one calculates (Marburg, 1961) that B is 1.4×10^{-10} cm^3/s. This value is consistent with recent measurements and calculations. Garbuzov (1982) described a simple quantum-mechanical calculation for direct-bandgap semiconductors and obtained the following expression for the B-coefficient:

$$B = 3 \times 10^{-10} \left(\frac{E_g}{1.5}\right)^2 \left(\frac{300}{T}\right)^{1.5} \quad \text{(cm}^3\text{/s)}. \tag{49}$$

Putting in the parameters appropriate to GaAs at room temperature, one gets $B \sim 2.7 \times 10^{-10}$ cm^3/s.

Quantum-mechanical calculations of the B-coefficient were performed by Stern (1976) and by Casey and Stern (1976). They included the doping concentration dependence of the energy gap and the energy-dependent matrix elements for the dipole transition. The methods used in quantum-mechanical calculations are described by Dutta in Chapter 1 of this volume. Stern (1976) calculated the B-coefficient for p-type ($p = 4 \times 10^{17}$ cm^{-3}) GaAs at electron injection levels ranging from $n = 0$ to $n = 2.5 \times 10^{18}$ cm^{-3}. The results of these calculations are shown in Chapter 1 (Fig. 8) of this volume. Stern finds $B \sim 2 \times 10^{-10}$ cm^3/s at the injected electron densities n less than 1×10^{18} cm^{-3}. Near $n = 2.5 \times 10^{18}$ cm^{-3}, the calculations show that B decreases to about 1×10^{-10} cm^3/s. Thus, B is not constant but varies weakly with carrier concentration. Casey and Stern (1976) calculated the B-coefficient for

p-GaAs over a range of doping levels. Their calculations indicated that B (p-GaAs) ranged from about 2×10^{-10} cm^3/s ($\sim 1 \times 10^{18}$ cm^{-3}) to 1×10^{-10} cm^3/s (1×10^{19} cm^{-3}). H. van Cong (1981) calculated the B-coefficient (77 K) for n-GaAs and predicted a minimum in the radiative lifetime at $n \sim 6 \times 10^{17}$ cm^{-3} that is the onset of degeneracy. These calculations predict that B decreases dramatically at degenerate doping levels because of the band-tailing effects. However, this decrease in B has not been experimentally observed, as is discussed in Section VI.

The work of van Roosbroeck and Shockley (1954) (vRS) related the emission spectrum of a semiconductor to the absorption coefficient α by thermodynamic detailed balance. This relationship between the absorption coefficient $\alpha(E)$ and the spontaneous emission spectrum $S(E)$ was derived in which E is the photon energy:

$$S(E) = \frac{8\pi n^2 E^2 \alpha(E)}{h^3 c^2 (\exp(E/kT) - 1)}. \qquad (50)$$

In this expression, the quantities specific to a particular semiconductor are the index of refraction n and the absorption coefficient $\alpha(E)$. Figure 2, curve A is a plot of the absorption coefficient of undoped GaAs (Sell and Casey, 1974). Curve B is a normalized emission spectrum calculated from the van Roosbroeck–Shockley relationship using Curve A and Eq. (50) with $kT = 0.025$ eV. Curve C is normalized PL emission data from an undoped epitaxial GaAs DH. The GaAs film thickness is about 1 μm and is grown between two epitaxial layers of GaInP for surface passivation purposes. The self-absorption effect (discussed in Section IV) minimally distorts the emission spectra in thin devices ($d < 1$ μm). One sees that the data agree well with the calculated spectra. Similar data are shown by Dutta (Chapter 1, Fig. 5) for GaAs doped n-type to 1.2×10^{18} cm^{-3}.

One may calculate the B-coefficient for a particular material by integrating Eq. (50) over the absorption spectra and dividing by n_i^2. Casey and Stern (1976) applied the vRS relationship to the absorption data for doped p-type GaAs and found $B = 3.2 \times 10^{-10}$ cm^3/s (1.2×10^{18} cm^{-3}) and 1.7×10^{-10} cm^3/s (1.6×10^{19} cm^{-3}). These B-values are somewhat larger than those calculated by quantum mechanics. However, these calculated B-values reflect the uncertainty in n_i for GaAs. This uncertainty is relatively large at high doping levels.

G. W. 't Hooft (1981) deduced B from measurements of turn-on delay and luminescence efficiency in the spontaneous regime of a GaAs laser. The experimental number found was $B = 1.3 \times 10^{-10}$ cm^3/s, which is within the range of calculated values. Hwang and Dyment (1973) also measured $B \sim 1.3 \times 10^{-10}$ cm^3/s from the measurements of turn-on delay

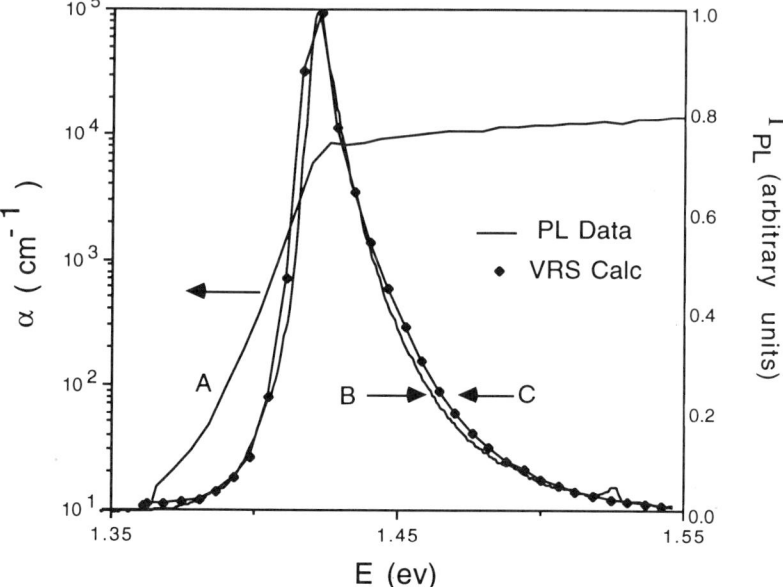

FIG. 2. Curve A is the absorption coefficient of undoped GaAs (Palik, 1985). Curve B is the normalized emission spectrum of a 1.0-μm GaAs thin film as measured at NREL. Curve C is the normalized emission spectrum calculated from the van Roosbroeck–Shockley relationship (1954) using the data from Curve A.

in LPE-grown laser diodes over the doping range of 2×10^{17} cm^{-3} to 3×10^{19} cm^{-3}. Thus, there is a great deal of consistency between the measurements and the calculations.

6. RADIATIVE LIFETIME

Equation (47) gives the light output per unit volume L of a semiconductor with n electrons and p holes per cubic centimeter. For a semiconductor under illumination, a density of $\rho(r)$ electron–hole pairs is generated, and the *general expression* for the net photoluminescence from the volume V is written in terms of an integral:

$$I_{PL} = B \int_V [(p_0 + \rho(r))(n_0 + \rho(r)) - n_i^2] \, dV$$

$$= B \int_V [\rho(r)^2 + \rho(r)(p_0 + n_0)] \, dV, \tag{51}$$

where $\rho(r)$ is the concentration of excess hole–electron pairs produced by the excitation. Here, Bn_i^2 is the equilibrium or blackbody luminescence. Under steady-state excitation, $\rho(r)$ remains constant in time, and therefore I_{PL} remains constant. Under pulsed excitation, $\rho(r)$ decreases because of recombination and I_{PL} decays accordingly. This is the physical process basic to measuring the minority-carrier lifetime by time-resolved photoluminescence.

Using the symmetric notation (N and ρ) and neglecting the steady state minority-carrier density, the recombination rate is given by Eq. (14) with B substituted for the generalized rate r. Assuming *spatially uniform* and instantaneous injection ρ_0 of electron–hole pairs per cm^3, the radiative decay of ρ is given by

$$\frac{d\rho}{dt} = -B(N\rho + \rho^2). \tag{52}$$

The solution to this differential equation is

$$\rho(t) = \frac{\rho_0 \exp(-t/\tau_R)}{1 + \frac{\rho_0}{N}(1 - \exp(-t/\tau_R))}. \tag{53}$$

Here, the radiative lifetime τ_R is defined as

$$\tau_R \equiv \frac{1}{BN}. \tag{54}$$

From Eq. (51), the PL intensity per unit volume can be written as a function of time:

$$I_{PL}(t) = B[N\rho(t) + \rho(t)^2]. \tag{55}$$

Here, $\rho(t)$ is given by Eq. (53).

a. *Low-Injection Lifetime*

At low-injection $\rho < N$, $I_{PL}(t)$ can be simplified to

$$I_{PL}(t) = BN\rho(t) = \frac{\rho(t)}{\tau_R}. \tag{56}$$

Thus, in low-injection, $I_{PL}(t)$ linearly tracks the excess-carrier density $\rho(t)$. If the excess carrier density is spatially nonuniform, then the total intensity $I_{PL}(t)$ comes from the volume integral

$$I_{PL}(t) = \frac{1}{\tau_R} \int_V \rho(r, t) \, dV. \tag{57}$$

Equation (57) ignores the self-absorption of the sample, which will be included in a later section. If a second, nonradiative recombination mechanism with lifetime τ_{nR} is active in the material, the rate equation is written using Eq. (36):

$$\frac{d\rho}{dt} = -\left(\frac{1}{\tau_R} + \frac{1}{\tau_{nR}}\right)\rho = -\frac{\rho}{\tau}. \tag{58}$$

Here, τ is the total lifetime and is given by the addition of the reciprocal lifetimes.

$$\frac{1}{\tau} = \frac{1}{\tau_R} + \frac{1}{\tau_{nR}}. \tag{59}$$

The solution for $I_{PL}(t)$ per unit volume in this case is

$$I_{PL}(t) = \frac{\rho_0 \exp(-t/\tau)}{\tau_R} \quad [\text{photons}/(\text{cm}^3 \text{ s})]. \tag{60}$$

b. *High-Injection Lifetime*

For the case in which ρ is not negligible compared to N, $I_{PL}(t)$ must be evaluated by integrating Eq. (55) over the volume. Figure 3 plots a normalized plot of $I_{PL}(t)$ versus time in units of τ_R. The injection levels ρ_0/N are 0.1, 1.0, and 10.0 in curves A, B, and C, respectively. It is obvious that the lifetimes in high injection are nonexponential for $t < \tau_R$. The instantaneous lifetime is given by the slope according to Eq. (20). To relate the slope of $I_{PL}(t)$ to the instantaneous excess carrier density, one may easily calculate the ratio

$$\frac{dI_{PL}/dt}{I_{PL}} = \frac{N + 2\rho}{\rho(\rho + N)} \frac{d\rho}{dt}. \tag{61}$$

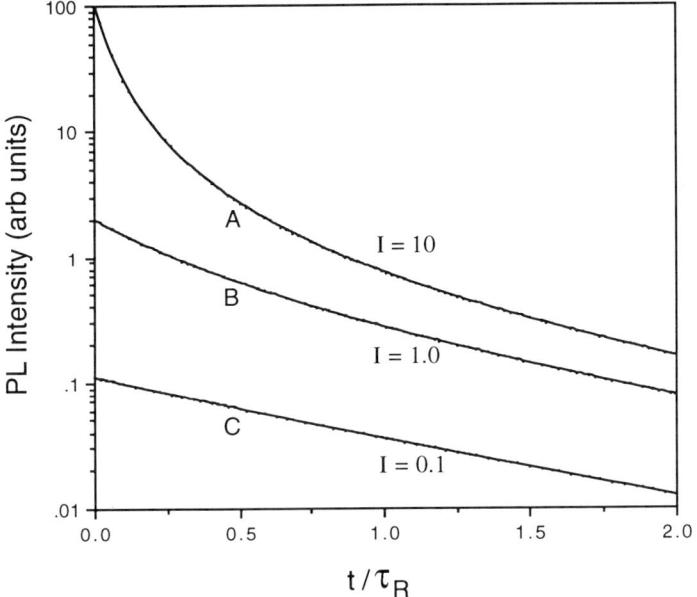

FIG. 3. A calculated plot of $I_{PL}(t)$ versus t/τ_R for three initial injection levels. The injection levels are 10, 1.0, and 0.1 for curves A, B, and C, respectively.

One can define the instantaneous lifetime from Eq. (61):

$$\frac{1}{\tau_{\text{eff}}} = -\frac{dI_{PL}/dt}{I_{PL}}. \tag{62}$$

The minimum value of τ_{eff} occurs at $t = 0$ and has the value

$$\frac{1}{\tau_{\text{eff}}} = B(N + 2\rho_0). \tag{63}$$

c. *Very High-Injection*

At very high injection ($\rho_0 \gg N$) and *short times* ($t \ll \tau_R$), the expansion of Eq. (53) will produce the very high-injection decay curve:

$$\rho(t) = \frac{\rho_0}{1 + B\rho_0 t}. \tag{64}$$

The PL intensity versus time in this case is

$$I_{PL}(t) = \frac{B\rho_0^2}{(1 + B\rho_0 t)^2} \quad (t \ll \tau_R). \tag{65}$$

The slope or instantaneous lifetime of the curve is obtained by applying Eq. (20) to Eq. (65):

$$\tau_{\text{eff}} = \frac{1}{2B\rho_0} + \frac{t}{2} \quad (t \ll \tau_R). \tag{66}$$

The initial and minimum value of τ_{eff} is $1/2B\rho_0$, and the effective lifetime increases linearly with time. When t becomes comparable to the radiative lifetime τ_R, Eqs. (53) and (55) must be used to describe the PL decay.

d. Temperature Dependence of the Radiative Lifetime

From Eqs. (48) and (49), one sees that the B-coefficient increases as $T^{-1.5}$ as the absolute temperature is lowered. One can calculate the radiative lifetime as a function of temperature from Eqs. (48) or (49) and (54). A generalized form of $B(T)$ can be written

$$B = A(E_g(T))^2 \left(\frac{300}{T}\right)^{1.5}. \tag{67}$$

Here, A is a constant and $E_g(T)$ is the bandgap as a function of temperature.

$$\tau_R = \frac{1}{A(E_g(T))^2 N(T)} \left(\frac{T}{300}\right)^{1.5}. \tag{68}$$

Here, $N(T)$ is the majority-carrier density as a function of temperature. At temperatures above donor or acceptor "freeze-out," $N(T)$ does not vary appreciably with temperature (Sze, 1981). One can estimate $E_g(T)$ with a linearized equation (Sze, 1981)

$$E_g(T) = E_g^0 - \gamma T. \tag{69}$$

Here, E_g^0 is the bandgap at 0 K and γ is a positive constant. Therefore, the radiative lifetime above "freeze-out" can be written as

$$\tau_R = \frac{1}{A(E_g^0 - \gamma T)^2 N(T)} \left(\frac{T}{300}\right)^{1.5}. \tag{70}$$

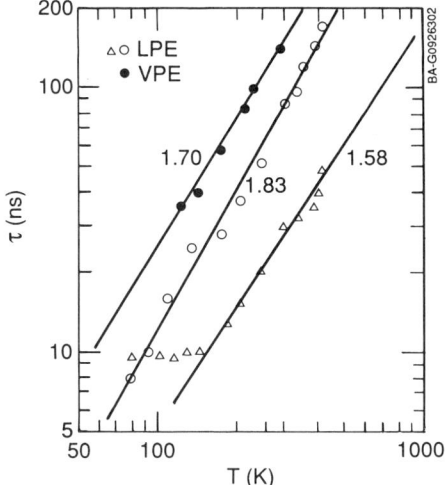

FIG. 4. Temperature dependence of the radiative lifetime for GaAs ('t Hooft and van Opdorp, 1983).

In cases in which radiative lifetime is the dominant mechanism, the photoluminescence lifetime varies according to Eqs. (68) and (70). The radiative lifetime varies with temperature at a rate slightly larger than $T^{1.5}$ because of the $E_g(T)$ dependence.

Figure 4 shows PL lifetime versus temperature measurements reported by 't Hooft and van Opdorp (1983) and Garbuzov and co-workers (1978) on epitaxial GaAs. The data show that the temperature dependence varies between $T^{1.58}$ and $T^{1.83}$. These structures had very large quantum efficiencies, and therefore the recombination was primarily radiative. A temperature-dependent lifetime τ, which varies approximately as $T^{1.5}$, is then indicative of dominant radiative recombination.

III. Lifetime Measurement Techniques

7. General Techniques

The minority-carrier lifetime of III–V semiconductors has historically been measured by transient or time-resolved photoluminescence (TRPL) rather than by the photoconductive decay techniques that are used for silicon. One advantage of the TRPL technique is that it is applicable to subnanosecond lifetimes. The relationship between the PL lifetime and the various mechanisms controlling minority-carrier lifetime is discussed in Section IV.

There are many publications describing the various methods of minority-carrier lifetime measurement in semiconductors. Two recent textbooks provide descriptions of a variety of minority-carrier lifetime measurement techniques. Comprehensive reviews are given in Chapter 8 of the recent book by Schroder (1990) and Chapters 3–6 of the volume by Orton and Blood (1990). The measurement techniques can be divided into contacting and noncontacting techniques. The former techniques require the preparation of a device with ohmic contacts to electrical wires that are connected to the measurement system. Contacting lifetime measurement methods include photoconductive decay and current or voltage decay in junction devices.

A number of contactless minority-carrier lifetime measurement techniques are described by Schroder (1990). These include photoluminescence decay, transient free-carrier absorption, and microwave reflection. Some of these requrie somewhat complicated multiple parameter fits to the data.

The photoluminescence decay methods include the phase-shift and time-resolved photoluminescence techniques. The phase shift technique was recently reviewed by this author (Ahrenkiel, 1988), and the limitations were discussed. In particular, phase shift data analysis assumes a pure exponential decay of the PL signal. The analysis produces inaccurate results if the decay is nonexponential. As will be shown, minority-carrier decay is often nonexponential and the phase shift analysis fails to accurately describe the lifetime. For a number of reasons, the preferred measurement method for compound semiconductors is the time-resolved photoluminescence technique called time-correlated single photon counting (Bachrach, 1972). The TRPL technique provides detailed features of nonexponential decay that contain many insights into the specific recombination mechanisms. The TRPL technique is very sensitive, and lifetimes at very low injection levels may be measured. Finally, the technique is contactless and may be applied to a very small sample.

8. Time-Correlated Single Photon Counting

I will describe the measurement apparatus that has been constructed at the National Renewable Energy Laboratory (NREL) for lifetime measurements. NREL has been heavily involved in minority-carrier lifetime measurements since the mid-1980s. This emphasis has developed because of the great importance of the minority-carrier lifetime in the development of photovoltaic materials.

The time-correlated single photon counting method is described in great detail in a textbook by Demas (1983). The photon counting technique was first applied to semiconductors by Bachrach (1972). The NREL facility has essentially the same configuration with modernized components (Ahrenkiel et al., 1988c). For use in time-correlated single photon counting, the photodetector must be able to detect single photons produced by free electron–hole recombination luminescence. There are a number of commercial photodetectors with sufficient gain for single photon counting. These include photomultiplier tubes (PMT) and microchannel plates (MCP). With these detectors, the experimental technique can only be used on larger bandgap materials with $E_g > 1.1$ eV. A recent paper (Louis, 1990) used a single photon avalanche photodiode (SPAD) to measure TRPL in semiconductors. Extension of the SPAD detector technology to the near-infrared region suggests the possibility of TRPL measurements on smaller bandgap semiconductors ($E_g < 1.1$ eV).

NREL has two TRPL measurement systems that will be described here. The first system uses an S-1 photomultiplier tube that has an impulse response of about 300 picoseconds (ps). The relatively long impulse response is caused by the transit-time dispersion of electrons as they cascade down the dynode chain. Transit-time dispersion is greatly reduced with a newer detector; the latter is called a microchannel plate detector. The newer and second system uses a Hamamatsu model R2809U microchannel plate detector that has a time resolution of 30 ps full-width half-maximum (FWHM). The time-correlated photon counting electronics are identical in the remainder of the detection system.

A schematic of the NREL experimental setup is shown in Fig. 5. The photons emitted by the sample are focused on the input slit of a scanning monochromator. The monochromator is tuned to a narrow band of wavelengths in the intrinsic or band-to-band emission spectra of the semiconductor. Using a beam splitter and a photodiode, a small fraction of the laser light pulse is deflected to a fast photodiode for initiating a timing pulse. The electrical output of the photodiode triggers the time-to-amplitude converter (TAC). The pulse-height discriminator is necessary to block electrical pulses that are produced by thermal and other multiple-photon events. The first collected photon initiates an electrical pulse in the photodetector that is amplified and passed by the amplitude discriminator. The electrical pulse produces a stop message at the TAC. The TAC output is a voltage pulse with an amplitude proportional to the time delay between the trigger pulse and the arrival of the first photon. The TAC signal is fed to a multichannel pulse height analyzer (MCA). Each count is stored in a channel appropriate to the time delay, and a maximum of one count is recorded for every laser

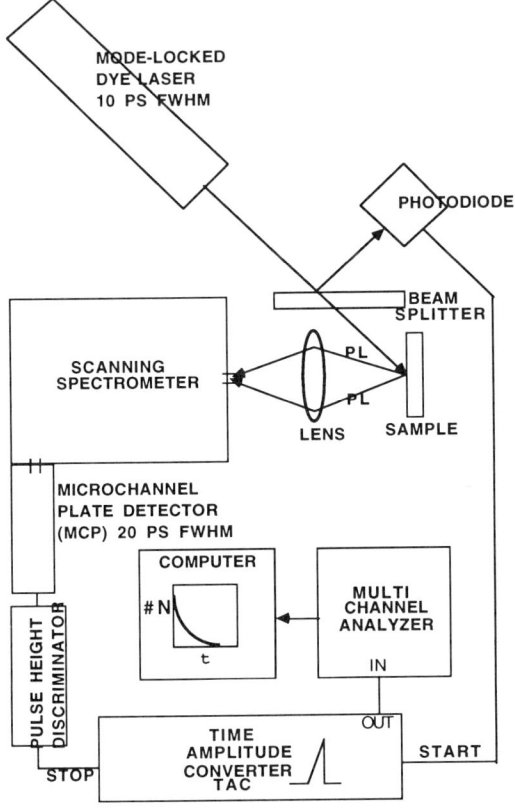

FIG. 5. Schematic of the NREL time-correlated single-photon counting apparatus (Ahrenkiel *et al.* 1988c).

pulse. In this way, a histogram of the PL decay is built up in terms of counts versus time.

Single photon counting is a very sensitive technique requiring very low light levels. PL decay measurements can be made at very low injection levels because of this sensitivity. Detectable measurements of PL decay can be made at excitation levels as low as $10^{-4}\,\mu\text{J/cm}^2$. Photon counting also has a wide dynamic range so that excitation levels can range from very low to very high injection. Most of the data described in this chapter have been obtained by this technique.

The samples are mounted in a cryostat so that the lifetime can be measured as a function of temperature. The variable-temperature cryostat provides a temperature range from 4.2 K to room temperature. As shown in Section II, the various recombination mechanisms have distinctive dependences on temperature. When the identification of mechanisms is strongly desired, temperature–lifetime measurements are very valuable.

9. Inductively Coupled Probe

A second, contactless pulse technique uses the inductive coupling between a high-frequency coil and a semiconductor wafer or film. The technique, first described by Miller and co-workers (1978), was recently reviewed by Yablonovitch and Gmitter (1992). The technique was recently described by Yablonovitch and co-workers (1986) as applied to lifetime measurements in silicon and germanium wafers. The technique was used to measure the minority-carrier lifetime in undoped GaAs thin films grown by different epitaxial techniques (Yablonovitch et al., 1987).

The detection system uses a 3–5-turn high-frequency coil to inductively couple a detector to the semiconductor under test. The radio-frequency magnetic field is the coupling mechanism, and the sample acts like a one-turn secondary in an rf transformer. Reported operating frequencies range from 70 to 500 MHz. The measurement circuit is essentially a radio-frequency bridge, which is balanced to produce a null signal. A change in sample impedance produces an rf signal in the bridge circuit. The latter is processed by the electronics and converted to a dc voltage that is proportional to the change in sample impedance. Excess minority carriers are generated in the sample by a pulsed light source, which is usually a laser. The decrease in sample impedance is proportional to the density of excess minority carriers. As the minority carriers decay, the dc offset voltage tracks the minority-carrier density. The offset voltage is usually connected to a transient digitizer or oscilloscope and averaged over a number of generation pulses.

The advantage of this system is that non–light-emitting materials and small-bandgap materials may be measured. Thus, besides silicon and germanium, the lifetime of small-bandgap ($E_g < 1.0$ eV) compound semiconductors may be measured. The disadvantage of the technique is that minimum measurable lifetime is large compared to that measurable by time-correlated single photon counting. Measurements at the NREL facility indicate that the minimum measurable lifetime is about 50 ns. Nevertheless, the technique will likely prove to be a valuable tool for lifetime measurement in many materials.

10. Photoluminescence Spectroscopy

Measurement of the photoluminescence spectra of semiconductors is a valuable component of semiconductor characterization. Photoluminescence spectroscopy is a quick and nondestructive technique to measure the bandgap E_g of a material of unknown composition. The derivation of the internal quantum efficiency result is given in Section IV. The quantum efficiency is expressed in terms of radiative and nonradiative recombination lifetimes. It is shown in Section IV that the total internal quantum efficiency is given by

$$\eta = \frac{\tau_{nR}}{\tau_{nR} + \tau_R}. \qquad (71)$$

As the nonradiative lifetime becomes long compared to the radiative lifetime, the internal efficiency approaches unity. Thus, measuring the quantum efficiency of a device is a diagnostic for detecting SRH or other nonradiative recombination mechanisms. This measurement is used by many researchers to make an assessment of material quality and *relative* lifetime.

Low-temperature photoluminescence spectra also produce information about chemical impurities and mechanical defects in the sample. These often produce sub-bandgap emission peaks. In addition, mechanical quality and strain affect the width of intrinsic band-to-band or exciton peaks.

The NREL photoluminescence spectroscopy system uses a Coherent Radiation krypton ion laser as the excitation source. The laser can be tuned to one of 12 different lines between the blue (~400 nm) and the near infrared (~800 nm). The laser intensity is adjusted by means of a continuous gradient neutral density filter. The sample or device is mounted in a Janis Super-Varitemp continuous-flow cryostat with four infrasil windows. This cryostat allows the changing of samples while the system is cold, which greatly decreases the time needed to change samples. This cryogenic system allows a much higher sample throughput for low-temperature measurements. As the samples are cooled by helium gas, thermal gradients between the sample and the temperature sensors are small. Therefore, the sensor temperature is very close to the actual sample temperature.

The light emission from the sample is collected by a lens, wavelength dispersed, and focused on light-sensing arrays. These arrays are components of an EG &G Model OMA III optical multichannel analyzer (OMA). Our system has a charge-coupled device (CCD) array, an InGaAs photodiode array, and a silicon diode array. These detectors covers the wavelength ranges from 400 to 1,700 nm. The OMA detector allows very rapid acquisition of spectral data, compared to conventional mechanical scanning methods.

IV. Time-Resolved Photoluminescence in Device Structures

11. MINORITY-CARRIER AND PHOTOLUMINESCENCE LIFETIME

The bulk minority carrier lifetime in III–V compounds is commonly found by measuring the time-resolved photoluminescence or TRPL. The technique works best with direct-bandgap materials that are strong light emitters. By using this technique, the experimentalist is able to measure the lifetime over a wide range of injection conditions, from very low injection to very high injection. In addition, the sample temperature may be varied over a wide range for additional diagnostic information.

An analysis of the data from transient PL experiments involves the solution of the transient diffusion equation applied to the particular device structure. The initial ($t = 0$) distribution of minority carriers deposited in the semiconductor by the pulsed light source is given by Beer's law:

$$\rho(x, 0) = \alpha I_0 e^{-\alpha x}. \tag{72}$$

Here, ρ is the excess minority-carrier density at $t = 0$. I_0 is the flux density of incident light in units of photons per square centimeter, and α is the absorption coefficient of the absorbing material at the wavelength of the light source. One can assume in this derivation that the light source has an impulse response $\delta(t)$. This approximation is accurate if the FWHM of the light pulse is short relative to the minority-carrier diffusion times and the recombination lifetimes. Equation (51) shows that the total light emission from the active layer at any instant of time is given by the volume integral over the active layer:

$$I_{PL}(t) = B \int_V p(r, t) n(r, t) \, dV. \tag{73}$$

Here, I_{PL} is the total emission rate in photons per second and V is the active volume of the device. The blackbody component of luminescence is neglected here. B is the radiative recombination coefficient specific to a particular semiconductor. Also, p and n are the instantaneous hole and electron concentrations, respectively. For an extrinsic or doped semiconductor (neglecting the equilibrium minority-carrier density, n_i^2/N), one can easily show

$$I_{PL} = B \int_V (N + \rho) \rho \, dV. \tag{74}$$

Using the symmetric notation, N is the majority-carrier density and ρ is the density of photogenerated electron–hole pairs. In the low injection limit ($\rho \ll N$), the integral simplifies to

$$I_{PL} = BN \int_V \rho \, dV = \frac{1}{\tau_R} \int \rho \, dV. \tag{75}$$

Here the radiative lifetime is substituted for $1/BN$. Thus, I_{PL} is proportional to the excess-carrier density ρ. By measuring $I_{PL}(t)$, one can measure $\rho(t)$. However, as will be shown, the density $\rho(t)$ may or may not decay with the minority-carrier lifetime.

12. CONFINEMENT STRUCTURES

The time evolution of the minority-carrier density $\rho(r, t)$ is found by solving the time-dependent diffusion equation. The latter is written as

$$\frac{\partial p(r, t)}{\partial t} = D\nabla^2 \rho(r, t) - \frac{\rho(r, t)}{\tau}. \tag{76}$$

Here, D is the minority-carrier diffusivity and τ is the total minority-carrier lifetime that includes radiative and nonradiative recombination lifetimes. The total lifetime τ is defined by combining Eqs. (36) and (58). By integrating Eq. (76) over the entire volume, one gets

$$\frac{\partial}{\partial t} \int_V \rho(r, t) \, dv = -\frac{1}{\tau} \int_V \rho(r, t) \, dv + D \int_V \nabla^2 \rho(r, t) \, dv. \tag{77}$$

By the application of Green's theorem, the second integral on the right can be related to the integral of a current J through the bounding surface

$$D \int_V \nabla^2 \rho(r, t) \, dv = \oint_S \bar{J} \cdot \hat{n} \, ds. \tag{78}$$

If the minority carriers are confined to the active volume, the surface integral is zero. A confinement structure is one in which the surface integral of Eq. (78) vanishes.

If one defines the total number of minority carriers in the active volume by $\rho_t(t)$, then

$$\rho_t(t) = \int_V \rho(r, t) \, dV. \tag{79}$$

2. MINORITY-CARRIER LIFETIME IN III–V SEMICONDUCTORS

Combining Eqs. (77) and (79), one can easily calculate the decay of the total minority-carrier population in a confinement structure:

$$\frac{\partial}{\partial t} \rho_t = -\frac{1}{\tau} \rho_t, \qquad \rho_t = \rho_0 \exp(-t/\tau). \tag{80}$$

Here ρ_0 is the integral of Eq. (79) at $t = 0$. Combining Eq. (80) with Eq. (75), one finds

$$I_{\text{PL}}(t) = \frac{1}{\tau_R} \int_V \rho(t) \, dV = \frac{\rho_0 \exp(-t/\tau)}{\tau_R}. \tag{81}$$

Hence, one sees that a confinement structure produces a PL emission decay time equal to the minority-carrier lifetime.

Integrating $I_{\text{PL}}(t)$ in Eq. (81) from $t = 0$ to infinity produces the well-known result that was stated in Eq. (71):

$$I_{PL}(\infty) = \frac{1}{\tau_R} \int_0^\infty \rho_0 \exp(-t/\tau) \, dt = \rho_0 \frac{\tau}{\tau_R}. \tag{82}$$

Or:

$$\eta = \frac{I_{PL}(\infty)}{\rho_0} = \frac{\tau_{nR}}{\tau_{nR} + \tau_R}. \tag{83}$$

Here, $I_{\text{PL}}(\infty)$ is the total number of emitted photons and τ_{nR} is the nonradiative lifetime. The ratio $I_{\text{PL}}(\infty)/\rho_0$ is the total internal quantum efficiency (emitted photons/absorbed photons). Equation (83) is the basis of the experimental technique described in Section III.

A more detailed analysis of the PL lifetime of the device under test requires a solution of Eq. (76) with the appropriate boundary conditions. A general solution can be found for nonconfinement structures. The PL decay is a function of the minority-carrier diffusivity and lifetime with the appropriate boundary conditions to describe all interfaces. For confinement layer devices, one may find that all interfaces can be described by a single recombination velocity S. By contrast, a reverse-biased p–n junction is described by an infinite recombination velocity with $\rho = 0$ at the edge of the space charge region.

Bulk recombination is described by a single minority-carrier lifetime τ where

$$\frac{1}{\tau} = \frac{1}{\tau_R} + \sum_i \frac{1}{\tau_i}. \tag{84}$$

Equation (84) is an extension of Eq. (36) and includes the addition of radiative recombination. Each τ_i is the individual nonradiative recombination lifetime related to the ith mechanism. The inverse lifetime relationship is correct in the case of nondegenerate semiconductors and low-injection. For degenerate semiconductors, the addition of recombination probabilities becomes more complicated (Landsberg, 1991).

The majority of diagnostic devices are epitaxial thin films for which the film cross-sectional dimensions are much larger than the thickness. The three-dimensional diffusion equation can be replaced by the one-dimensional diffusion equation. Here one solves for $\rho(x, t)$, where x is the dimension perpendicular to the plane of the film. The one-dimensional time-dependent diffusion equation is written as

$$\frac{d\rho(x, t)}{dt} = D\frac{d^2\rho(x, t)}{dx^2} - \frac{\rho(x, t)}{\tau}. \tag{85}$$

The next section will discuss the solution of Eq. (76) for the isotype double heterostructure.

13. Transient Analysis of the Isotype Double Heterostructure

The isotope DH has been widely used as a diagnostic structure for time-resolved PL measurements. As will be shown, by fabricating and measuring two devices that are identical except for the active layer thicknesses, one can determine the bulk minority-carrier lifetime and the interface recombination velocity (S). As a confinement layer of the same composition is often a component of the actual working device, a measurement of the interfacial S in the diagnostic device is very valuable. A schematic of a generic isotype DH structure is shown in Fig. 6. This structure has an active layer sandwiched between two confinement or window layers. The device structure is characterized by the bandgaps $E_w/E_g/E_w$, where $E_w > E_g$. In the device, injected minority carriers are confined to the active region by deflection at the heterointerface, because of a barrier presented by the bandgap discontinuities ΔE_c and ΔE_v. A number of combinations of III-V materials may be used to make DH structures for the various binary, ternary, and quaternary combinations. Most work described in the literature describes lattice-matched confinement layers. Lattice matching greatly reduces the dislocation densities at the interface and in the active layer. This requirement greatly limits the choice of confinement layer composition for binary and ternary alloys.

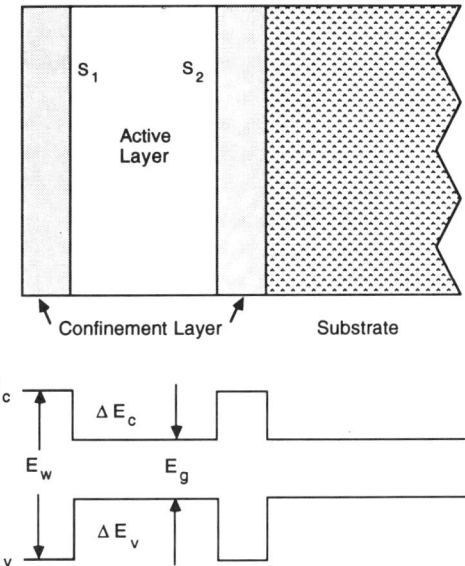

FIG. 6. Schematic of an isotype double heterostructure with an active layer bounded by two confinement layers. A simplified band diagram is also shown.

The exception to this exists in the GaAs-$Al_xGa_{1-x}As$ series, for which the whole compositional range is well matched. The lattice-constant matching results from the nearly equal ionic radii of the Al and Ga ions.

A standard structure for confining $Al_xGa_{1-x}As$ active layers has the composition $Al_yGa_{1-y}As/Al_xGa_{1-x}As/Al_yGa_{1-y}As$, where $y > x$. Diagnostic devices with a GaAs active layer have the composition of $Al_yGa_{1-y}As/GaAs/Al_yGa_{1-y}As$, where y varies from 0.25 to 0.90. The choice of alloy composition is determined by the grower's preference, and the lower aluminum values ($x > 0.30$) adequately confine minority carriers. The higher aluminum compositions have larger bandgap discontinuities, but have the adverse effect of greater susceptibility to atmospheric gases such as oxygen and water vapor. Crystal growers often deposit a moisture-resistant film, such as silicon nitride or a thin layer of GaAs, on the exposed $Al_yGa_{1-y}As$ layer to prevent degradation by atmospheric gas adsorption.

For TRPL measurements, a pulsed, monochromatic light source or laser generates electron–hole pairs in the active layer. Optical absorption in the window layer is not usually a limitation because the layer thickness is usually kept to several hundred angstroms. On the other hand, the window thickness must be large enough to prevent minority-carrier tunneling out of the active layer. The preferable wavelength of excitation is chosen so that it is absorbed by the active layer but transmitted by the window layer.

a. Analysis of TRPL in DH structures

The solution of the time-dependent diffusion Eq. (85) usually starts with the following substitution:

$$\rho(x, t) \equiv U(x, t) \exp(-t/\tau). \tag{86}$$

This solution implies that there is a unique, bulk minority-carrier lifetime τ. On substitution into Eq. (85), the exponential term $\exp(-t/\tau)$ cancels, and the differential equation involving $\rho(x, t)$ is transformed into the well-known equation characteristic of heat flow:

$$\frac{\partial U(x, t)}{\partial t} = D \frac{\partial^2 U(x, t)}{\partial x^2}. \tag{87}$$

A number of researchers have used the time-dependent diffusion equation to analyze the transient PL decay in DH structures. The solutions to this boundary value equation subject for a variety of geometries are well documented in the literature on heat flow (Crank, 1975; Carslaw and Jaeger, 1978). The mathematics of heat flow has been carried over to the minority-carrier diffusion problem by several authors (Boulou and Bois, 1977; 't Hooft and van Opdorp, 1983; Ahrenkiel and Dunlavy, 1989). Boulou and Bois (1977) were among the first workers to calculate solutions to this problem and used the heat-flow mathematical techniques to analyze cathodoluminescence decay. Early calculations by van Opdorp and co-workers (1977) used the time-dependent diffusion equation to calculate minority-carrier decay in structures with various geometries. Ioannou and Gledhill (1984) calculated solutions for a finely focused laser beam as the excitation source. More recent formulations of the problem were given by 't Hooft and van Opdorp (1986) and Ahrenkiel and Dunlavy (1989).

For symmetric structures, such as the double heterostructure, one may use the formalism developed by Carslaw and Jaeger (1978). The mathematics was developed for heat flow in thin, infinite planes of thickness d. A Fourier series is a solution of Eq. (87):

$$U_n(x, t) = A(\cos \alpha_n x + b_n \sin \alpha_n x) e^{-\beta_n t}, \tag{88}$$

where

$$U(x, t) = \sum_n U_n(x, t). \tag{89}$$

Here, the parameters α_n are determined by the boundary conditions.

Substitution of $U(x, t)$ into Eq. (87) requires term-by-term equality producing a set of relationships between the α_n and β_n:

$$\beta_n = D\alpha_n^2. \tag{90}$$

The boundary conditions at $x = 0$ and $x = d$ equate the diffusion currents to the recombination currents at the two interfaces. Here, S_1 and S_2 are the recombination velocities at the first and second interface, respectively, of Fig. 6. Therefore,

$$\left| qD \frac{\partial U_i}{\partial x} = qU_iS_1 \right|_{x=0} \tag{91}$$

and

$$\left| qD \frac{\partial U_i}{\partial x} = -qU_iS_2 \right|_{x=d}. \tag{92}$$

The boundary conditions of Eq. (91) produce the relationship $b_n = S_1/(\alpha_n D)$. The boundary condition at $x = d$ produces an equation containing the recombination velocities S_1 and S_2 and contains the unknown quantity α_n:

$$\tan \alpha_n d = \frac{\alpha_n d(S_1 + S_2)d/D}{(\alpha_n d)^2 - S_1 S_2 (d/D)^2}. \tag{93}$$

A series of discrete values α_n are solutions to Eq. (93); i.e., $n = 1, 2, 3$, etc. These values of α_n are then substituted back into Eqs. (88) and (89). It can be shown that the functions $U_n(x, t)$ form an orthogonal basis set for describing $\rho(x, t)$. These functions $U_n(x, t)$ are mathematically similar to the wave functions used in quantum mechanical problems. The set of values α_n are then eigenvalues of Eq. (93), and the latter is called an eigenvalue equation.

For the case of equal interface recombination velocities (or $S_1 = S_2 = S$), the equation simplifies to

$$\tan(\alpha_n d) = \frac{2(\alpha_n d)(Sd/D)}{(\alpha_n d)^2 - (Sd/D)^2}. \tag{94}$$

One can write this equation in more compact form by defining the dimensionless quantities $\theta_n = \alpha_n d$ and $z = Sd/D$. Then, in terms of these new parameters,

$$\tan(\theta_n) = \frac{2z\theta_n}{\theta_n^2 - z^2}. \tag{95}$$

The solutions, $U_n(x, t)$, of this transcendental equation are discussed by Carslaw and Jaeger (1978). The eigenvalues or roots, θ_n, of the equation are seen by plotting the two sides of Eq. (95) versus θ and finding the intersections. Figure 7 shows a plot of $\tan \theta$ (Curve B) and the function $2z\theta/(\theta^2 - z^2)$ (Curve A), with the points of intersection indicated by a

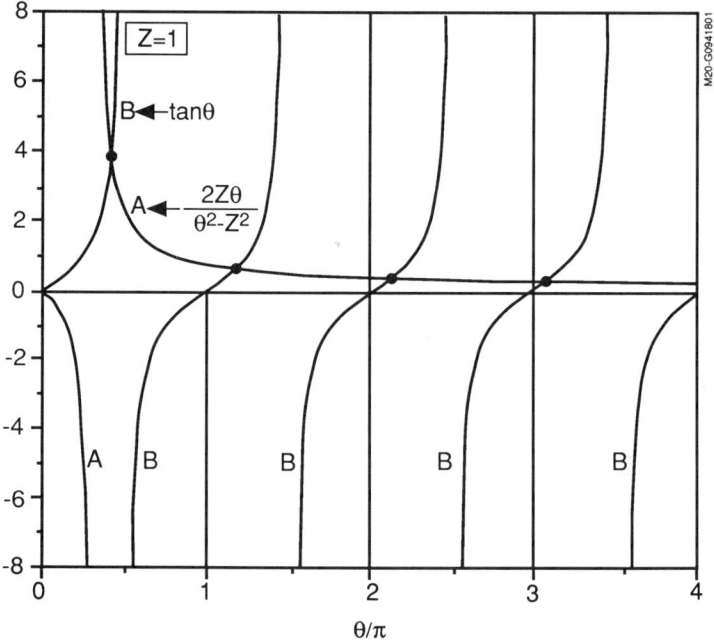

FIG. 7. A plot of the transcendental Eq. (95). The circles are the points of intersection that give the eigenvalues θ_n.

2. MINORITY-CARRIER LIFETIME IN III–V SEMICONDUCTORS 79

solid circle. A value of $z = 1$ was used here to illustrate the solutions for the lower range of z. The function $\tan(\theta)$ intersects with the hyperbolic function (the right side of Eq. (94)) once in each pair of quadrants, i.e., between 0 and π, π and 2π, and $n\pi$ and $(n+1)\pi$, etc. These intersections provide the unique solutions (θ_n) that are the eigenvalues of Eq. (95). For this example of $z = 1$, the value of θ_1 is 0.30π. As can be seen from Fig. 7, the value of θ_1 is 0.30π. As can be seen from Fig. 7, the higher-order solutions $(n > 1)$ are $\theta_n \sim (n-1)\pi$, where n is the index number. In the limit of S (or z) $= 0$, the eigenvalues are, for $n = 1, 2, 3$, etc.,

$$\theta_n = \alpha_n d = 0, \pi, 2\pi, \ldots,$$
$$\beta_n = 0, D\pi^2/d^2, 4D\pi^2/d^2, \ldots \quad (96)$$

The optical excitation produces an initial excess minority-carrier density given by Beer's law. If one assumes that the duration (FWHM) of the laser pulse is instantaneous compared to diffusion processes, the generation rate can be represented by a delta function $\delta(t)$. Therefore,

$$\rho(x, 0) = I_0 \alpha e^{-\alpha x} \delta(t), \quad (97)$$

where α is the absorption coefficient of the incident monochromatic light at the laser wavelength. Also, I_0 is the incident light intensity in units of photons per square centimeter. The dye laser described in the Section III has a pulse width of about 10 ps FWHM, and for most measurements the delta function approximation is adequate.

By substitution into Eqs. (88) and (97), one finds A_n to be

$$A_n = I_0 \frac{2\alpha_n^2}{\left[\alpha_n^2 + \left(\frac{S}{D}\right)^2\right]d + 2\frac{S}{D}} \frac{\alpha\left(\alpha + \frac{S}{D}\right)}{[\alpha_n^2 + \alpha^2]}. \quad (98)$$

Finally, one finds $\rho(x, t)$ equal to

$$\rho(x, t) = e^{-t/\tau} \sum_{n=1}^{\infty} e^{-\beta_n t} A_n \left(\cos \alpha_n x + \frac{S}{\alpha_n D} \sin \alpha_n x\right). \quad (99)$$

In low injection, $I_{\mathrm{PL}}(t)$ is the integral of $\rho(x, t)/\tau_R$ over the active volume. One integrates Eq. (99) from $x = 0$ to $x = d$.

$$I_{PL}(t) = \frac{1}{\tau_R} \sum_{n=1}^{\infty} A_n C_n \exp(-t/\tau_n). \tag{100}$$

Here, τ_n is the lifetime of the nth term, which is given by

$$\frac{1}{\tau_n} = \frac{1}{\tau} + \beta_n = \frac{1}{\tau} + D\alpha_n^2. \tag{101}$$

Also, C_n comes from the integration of Eq. (99) over the active region that has a thickness d:

$$C_n = 1/\alpha_n \left[\sin \alpha_n d + \frac{S}{\alpha_n D} (1 - \cos \alpha_n d) \right]. \tag{102}$$

The function $I_{PL}(t)$ is the sum of a series of exponential decays of individual modes. Each mode has a specific time constant $\beta_n + 1/\tau$ and is weighted by a coefficient $A_n \cdot C_n$. Thus, the general solution for the total decay of $I_{PL}(t)$ is nonexponential, and only an instantaneous lifetime can be defined. The coefficients β_n increase with n and therefore $\beta_1 < \beta_2 < \beta_3$, etc., as will be shown.

b. *Low Recombination Velocity: $z < 1$*

For $S = 0$, Eq. (96) shows the following difference between successive values of β_n:

$$\beta_{n+1} - \beta_n = (2n - 1) \frac{D\pi^2}{d^2}. \tag{103}$$

The diffusion transit time τ_D is defined as

$$\tau_D = \frac{d^2}{D\pi^2}. \tag{104}$$

Thus,

$$\beta_{n+1} - \beta_n = \frac{(2n - 1)}{\tau_D}. \tag{105}$$

At $z = 0$, the values of β are $\beta_1 = 0$, $\beta_2 = 1/\tau_D$, $\beta_3 = 4/\tau_D$, $\beta_4 = 9/\tau_D$, etc. The term with the smallest value of β_n dominates at large values of time. The higher modes decay very rapidly as β_n increases rapidly with the index n. For nonzero but small values of z, the first ($n = 1$) mode dominates at times longer than τ_D. At longer times (i.e., $t > \tau_D$), $I_{PL}(t)$ can be approximated by a single exponential decay process:

$$I_{PL}(t) \simeq \frac{1}{\tau_R} A_1 C_1 e^{-\beta_1 t} e^{-t/\tau}. \tag{106}$$

The long-term lifetime is clearly $1/(\beta_1 + 1/\tau)$. If $z = 0$, then $\beta_1 = 0$ and the excess density decays with the bulk lifetime τ. The term β_1 is related to the surface/interface recombination rate, and the surface lifetime is defined as $1/\beta_1$:

$$\frac{1}{\tau_s} \equiv \beta_1 = D \frac{\theta_1^2}{d^2}. \tag{107}$$

The range of values of z for $Al_xGa_{1-x}As/GaAs$ DHs is less than unity for reasonably good interfaces. Equation (94) for θ may be solved as a function of z with a nonlinear algorithm such as the secant procedure (Press et al., 1988, p. 248). As shown earlier, β varies as α^2 and therefore varies with θ^2. One may write the surface recombination rate $1/\tau_s$ (Eq. (107)) in a dimensionless form as

$$\frac{d^2}{D\tau_s} = \theta_1^2. \tag{108}$$

Of course, τ_s and θ_1 are both functions of the dimensionless parameter z. Figure 8, Curve A, comes from the exact, numerical solution to Eq. (95). Curve A plots the exact numerical solution for $d^2/D\tau_s$ versus $z(Sd/D)$ over the range $z = 0$ to 10. One sees from the plot that $d^2/D\tau_s$ (or θ_1^2) is approximately equal to $2z$ for $z < 1.0$. In this range of z, β_1 varies linearly with z, as will be shown later. For small values of z, an approximate solution to the eigenvalue equation is found by assuming $\tan \theta_1 \sim \theta_1$. Then it is easy to show that

$$\theta_1^2 = \frac{d^2}{D\tau_s} \cong 2z. \tag{109}$$

In this range of z, one may easily prove that

$$\beta_1 \cong \frac{2S}{d}. \tag{110}$$

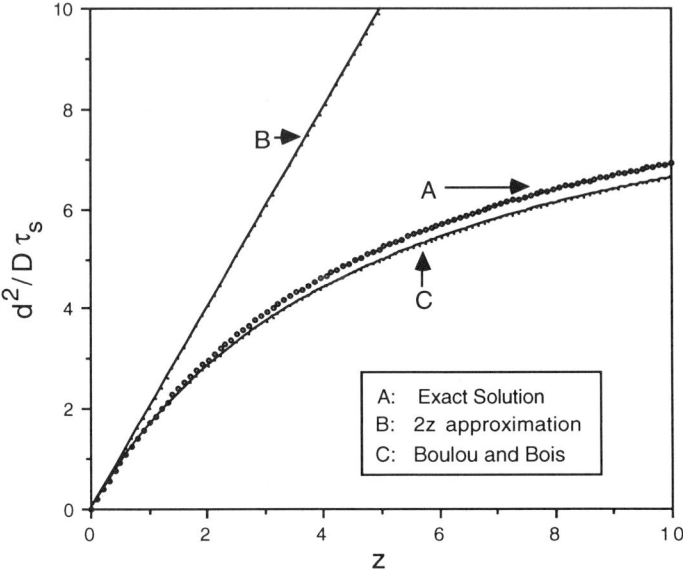

FIG. 8. Curve A is a plot of $\theta_1^2/2$ versus Sd/D (or z). Curve B is the linear approximation that $\theta_1^2/2$ equals Sd/D.

Figure 8, Curve B, is a plot of the linear approximation of Eq. (109). This approximation becomes very inaccurate for $z > 1$.

Thus, the PL lifetime for the DH structure in the range $z < 1$ is given by

$$\frac{1}{\tau_{PL}} = \frac{1}{\tau} + \frac{2S}{d}, \quad (111)$$

where S in the recombination velocity of the two interfaces.

c. *High Recombination Velocity: $z > 1$*

At large values of z, θ approaches the value π, will be shown here. For very large values of z ($z \gg 1$), the hyperbolic function is a small, negative number for $\theta < z$, and therefore the intersection points are near the zeros of tan θ. The first solution θ_1 is slightly less than π, θ_2 is slightly less than 2π, etc. For z approaching infinity, the solutions of θ_n are the zeros of tan θ. These solutions are for $n = 1, 2, 3$, etc.:

$$\theta_n = \alpha_n d = \pi, 2\pi, 3\pi \ldots ,$$
$$\beta_n = D\pi^2/d^2, 4D\pi^2/d^2, \ldots \quad (112)$$

Thus, β_n may be written as n^2/τ_D, and the difference of successive terms goes as

$$\beta_{n+1} - \beta_n = \frac{(2n+1)}{\tau_D}. \tag{113}$$

The successive values of β increase as $\beta_1 = 3/\tau_D$, $\beta_2 = 5/\tau_D$, $\beta_3 = 7/\tau_D$, etc., and the higher modes do not decay as quickly as for the low-z case. Longer times are required for the first mode (i.e., $n = 1$ term) to become dominant.

Boulou and Bois (1977) suggested an approximate solution to Eq. (95) that is accurate to within 5% ('t Hooft, 1992) over a wide range of z. This formula for the surface lifetime is

$$\tau_s = \frac{d^2}{\pi^2 D} + \frac{d}{2S}. \tag{114}$$

By rearrangement, one may define the dimensionless surface recombination parameter $d^2/(D\tau_s)$ as a function of z:

$$\frac{d^2}{D\tau_s} = \frac{2\pi^2 z}{\pi^2 + 2z}. \tag{115}$$

Equation (115) is plotted in Fig. 8, Curve C and good agreement with the exact solution (Curve A) is very obvious.

The expression for $I_{PL}(t)$ at long times ($t > \tau_D$) is now given by

$$I_{PL}(t) \cong \frac{1}{\tau_R} A_1 C_1 \exp\left(\frac{-t}{\frac{d^2}{\pi^2 D} + \frac{d}{2S}}\right) \exp(-t/\tau). \tag{116}$$

For the case of $z \gg 1$, the minority-carrier lifetime is an exponential function of the minority-carrier diffusivity. The PL lifetime is limited by the diffusion transit time to the interfaces. From Eq. (116) one can write the total PL lifetime as

$$\frac{1}{\tau_{PL}} = \frac{1}{\tau} + \frac{D\pi^2}{d^2} = \frac{1}{\tau} + \frac{1}{\tau_D} \quad (z \gg 1). \tag{117}$$

For intermediate values of z, Eq. (93) may be solved numerically to obtain the eigenvalue θ_n. The exact solutions for the θ_n are obtained using standard numerical procedures for the solution of nonlinear or transcendental equations. Alternatively, the Boulou and Bois approximation (Eq. (114)) may be used.

By making two structures that are identically processed except for the active-layer thickness d, one can obtain unique values of bulk lifetime and S. This information is, of course, very useful for process monitoring and development. It is also vital information in device-modeling calculations.

14. Self-Absorption and Photon Recycling

The distinction between self-absorption and photon recycling is not always made clear, and the two terms are often used interchangeably. Self-absorption is a simple consequence of the overlap of the absorption spectrum $\alpha(E)$ with the internal emission function $S(E)$, as seen in Fig. 2. The van Roosbroeck–Shockley relationship (1954) indicates that these two spectra are intimately related and must overlap. For direct-bandgap semiconductors, the absorption rises steeply to above $1 \times 10^4 \text{ cm}^{-1}$ for $h\nu$ greater than E_g. Therefore, the self-absorption effect is very strong in direct semiconductors. The term photon recycling is used to describe the generation of a new electron–hole pair by the self-absorbed photon. The secondary photon may escape from the material, or it may be again self-absorbed, producing a new electron–hole pair.

The self-absorption and re-emission of radiative recombination were proposed by Dumke (1957) many years before significant experimental data were available. Moss (1957) calculated the effects of self-absorption on the emission spectra of InSb. He also proposed that the radiative decay time increases because of self-absorption and re-emission of photons. Kameda and Carr (1973) proposed that self-excited luminescence or photon recycling is an important effect in GaAs. Calculations by Stern and Woodall (1974) indicated that photon recycling lowered the threshold current density in GaAs DH diode lasers. Early measurements using TRPL in $Al_xGa_{1-x}As/GaAs$ DH devices showed that the lifetime increased with active layer thickness (Ettenberg and Kressel, 1976). In this work, the experimental lifetimes were five times larger than the calculated radiative lifetime in a 10 μm DH device. Calculations by Asbeck (1977) explained this anomaly in terms of the photon recycling effect. Calculations by Kuriyama and co-workers (1977) and Kamiya and co-workers (1979) used the photon recycling model to explain photoluminescence lifetimes that are larger than the radiative lifetime. Garbuzov and co-workers (1977, 1982) also calculated the photon recycling effect in AlAs/GaAs DHs. The latter calculation, and some experimental data, indicate that the recycling effect is greatly enhanced by substrate removal. Recent data, presented in Section VI, describe the increase in the recycling effect caused by substrate etching (Lush et al., 1992b). In all of these calculations, the radiative lifetime is replaced by an effective radiative lifetime. The effective lifetime can be written in terms of a photon recycling factor ϕ that multiplies the radiative lifetime:

$$\frac{1}{\tau_R} \to \frac{1}{\phi \tau_R}. \tag{118}$$

The factor ϕ is calculated for specific device structures. For the DH structure, the factor ϕ is a function of the active layer thickness d, the window layer composition, and the doping concentration. The low-injection photoluminescence lifetime of a DH structure can then be written as

$$\frac{1}{\tau_{PL}} = \frac{1}{\tau_{nR}} + \frac{1}{\phi \tau_R} + \frac{1}{\tau_S}. \tag{119}$$

Here τ_s, the surface lifetime, is determined from the techniques described earlier.

a. One-Dimensional Self-Absorption model

The integration of $\rho(x, t)$ (Eq. (99)) to give I_{PL} was carried out assuming that all of the internally generated photons are able to escape the active volume. That assumption is incorrect because self-absorption and total internal reflection prevent the escape of a large fraction of the photons. One can use a simple, one-dimensional model of self-absorption that modifies the quantity C_n of Eq. (102). The only photons that are included in the calculation are those propagated along the x-axis. This approximation is reasonably accurate for most experimental configurations in which the active photodetector area subtends a small solid angle relative to the sample. The self-absorption effect increases from the low-energy side to the high-energy side of the internal emission spectrum, which is given by Eq. (50). The mechanism of self-absorption in GaAs is very obvious from an examination of Fig. 2. Strong band-to-band absorption ($\alpha > 1 \times 10^4$ cm^{-1}) overlaps the internal or vRS spectra for which $h\nu > E_g$ or 1.42 eV. The external emission spectrum (Curve C) matches the VRS spectra in Fig. 2. The external emission spectrum is modified by the self-absorption effect as d increases. The distortion is minimal here because d is 1.0 μm and self-absorption effects are small.

The emission radiating from the differential volume bounded by x and $x + \Delta x$ and attenuated by self-absorption is given by

$$\Delta I_{PL} = \rho(x, t) \exp(-\beta x) \Delta x. \tag{120}$$

Here, $\beta(E)$ is obtained from the appropriate absorption curve. For undoped GaAs, one would use Curve A from Fig. 2. These are published absorption data for undoped GaAs by Casey and co-workers (1975). As $\beta\,(h\nu > E_g)$ is greater than $10^4\,\text{cm}^{-1}$, self-absorption is significant for GaAs films thicker than about 1.0 μm.

Data are shown in Fig. 9 that illustrate the strong self-absorption effects in GaAs. The emission spectra of two GaAs DH devices with very different active layer thicknesses are shown here (Ahrenkiel et al., 1992a). The active layer thicknesses are 0.25 μm (Curve A) and 10.0 μm (Curve B). The DH structures are $Al_{0.3}Ga_{0.7}As/GaAs/Al_{0.3}Ga_{0.7}As$ and

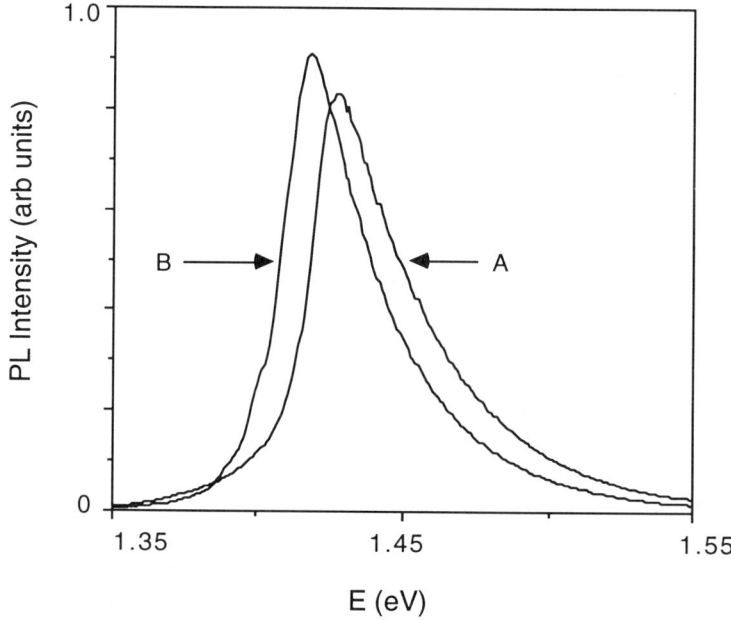

FIG. 9. The emission spectra of two GaAs DH devices as measured at NREL (Ahrenkiel et al., 1992a). The thickness of device A is 0.25 μm and that of device B is 10.0 μm.

are selenium-doped to 1.3×10^{17} cm^{-3}. A marked "red shift" of the emission band is obvious for the 10-μm DH device of Curve B relative to the 0.25-μm device of Curve A. This red shift is easily simulated on the computer by inserting Eq. (120) into Eq. (57) and integrating across the active region. Thus, the self-absorption effects are readily observable from conventional, steady-state spectroscopy.

The self-absorption effects are also seen in time-resolved photoluminescence spectroscopy. For the TRPL experiment, the monochromator may be tuned to various wavelengths in the emission spectrum. Each monitored wavelength has a corresponding photon energy, $h\nu$, for which the self-absorption coefficient β may be determined from Curve A, (Fig.2). After the choice of the measurement wavelength (or photon energy $h\nu$) to which the monochromator is tuned, the one-dimensional calculation is modified by the self-absorption effect as

$$C_n(h\nu) = \int_0^d \left(\cos \alpha_n x + \frac{S}{\alpha_n D} \sin \alpha_n x \right) \exp(-\beta(h\nu)x) \, dx. \quad (121)$$

The integration is somewhat laborious, but the final result is

$$C_n(h\nu) = \frac{\beta + S/D}{\beta^2 + \alpha_n^2} - \frac{e - \beta d}{\beta^2 + \alpha_n^2}$$

$$\times \left\{ (\beta + S/D) \cos(\alpha_n d) + \left(\frac{S\beta}{\alpha_n D} - \alpha_n \right) \sin(\alpha_n d) \right\}. \quad (122)$$

The β in Eq. (121) is equal to $\beta(h\nu)$ of Eq. (120).

Figure 10 shows the PL decay curves by inserting C_n from Eq. (122) into Eq. (100). The simulation describes an experiment in which the monochromator is tuned to various energies in the emission band as indicated. In these curves, the values of the semiconductor parameters are $D = 5$ cm^2/s, $S = 0$ cm/s, $\tau = 500$ ns, and $d = 10$ μm. The τ-value is typical of the PL lifetime measured for high-quality n-type DHs with N equal to about 1×10^{17} cm^{-3} and a 10-μm active layer. For curve A, the simulation assumes that $h\nu$ equals 1.38 eV, corresponding to a β-value of 12 cm^{-1} from the data for GaAs. For curves B and C, the photon energies $h\nu$ are 1.42 and 1.55 eV, corresponding to β-values of 0.577×10^4 and 1.35×10^4 cm^{-1}. A self-absorption transient is obvious from the calculations. This transient increases as the monochromator is tuned to higher-energy photons in the emission spectra.

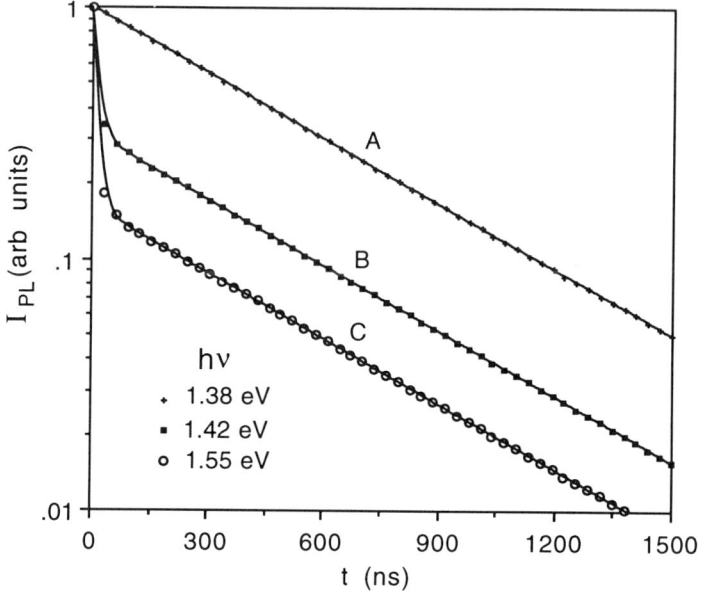

FIG. 10. The calculated PL decay curves using Eqs. (100) and (121) and illustrating the self-absorption transient. The model assumes that $S = 0$ cm/s, $\tau = 500$ ns, $D = 5$ cm/s, and $d = 10.0\,\mu$m. The self-absorption coefficients for curves A, B, and C are 12, 0.577×10^4, and 1.35×10^4 cm^{-1}, respectively.

The origin of the transient is the diffusion of excess carriers out of the initial Beer's law distribution. The function $\rho(x, t)$ is calculated at various times after the injection pulse. The initial distribution has a Beer's law profile, but the minority carriers diffuse into the active volume very rapidly. A "snapshot" of the function $\rho(x, t)$ is shown in Fig. 11 at various times after the pulsed generation of carriers. The excitation absorption coefficient chosen for the calculation was $\alpha = 5 \times 10^4$ cm^{-1}, corresponding to $\lambda \sim 600$ nm for GaAs. From Beer's law, one knows that at $t = 0$, the excess carriers lie in a thin layer at the front surface. The thickness of the layer is about $1/\alpha$ or $0.5\,\mu$m in this case. Photons generated from the surface layer escape the front surface with high probability. Curve A shows the $\rho(x, t)$ distribution at $t = 0.5$ ns, and diffusion has spread the distribution out to about $1.5\,\mu$m. Curve B is calculated with $t = 5$ ns, and Curve C is calculated with $t = 50$ ns. Curve C shows that the excess carrier density has become uniform by diffusion. The calculated diffusion transit time τ_D is about 20 ns from Eq. (104). This corresponds to the time between the excitation pulse and the nearly "flat band" condition. At times $t > \tau_D$, the ρ distribution remains "flat"

2. MINORITY-CARRIER LIFETIME IN III–V SEMICONDUCTORS

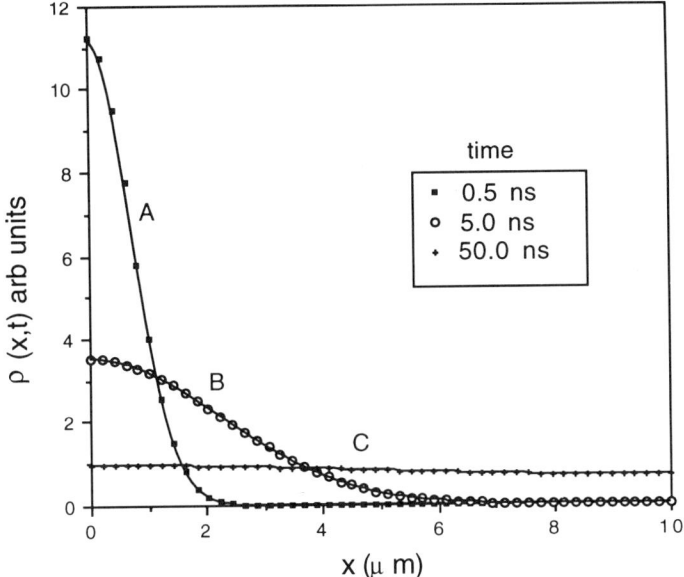

FIG. 11. Snapshots of $\rho(x, t)$ for the model used in Fig. 10. The times for curves A, B, and C are 0.5, 5.0, and 50.0 ns, respectively.

and decreases in magnitude because of recombination. At "flat band," only the low-energy photons ($h\nu < E_g$) are able to escape from the entire active volume. The only escaping higher-energy photons ($h\nu < E_g$) are those generated within $1/\alpha$ from the front surface when $t > \tau_D$. Consequently, the PL intensity drops precipitously at $t > \tau_D$ for higher-energy photons ($h\nu > E_g$).

TRPL data are shown in Fig. 12 for an n-type $Al_xGa_{1-x}As/GaAs/Al_xGa_{1-x}As$ DH structure with active layer doping of 1.3×10^{17} cm^{-3} and $d = 10$ μm. Three data sets were obtained by tuning the detecting monochromator across the PL emission band. These are shown with the monochromator adjusted to wavelengths corresponding to the same three photon energies 1.38 eV (Curve A), 1.42 eV (Curve B), and 1.55 eV (Curve C), as in the calculated curves of Fig. 10. The long-term slopes of all decay curves of Fig. 12 are identical and indicate a decay time of 465 ns. A diffusion transient is observed in curves B and C but is not present in curve A. This transient is observed for all monochromator settings from about the emission peak at 1.42 eV to the high-energy edge of the band at about 1.57 eV. The shapes of these TRPL curves agree quite well with the calculated curves of Fig. 10. The self-absorption effect is very obvious from measurements of this type.

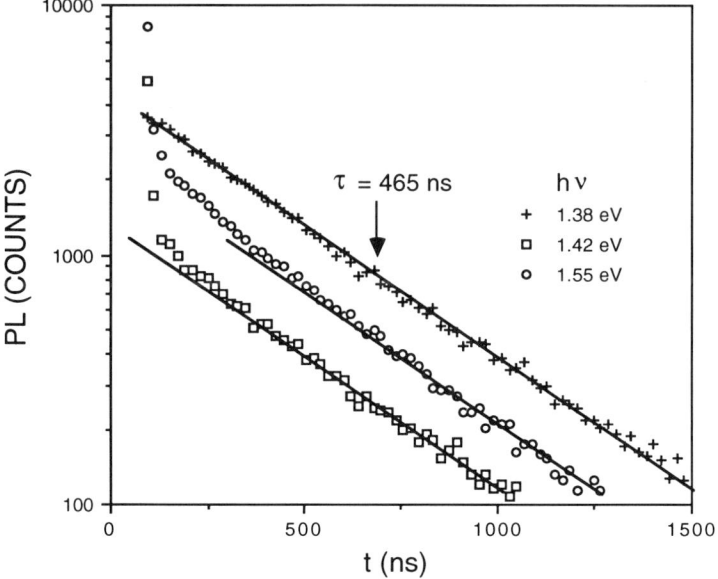

FIG. 12. TRPL data for a 10.0-μm DH doped n-type 1.3×10^{17} cm^{-3} as measured at NREL. The three data sets correspond to the photon energy to which the detection apparatus is tuned. The self-absorption transient is observed for $h\nu > 1.41$ eV.

b. Calculation of the Photon Recycling Factor

The photon recycling effect results in the generation of new minority carriers in the device active region. The estimation and calculation of the photon recycling factor is simple in principle but is numerically complicated. The time-dependent diffusion equation (Eq. (76)) must be modified to include this generation effect:

$$\frac{\partial \rho(r, t)}{\partial t} = D\nabla^2 \rho(r, t) - \frac{\rho(r, t)}{\tau} + G(\rho(r)). \qquad (123)$$

Here, $G(r)$ is the generation of electron–hole pairs by self-absorption. For the one-dimensional model that has been used for DHs, Eq. (123) becomes

$$\frac{d\rho(x, t)}{dt} = D\frac{d^2\rho(x, t)}{dx^2} - \frac{\rho(x, t)}{\tau} + G(\rho(x)). \qquad (124)$$

To find $G(\rho(x))$, one must calculate the probability of self-absorption

2. MINORITY-CARRIER LIFETIME IN III–V SEMICONDUCTORS

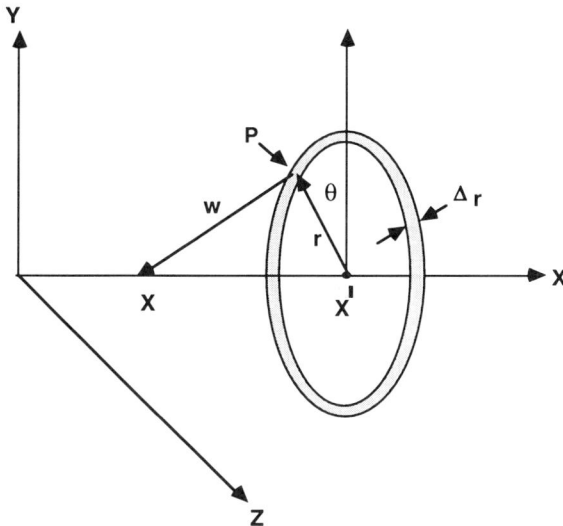

FIG. 13. The cylindrical coordinate system used for calculating the photon recycling factor.

for emission from each volume element $\rho(r)\Delta V$. The emission is usually assumed to be isotropic, and therefore this problem is inherently three-dimensional. For planar structures such as DHs, cylindrical coordinates are usually used. Integration over the polar angle θ reduces the problem to a one-dimensional problem involving the cylindrical axis x. The solution also involves an integrated average over the internal emission spectrum, which is convoluted with the absorption spectra $\alpha(h\nu)$. The emission spectra are given by the Shockley–van Roosbroeck relationship (Eq. (50)).

Figure 13 shows a cylindrical coordinate system, similar to that used by Kuriyama and co-workers (1977), for calculating the photon recycling factor. For consistency with the previous one-dimensional models, the x-coordinate is the cylindrical axis that is normal to the plane of the DH film. One assumes that the film is infinite in extent in the y–z plane. Also, the excess carrier density is uniform in the y–z plane. Then

$$\rho(r) = \rho(r, \theta, x) = \rho(x). \tag{125}$$

Consider the emission from the annular ring of minority-carrier charge at point x' and radius r. The probability of a photon $h\nu$ at point $P(x', r, \theta)$ reaching the cylindrical axis at point x is

$$P = \exp(-w\alpha(h\nu)) = \exp(-\alpha(h\nu)\sqrt{(x'-x)^2 + r^2}). \tag{126}$$

As the radiation at P is isotropically emitted into a full solid angle, the intensity is reduced by $1/4\pi w^2$ at point x. The number of monochromatic photons of energy $h\nu$ emitted from the ring of width Δr and thickness Δx is

$$\Delta G(x, x') = \frac{\rho(x')}{2\tau_R} S(h\nu) \frac{r \exp\{-\alpha(h\nu)\sqrt{(x'-x)^2 + r^2}\}}{(x'-x)^2 + r^2} \Delta r \, \Delta x. \quad (127)$$

The number of monochromatic photons emitted from the semi-infinite plane at x' and thickness Δx is found by integrating r from 0 to infinity. The total photon intensity at x from the plane at x' is obtained by integrating over the normalized emission spectra $S(h\nu)$:

$$G(x, x') = \frac{\rho(x')}{2\tau_R} \int_0^\infty d(h\nu) \int_0^\infty dr \, S(h\nu) \frac{r \exp(-\alpha(h\nu)\sqrt{(x'-x)^2 + r^2})}{(x'-x)^2 + r^2} \Delta x. \quad (128)$$

The total $G(x, x')$ is now found by integrating Eq. (128) over the active volume or from 0 to d:

$$G(d) = \int_0^d dx \int_0^x dx' G(x', x). \quad (129)$$

This derivation has included the primary self-absorption effects but has not included the effects of total internal reflection at the interface. When the external emission or PL probability is calculated, interfacial reflection must be included. The schematic of Fig. 14 shows some possible optical and transport events that are involved in external photon emission. For example, a minority carrier is generated at point A_0 and diffuses to point A_1. There, electron–hole recombination occurs, producing a photon of energy $h\nu$. This photon propagates through the active layer at an angle θ relative to the normal. The probability P_{ext} of the photon reaching the left-hand $Al_xGa_{1-x}As/GaAs$ interface is simply

$$P_{ext} = \exp\left(-\frac{\alpha(h\nu)x}{\cos \theta}\right). \quad (130)$$

As before, x is the distance of the generation event from the left-hand interface and θ is the angle of propagation with respect to the normal.

2. MINORITY-CARRIER LIFETIME IN III–V SEMICONDUCTORS 93

One can define an average absorption length L for photon with energy $h\nu$:

$$L_\alpha = 1/\alpha(h\nu). \tag{131}$$

The index of refraction step (Δn) at the $Al_xGa_{1-x}As/GaAs$ interface is $n_1(Al_xGa_{1-x}As) - n_2(GaAs)$. The critical angle θ_c for total internal reflection is

$$\theta_c = \sin^{-1}\left(\frac{n_1}{n_2}\right). \tag{132}$$

The reflection coefficients for transverse electrical (TE) and transverse magnetic (TM) modes have been taken into account by numerous authors. The standard approximation for the reflection coefficient $R(\theta)$, after averaging over polarization effects, is:

$$R(\theta) = 0, \quad \theta < \theta_c, \quad \text{and} \quad R(\theta) = 1, \quad \theta > \theta_c. \tag{133}$$

Thus, photons can only escape to the window layer when emitted within the cylindrical cone $\theta < \theta_c$.

Other possible events for photon self-absorption and reflection in a planar DH are shown in Fig. 14. For the photon originating at point A_1, the propagation angle θ is less than θ_c. The probability for emission into the window layer is

$$P_{ext} = \exp(-x/L_\alpha)(1 - R(\theta)). \tag{134}$$

If the distance $x < L_\alpha$ and as $\theta < \theta_c$, the photon has a high probability of escaping from the active region.

Other possible paths for the energy transport are indicated in the schematic. A more complex mode of energy transport is indicated by the incident photon absorbed at point B_0. The photogenerated minority carrier diffuses to point B_1 and recombines, producing a second photon. The second photon is self-absorbed at B_2, producing a new electron–hole pair. The minority carrier diffuses to B_3 and recombines, generating a new photon. This third photon is generated at an angle θ that is greater than the critical angle θ_c. A reflection at the interface directs the photon back to B_4, where it is absorbed. The minority carrier diffuses to B_5, where it recombines, producing the third photon in this particular cycle. Finally, the photon generated at B_5 is able to escape from the active layer because θ is less than θ_c. Although an energy transport process such as

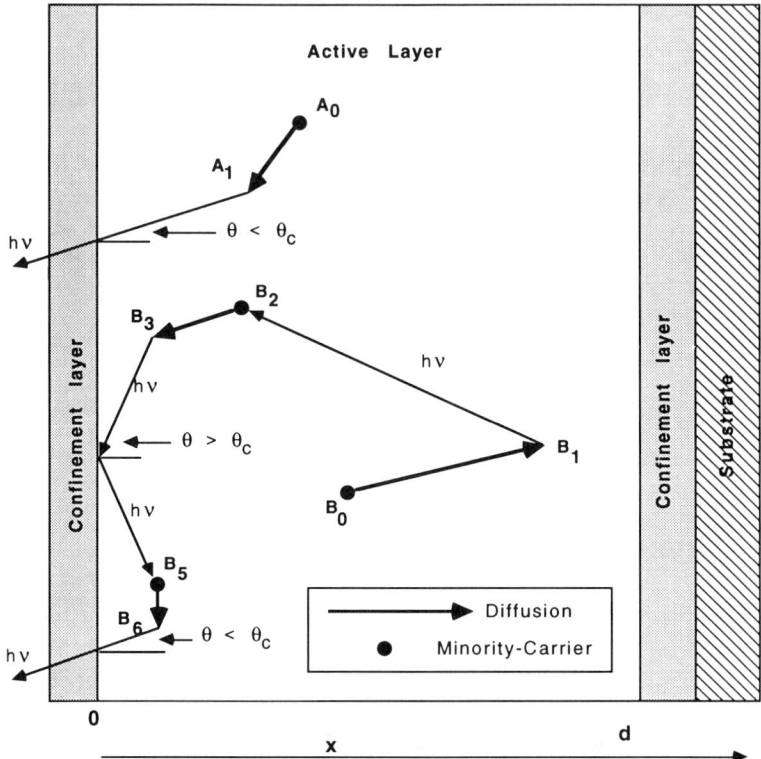

FIG. 14. Possible paths of energy transfer in a DH structure. Energy may be transferred by minority-carrier diffusion of photon emission ($h\nu$).

this are realistic, the mathematics are complex. This problem is treated by statistical methods using simulation techniques on a large computer. A detailed description of the simulation technique has been recently given (Miller, 1989).

A variety of approximations have been used to calculate $G(x)$. The calculations of Asbeck (1977) and Nelson and Sobers (1978a) used a "flat band" approximation, i.e., $\rho(x)$ is a constant. Kuriyama and co-workers (1977) used a technique that allows an arbitrary carrier distribution to be used. Recent work by Bensaid and co-workers (1989) treats the problem using the functions $U_n(x)$ as a complete set of basis functions. Secular equations are developed using the matrix elements of $G(x)$. The matrix $(U_n |G| U_m)$ is diagonalized to produce solutions of Eq. (123).

After calculating $G(d)$ and including the reflection effects, one defines a quantity $F(d, \theta_{cr})$ that is the fraction of emitted photons generating electron–hole pairs by virtue of self-absorption. The lifetime can then be written as

$$\frac{1}{\tau_{PL}} = \frac{1}{\tau_{nR}} + \frac{2S}{d} + \frac{1}{\tau_R} - \frac{F(d, \theta_{cr})}{\tau_R}. \tag{135}$$

Comparing Eq. (135) with Eq. (119), it is obvious that F is related to ϕ by the relationship

$$\phi(d) = \frac{1}{1 - F(d, \theta_{cr})}. \tag{136}$$

Thus, as $\phi(d)$ becomes very large, F approaches unity.

The calculation of $\phi(d)$ is very sensitive to the device structure. The calculations of Asbeck (1977) were for DH devices with $Al_xGa_{1-x}As$ confinement layers of two compositions, $x = 0.25$ and $x = 0.50$. Asbeck also added a top layer of epitaxial GaAs on the $Al_xGa_{1-x}As$ window layer in his calculation. This layer is often used to protect $Al_xGa_{1-x}As$ from atmospheric gas contamination. Garbuzov and co-workers (1977) calculated $\phi(d)$ for a "free-standing" DH, i.e., one for which the substrate had been etched away. These calculations indicated a large increase in $\phi(d)$ as light trapping occurs at both interfaces in this case.

Asbeck's calculations of $\phi(d)$ for DH structures with different p- and n-type doping levels and $Al_xGa_{1-x}As$ window layer compositions are shown in Fig. 15. The recycling factor $\phi(d)$, calculated for the AlGaAs/GaAs DH structure, increases with the active layer thickness. The calculations of Fig. 15 also show that $\phi(d)$ depends on the doping type and doping density. These calculations indicate that ϕ is greater than 10 for undoped GaAs active layers with a 10 μm thickness. A later section shows experimental results and compares the measurements with Asbeck's calculations.

As the doping level increases, the absorption edge becomes less "steep" because of the perturbation effects of ionized impurities on the band structure (Casey et al., 1975). At values of d greater than 10 μm, $\phi(d)$ decreases rapidly for both n and p concentrations greater than 1×10^{18} cm^{-3}. At small values of d (less than 0.5 μm), $\phi(d)$ is 1.5 to 2.0 and has no doping dependence. The photon recycling factor, in this range of d, is dominated by total internal reflection. The internal reflection mechanism is verified by calculations of the lower graph of the figure. Here, θ_{cr} is a parameter in the plot of $\phi(d)$ versus d. As θ_{cr} increases from 66° to 72°, $\phi(d)$ decreases from about 1.9 to 1.6. The total internal

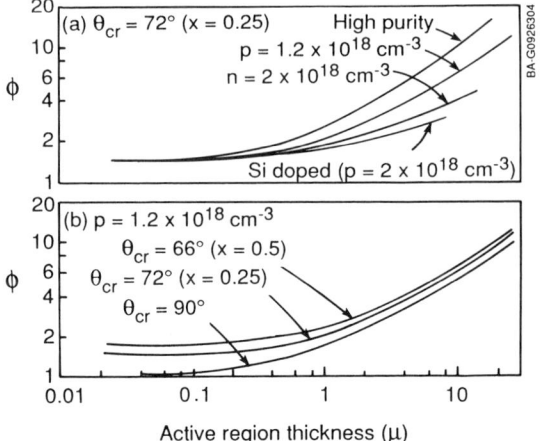

FIG. 15. Calculation of the photon recycling factor ϕ as a function of active layer thickness, doping type and concentration, and critical angle (Asbeck, 1977, reprinted with permission).

reflection disappears when the refractive indices on both sides of the boundary become identical.

Fits to the Asbeck calculations for *n*-type DHs indicate that $\phi(d)$ is approximately linear for $d < 3\,\mu$m. My fit of the calculations gave the approximate values

$$\phi(d) = 1.40 + 1.1d\ (\mu\text{m}) \quad \text{(undoped)},$$
$$\phi(d) = 1.43 + 0.4d\ (\mu\text{m}) \quad \text{(2E18)}. \tag{137}$$

As the effective radiative lifetime ($\phi\tau_R$) increases, the PL lifetime may become dominated by the nonradiative recombination processes. Using Eq. (119), the PL lifetime approaches a limit as $\phi\tau_R$ becomes very large:

$$\frac{1}{\tau_{PL}} = \frac{1}{\tau_{nR}} + \frac{2S}{d}. \tag{138}$$

This is often the limiting case for undoped GaAs, as τ_R is $1/BN$ and N can be as low as $10^{15}\,\text{cm}^{-3}$ by epitaxial growth. When d is made sufficiently large, $\tau_{PL} \sim \tau_{nR}$ as bulk SRH recombination dominates the measurement. For smaller values of d, the surface lifetime may dominate the measurement and $\tau_{PL} \sim d/2S$. The fabrication of undoped DH structures is a common method of testing for SRH recombination centers that are introduced during growth. Measurement of the photon recycling effect is discussed further in Section V.

15. THE PL LIFETIME IN JUNCTION DEVICES

The isotype DH is a powerful diagnostic tool that avoids many of the processing steps that are involved in commercial devices. These include contact formation, mesa etching, and other processing steps that might affect the minority-carrier lifetime. The homojunction or heterojunction is a basic structure in devices ranging from diode lasers to photovoltaic cells. Direct measurement of the minority-carrier lifetime in a junction device is often desirable for analysis of the device performance. The interpretation of the data is more difficult for nonconfinement structures. However, models of junction PL decay have been made using the mathematical techniques described earlier here.

Figure 16 shows a schematic diagram of a device structure that will be used for this calculation. The figure shows an emitter and a base of a junction device. For GaAs devices, the emitter is usually passivated by epitaxial $Al_xGa_{1-x}As$. The junction is a component of a photovoltaic cell, a diode laser, or a transistor. Several papers have developed solutions for the PL decay from the emitter of a p–n junction such as that of Fig. 16. One can develop a solution of the PL decay in this structure bounded by a window layer with an interface recombination velocity S_w and by a p–n junction. Recent work by Ehrhardt and co-workers (1991) has provided an elegant mathematical solution for PL decay in p–n junctions. This solution applies a Fourier transform to the one-dimensional, time-dependent diffusion equation. A generation function of the form

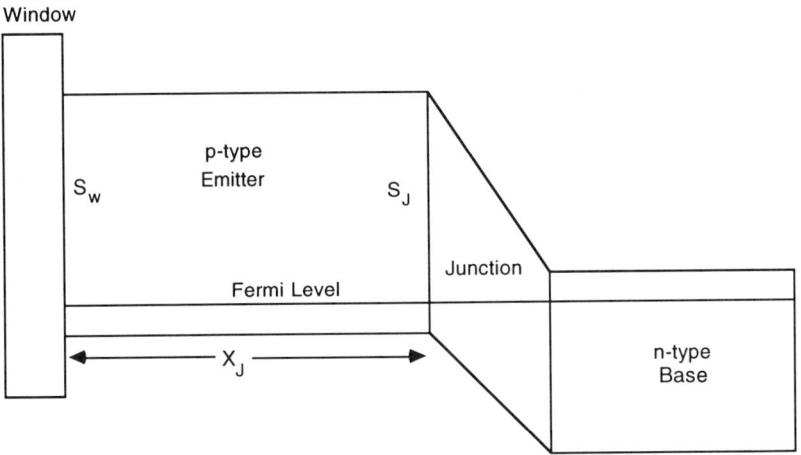

FIG. 16. Schematic of a confined or passivated p–n junction used for the PL decay calculations of Section IV.

$f(t) \propto \exp(-\alpha x)$ is used to provide a solution PL(t) that, in general, is mathematically quite complex. The Fourier transform converts the partial differential diffusion equation into an ordinary differential equation of the spatial variable x. Parameters are the window recombination velocity S_w, the junction recombination velocity S_J, and the junction depth X_J. The value of S_J depends on the bias voltage across the junction. Under reverse bias, S_J approaches infinity. The boundary condition then is that $\rho(X_J)$ is 0. Under forward bias, $\rho(X_J)$ is an exponential function of the voltage, and the solutions are more mathematically complicated. The solutions for $f(t) = \delta(t)$ and S_J approaching infinity were developed by Ehrhardt et al. (1991). The eigenvalue equation is identical to Eq. (93) for the DH device with unequal recombination velocities at the two interfaces:

$$\tan \alpha_n X_J = \frac{\alpha_n X_J (S_w + S_J) X_J / D}{(\alpha_n X_J)^2 - S_w S_J (X_J/D)^2}. \quad (139)$$

In the limit of S_J approaching infinity, the equation becomes

$$\tan \alpha_n X_J = -\frac{\alpha_n X_J}{S_w X_J / D}. \quad (140)$$

Letting $\alpha_n X_J$ equal θ_n and $S_w X_J / D$ equal to z, this equation becomes

$$\tan \theta_n = -\frac{\theta_n}{2}. \quad (141)$$

Ehrhardt et al., evaluate these solutions for specific device parameters. The time decay is written as a sum of modes with a specific lifetime for each mode. The mode lifetime is identical to Eq. (101):

$$\frac{1}{\tau_n} = \frac{1}{\tau} + \beta_n = \frac{1}{\tau} + D\alpha_n^2. \quad (142)$$

Here, of course, the α_n values are closer to those of Eq. (114) because of the infinite value of S_J. For example, the first root of Eq. (141) lies in the range $\pi/2 < \theta_n < \pi$; the second root lies in the range $3\pi/2 < \theta_n < 2\pi$, etc.

Figure 17 shows the spatial distribution of $\rho(x, t)$ in a model system at various times after a $\delta(t)$ excitation of minority carriers (Ehrhardt et al., 1991). The parameters used in the calculation are $X_J = 4.0\,\mu\text{m}$, $\tau = 3$ ns,

2. MINORITY-CARRIER LIFETIME IN III-V SEMICONDUCTORS

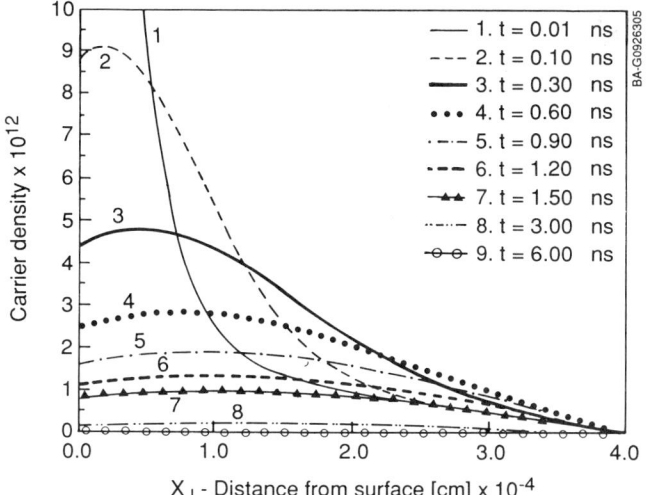

FIG. 17. Calculated spatial distribution of minority carriers in the emitter of the $p-n$ junctions at the indicated times (Ehrhardt et al., 1991, reprinted with permission). The parameters used in the calculation are given in the text.

and $S_w = 10^5$ cm/s. As S_w is large, $\rho(0, t)$ bends downward because of minority-carrier recombination at the interface. At X_J, $\rho(x_J, t)$ is maintained at 0 as S_J is set to the very large value of 10^7 cm/s. Setting S_J to infinity does not noticeably change the solution.

This author (Ahrenkiel, 1987a) developed solutions to this problem for the case $S \approx 0$ and $S_J \to \infty$. In this limit, the eigenvalues of θ_n are

$$\theta_n = \pi/2, \, 3\pi/2, \, 5\pi/2, \ldots \quad \text{for } n = 1, 2, 3, \ldots. \quad (143)$$

The corresponding values of β_n are, for $n = 1, 2, 3$, etc.,

$$\beta_n = D \frac{(2n - 1)^2 \pi^2}{4x_J^2}. \quad (144)$$

The one-dimensional function $U(x, t)$, as defined in Eq. (86), is developed as

$$U_n(x, t) = A_n \left(\cos \alpha_n x + \frac{S_w}{\alpha_n D} \sin \alpha_n x \right) e^{-\beta_n t}. \quad (145)$$

Therefore, for the boundary condition $S_w = 0$ and the values of θ_n from

Eq. (143), the solution for $U(x, t)$ becomes

$$U(x, t) = \sum_{n=1}^{\infty} A_n \cos\left[(n - 1/2)\frac{\pi x}{X_J}\right] e^{-\beta_n t}. \tag{146}$$

For an initial Beer's law distribution, the coefficients A_n are found by substituting Eq. (146) into Eq. (97) to find

$$A_n = I_0 \frac{2\alpha^2 x_J + 2\pi^2 \alpha (-1)^{n+1}(n - 1/2) \exp^{-\alpha X_J}}{\alpha^2 X_J^2 + \pi^2 (n - 1/2)^2} \tag{147}$$

In the limit of $\alpha x_J \gg 1$, one finds $A_n = 2I_0/x_J$. Making this approximation, one finds the total excess charge density by integrating $\rho(x, t)$ from 0 to x_J:

$$\rho_t(t) = \exp(-t/\tau) \int_0^{x_n} U(x, t)\, dx = \frac{4I_n}{\pi} \sum_{n=1}^{\infty} \frac{(-1)^{n-1}}{2n - 1} \exp\left[-\left(\beta_n + \frac{1}{\tau}\right)t\right]. \tag{148}$$

Thus,

$$I_{\text{PL}}(t) = \frac{4I_0}{\pi \tau_R} \sum_{n=1}^{\infty} \frac{(-1)^{n+1}}{2n - 1} \exp\left[-\left(\beta_n + \frac{1}{\tau}\right)t\right]. \tag{149}$$

This series converges rapidly for $t \gg X_J^2/(\pi^2 D)$ as the first mode ($n = 1$ term) dominates:

$$I_{\text{PL}}(t) = \frac{4I_0}{\pi \tau_R} \exp\left[-\left(\frac{\pi^2 D}{4X_J^2} + \frac{1}{\tau}\right)t\right]. \tag{150}$$

The term equivalent to $X_J^2/(D\pi^2)$ was defined in Eq. (104) as the diffusion transit time τ_D. Therefore,

$$\frac{1}{\tau_{\text{PL}}} = \frac{1}{\tau} + \frac{1}{4\tau_D} = \frac{1}{\tau} + \frac{\pi^2 D}{4X_J^2}. \tag{151}$$

Calculated PL decay curves, as a function of X_J, are shown in the curves of Fig. 18 (Ehrhardt et al., 1991). These were calculated using the

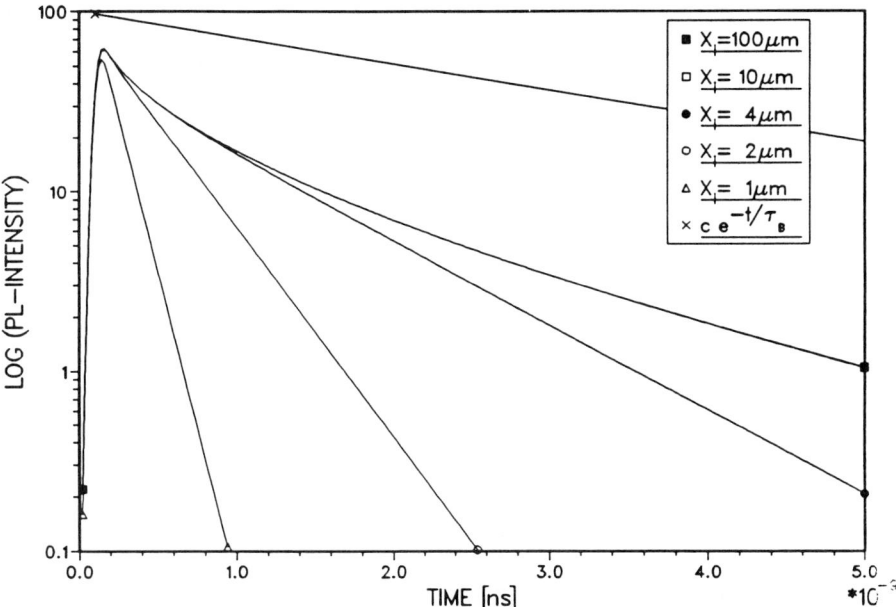

FIG. 18. Calculation of the PL decay of $p-n$ junctions versus time with the emitter thickness, x_J, as a parameter (Ehrhardt et al., 1991, reprinted with permission).

complete theory developed by the authors. The parameters used here are $\tau_B = 3$ ns, $D = 25$ cm^2/s, $S_w = 1 \times 10^5$ cm/s, and $S_J = 1 \times 10^7$ cm/s. The junction depth is varied from 1.0 μm to 100 μm in these calculations. The PL decay is approximately described by Eq. (151), and at $X_J = 4.3$ μm, the bulk lifetime equals the diffusion lifetime τ_D. Thus, for "thinner" emitters, the PL decay time becomes dominated by X_J.

By estimating the minority-carrier diffusivity D and measuring the junction depth X_J, one may be able to obtain the minority-carrier lifetime, τ, from Eq. (151). Determination of bulk lifetime is best done when $\tau < \tau_D$. As with the DH structures, one can fabricate two junction devices that are identical except for the active layer thickness X_J. After τ_{PL} is measured in both devices, the bulk lifetime τ and diffusivity D can be calculated. If $\tau \gg \tau_D$, the bulk minority-carrier lifetime is insignificant in PL decay kinetics and depends only on D:

$$I_{PL}(t) = \frac{4I_0}{\pi \tau_R} \exp\left(-\frac{\pi^2 D t}{4 X_J^2}\right). \tag{152}$$

In this case, the photoluminescence lifetime is completely unrelated to the recombination lifetime and depends only on the junction depth and minority-carrier diffusivity.

The photon recycling effect will also be significant for thick junction devices, but no calculations have been reported in the literature. Recent diffusion length measurements (Partain et al., 1990) on GaAs homojunctions produced values that were significantly greater than expected. This work estimated minority-carrier diffusivities from the majority-carrier mobility at the given doping levels. Using the radiative lifetime as the bulk lifetime, a diffusion length was computed that was smaller than the experimental values. The photon recycling effect explains the inconsistency. One can define a recycling factor as in Eq. (118) and apply it to the junction device lifetime as

$$\frac{1}{\tau_{PL}} = \frac{1}{\tau_{SRH}} + \frac{1}{\phi(d)\tau_R} + \frac{1}{4\tau_D} = \frac{1}{\tau_{SRH}} + \frac{1}{\phi(d)\tau_R} + \frac{\pi^2 D}{4X_j^2}. \quad (153)$$

Here the $\phi(d)$ could be calculated from Eqs. (127)–(129) using the $\rho(x)$ calculated for the junction device. These values will depend on the boundary conditions and therefore on the voltage across the junction. For the general condition that $S_w > 0$, Eq. (142) is modified as

$$\frac{1}{\tau_{PL}} = \frac{1}{\tau_{SRH}} + \frac{1}{\varphi(d)\tau_R} + \beta_1. \quad (154)$$

At long times, the first mode dominates the decay process. In Eq. (154), β_1 comes from the solution of the eigenvalue equation (Eq. (139)).

16. THICK FILMS AND BULK CRYSTALS

The solution of the time-dependent diffusion equation for bulk crystals is best developed with the Fourier or Laplace transform techniques. In this approach, the entire diffusion equation is transformed from the time domain to the frequency domain. This procedure transforms the partial differential equation into an ordinary differential equation involving spatial variables. The spatial differential equation is solved using standard techniques. The inverse transform is applied, giving a solution in terms of space and time. The Laplace transform technique has been frequently applied in the literature to the solution of transients in semiconductors. Ahrenkiel (1988) used the Laplace transform to analyze the PL decay in bulk wafers. On the other hand, Ehrhardt and co-workers (1991) used a Fourier transform technique to analyze PL transients in bulk materials.

Another analysis of the PL lifetime in thick active layers was described by Tyagi and co-workers (1982). Earlier solutions to PL decay in a semi-infinite crystal were reported by Vaitkus (1976) and Boulou and Bois (1977). All of these calculations included the self-absorption effect on the transient waveform. However, none of the calculations include the photon recycling effect. The latter will be a function of time here as the minority carriers diffuse away from the front surface.

A mathematical solution was developed by Ioannou and Gledhill (1984) for a planar distribution of minority carriers diffusing from a finely focused laser beam. This solution was an application of the Green's function formalism developed by van Roosbroeck (1955). Recent work by 't Hooft and van Opdorp (1986) showed that a nonuniform distribution of generated carriers was inconsequential in the transient decay characteristics.

Here I will derive solutions for the bulk or thick film problem by using the method of Laplace transforms. An instantaneous Beer's law generation of minority carriers at the front surface will be assumed. This analysis will use a delta function, $\delta(t)$, to approximate the pulsed generation of electron–hole pairs. Thus,

$$G(x, t) = \alpha I_0 \exp(-\alpha x)\delta(t). \tag{155}$$

With this generation function $G(x, t)$, the one-dimensional diffusion equation is written

$$D\frac{d^2\rho(x, t)}{dx^2} = \frac{d\rho(x, t)}{dt} + \frac{\rho(x, t)}{\tau} - G(x, t). \tag{156}$$

The one-dimensional, time-dependent continuity equation (Eq. (85)) is Laplace transformed from the time domain to the frequency domain according to the standard procedure in which p is the Laplace frequency variable:

$$F(x, p) = LF(x, t) \equiv \int_0^\infty e^{-pt} F(x, t)\, dt. \tag{157}$$

The inverse transform is defined as

$$L^{-1}F(x, p) = F(x, t) = \frac{1}{2\pi i}\int_{a-i\cdot\infty}^{a+i\cdot\infty} e^{pt} F(x, p)\, dp. \tag{158}$$

Extensive tables exist for Laplace transform pairs of various functions (Roberts and Kaufman, 1966; Oberhettinger and Baddi, 1973). Substituting $U(x, t)\exp(-t/\tau)$ for $\rho(x, t)$ as in Eq. (86), the diffusion equation becomes

$$D\frac{d^2U(x, t)}{dx^2} = \frac{dU(x, t)}{dt} - G(x, t)\exp(t/\tau). \tag{159}$$

The individual terms of Eq. (158) are now transformed:

$$D\frac{\partial^2}{\partial x^2}\bar{U}(x, p) = D\bar{U}(x, p) + U(x, 0^-) - \alpha I_0 e^{-\alpha x}. \tag{160}$$

The transform of the time derivative is well known, in which $U(x, 0^-)$ is the excess charge density prior to excitation. Here the value is 0. The partial differential equation is now transformed into an ordinary differential equation of the spatial variable x. Defining $\lambda^2 = p/D$, the solutions are

$$\bar{U}(x, p) = Ae^{-\lambda x} + Be^{\lambda x} + \frac{\alpha I_0}{p - \alpha^2 D}e^{-\alpha x}. \tag{161}$$

The third term on the right is the particular solution to the differential equation. As $U(\infty, p)$ must vanish, the coefficient B must equal zero. To evaluate the constant A, one equates the surface recombination current and the diffusion current at the front surface:

$$D\left[\frac{d\bar{U}}{dx}\right]_{x=0} = S\bar{U}(0, p). \tag{162}$$

After some elementary algebra, one finds

$$\bar{U}(x, p) = \frac{\alpha I_0}{p - \alpha^2 D}\left[e^{-\alpha x} - \frac{S + \alpha D}{S + \lambda D}e^{-\lambda x}\right]. \tag{163}$$

The algebra is simplified by first integrating $U(x, p)$ over the range of x (0 to ∞) to get the total charge density:

$$U_t(x, p) = \int_0^\infty U(x, p)\,dx = \frac{I_0}{p - \alpha^2 D}\left[1 - \frac{\alpha(S + \alpha D)}{\lambda(S + \lambda D)}\right]. \tag{164}$$

The inverse transform is found in the tables of Roberts and Kaufman (1966). The inverse transform of the first and second terms are, respectively:
First term:

$$L^{-1}\left(\frac{I_0}{p - \alpha^2 D}\right) = I_0 e^{\alpha^2 Dt}. \qquad (165)$$

Second term:

$$L^{-1}\left(\frac{1}{\sqrt{p}(\sqrt{p} + S/\sqrt{D})(p - \alpha^2 D)}\right) = \frac{D \exp\left(\frac{S^2 t}{D}\right) \operatorname{erfc}(S\sqrt{t/D})}{S^2 - \alpha^2 D^2}$$

$$+ \frac{S \exp(\alpha^2 Dt) \operatorname{erf}(\alpha\sqrt{Dt})}{\alpha(S^2 - \alpha^2 D^2)}$$

$$- \frac{\alpha D}{S - \alpha D} \exp\left(\frac{S^2 t}{D}\right) \operatorname{erfc}(S\sqrt{t/D}). \qquad (166)$$

Combining, simplifying, and calculating $I_{PL}(t) = \rho_t(t)/\tau_R$;

$$I_{PL}(t) = \frac{I_0}{\tau_R} \exp\left(-\frac{t}{\tau}\right) \left[\frac{S}{S - \alpha D} \exp(\alpha^2 Dt) \operatorname{erfc}(\alpha\sqrt{Dt})\right.$$

$$\left. - \frac{\alpha D}{S - \alpha D} \exp\left(\frac{S^2}{D}\right) \operatorname{erfc}\left(S\sqrt{\frac{t}{D}}\right)\right]. \qquad (167)$$

This expression for PL decay assumes low injection and neglects self-absorption.

From Eq. (167), one obtains several limiting expressions. If $S = 0$, the decay is simple exponential:

$$I_{PL}(t) = \frac{I_0}{\tau_R} \exp\left(-\frac{t}{\tau}\right). \qquad (168)$$

For the case of a very high recombination velocity ($S \gg \alpha D$), one can easily show that

$$I_{PL}(t) = \frac{I_0}{\tau_R} \exp(\alpha^2 Dt) \operatorname{erfc}(\alpha\sqrt{Dt}) \exp\left(-\frac{t}{\tau}\right). \qquad (169)$$

For longer times ($t \gg 1/\alpha^2 D$), one can use the asymptotic form of the complementary error function to show that

$$I_{\text{PL}}(t) = \frac{I_0}{\alpha \tau_R \sqrt{\pi D t}} \exp\left(-\frac{t}{\tau}\right). \tag{170}$$

For $t > 1/\alpha^2 D$, the time dependence of the PL decay rate is $\exp(-t/\tau)/\sqrt{t}$. In this situation, one can fit Eq. (170) to the data with D and τ as parameters.

The self-absorption is added as in Eq. (120) by inserting the Beer's law term $\exp(-\beta x)$ appropriate to the monitored wavelength. Thus, $I_{\text{PL}}(p)$ is produced by integrating $\rho(x, p)$ over x

$$I_{\text{PL}}(p) = \frac{1}{\tau_R}\left(\frac{\alpha I_0}{p - \alpha^2 D}\right) \int_0^\infty \left[e^{-\alpha x} - \frac{S + \alpha D}{S + \lambda D} e^{-\lambda x}\right] \exp(-\beta x)\, dx. \tag{171}$$

The inverse transform gives $I_{\text{PL}}(t)$[1]:

$$I_{\text{PL}}(t) = \frac{I_0}{\tau_R} \exp\left(-\frac{t}{\tau}\right) \{A \exp(\alpha^2 D t)\, \text{erfc}(\alpha \sqrt{Dt})$$
$$+ B \exp\left(\frac{S^2 t}{D}\right) \text{erfc}(S\sqrt{t/D}) + C \exp(\beta^2 D t)\, \text{ercf}(\beta \sqrt{Dt})\}. \tag{172}$$

The constants[1] A, B and C are

$$A = \frac{2\beta}{\beta^2 - \alpha^2} + \frac{2S/D}{(S/D - \alpha)(\alpha - \beta)},$$

$$B = \frac{2S/D}{(S/D - \alpha)(\beta - S/D)}, \tag{173}$$

$$C = \frac{2}{\alpha + \beta} - (A + B).$$

This expression includes self-absorption, but it does not include photon recycling effects.

Figure 19 shows a plot of Eq. (168) with $\alpha = 5 \times 10^4\,\text{cm}^{-1}$, $D = 25\,\text{cm}^2/\text{s}$, and $\tau = 25\,\text{ns}$. These values are typical room-temperature electron parameters in moderately doped p-type GaAs or InP. The

---

[1] A printing error resulted in an incorrect value for this expression in a previous publication (Ahrenkiel, 1988).

2. MINORITY-CARRIER LIFETIME IN III–V SEMICONDUCTORS

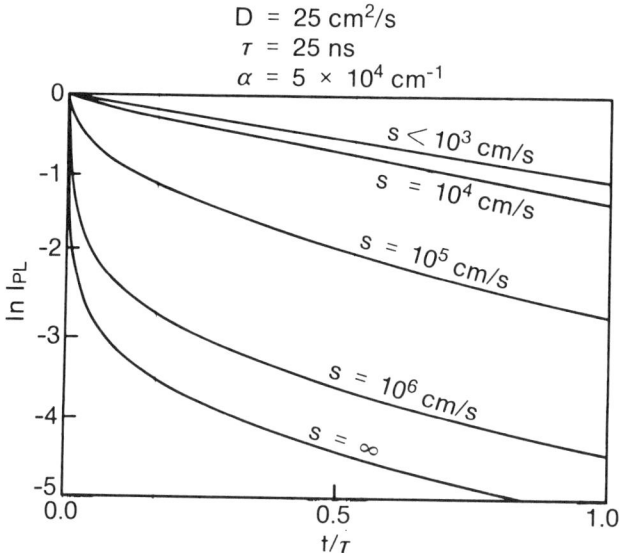

FIG. 19. Calculation of the PL decay of a bulk crystal versus time with the surface recombination velocity (S) as a parameter. The parameters shown are the minority-carrier diffusivity (D), the minority-carrier lifetime (τ), and the absorption coefficient of the incident light (α). (Ahrenkiel, 1988, reprinted with permission.)

curves show calculated decays with values of S ranging from 10^3 cm/s to infinity. For $S < 10^3$ cm/s, the PL decay is not noticeably affected by surface effects. For $S > 10^6$ cm/s, the decay is approximately described by Eq. (169) and does not increase appreciably as S approaches infinity. The rapid change in decay rates occurs in the range 10^4 cm/s $< S < 10^6$ cm/s. This is a range of S often found for the bare surfaces (unpassivated) of III–V semiconductors. The fast decay near $t = 0$ for $S > 10^5$ cm/s is apparent here. One can define an instantaneous lifetime from the slope as indicated before:

$$\frac{1}{\tau_{\text{eff}}} = -\frac{d}{dt}\ln(I_{\text{PL}}(t)). \tag{174}$$

One can show from Eq. (167) that at $t = 0$ ('t Hooft and van Opdorp, 1986)

$$\tau_{\text{eff}} = \frac{\tau}{1 + \alpha S \tau}. \tag{175}$$

The steep initial decay rates (short effective lifetimes) are found with strong absorption of the incident light and $S > 10^5$ cm/s. Large values of both α and S are necessary to produce the steep initial slopes evident in Fig. 19 (Ahrenkiel, 1988). The surface recombination effect can be greatly reduced by using weakly absorbed wavelengths. Finding an appropriate wavelength for direct-bandgap semiconductors is not always easy with pulsed laser light sources. The chosen wavelength must match the weakly absorbing "tail" that exists in the spectral region $h\nu < E_g$. For indirect semiconductors, such as silicon, the choice of suitable light sources is somewhat less exacting.

V. High-Injection Effects in Double Heterostructure

17. Radiative Lifetime

The radiative lifetime under high-injection conditions was described in Section II. Equation (53) describes the time-dependent radiative decay at an arbitrary injection density $\rho(t)$. Under high-injection, the decay process is nonexponential and is called bimolecular decay. The equations in Section II were derived with no consideration of the self-absorption/photon recycling process. Figure 15 shows the calculated effects of photon recycling on the effective radiative lifetime in low injection. Photon recycling effects influence radiative decay at all injection levels, and the mathematical expressions must be modified accordingly. Assuming for the moment that radiative recombination is the only active process, Eq. (52) is rewritten

$$\frac{d\rho}{dt} = -B(\rho N + \rho^2) + G(\rho(r)). \tag{176}$$

Here, $G(\rho(r))$ is the photon recycling generation term that arises from self-absorption. After calculating the self-absorption/photon recycling fraction $F(\theta_{cr}, d, \rho)$, the recycling factor $\phi(d, \rho)$ is calculated by Eq. (136). The excess-carrier distribution ρ is retained as an argument, as the recycling factor ϕ depends on the spatial distribution. As shown previously, one calculates the factor ϕ by the volume integration of $\rho(r)S(h\nu)$ over the full solid angle. Finally, Eq. (176) can be rewritten as

$$\frac{d\rho}{dt} = -\frac{B}{\phi(d)}(\rho N + \rho^2). \tag{177}$$

Implicit in the calculation of $\phi(d)$ is the instantaneous charge distribution ρ.

2. MINORITY-CARRIER LIFETIME IN III–V SEMICONDUCTORS

If one uses the one-dimensional model for the DH structure, ρ is a function of x only. Many calculations of ϕ have assumed that the excess charge fills the DH potential well as in Fig. 11, Curve C. Diffusion removes the large gradients in $\rho(x)$ during the time period $0 < t < \tau_D$. Therefore, this assumption generally applies to times $t > \tau_D$. Assuming that diffusion does not change the spatial profile of $\rho(t)$ with time, the solution of Eq. (177) is

$$\rho(t) = \frac{\rho_0 \exp(-t/\phi\tau_R)}{1 + \frac{\rho_0}{N}(1 - \exp(-t/\phi\tau_R))}. \tag{178}$$

As in the low-injection case, the radiative lifetime is multiplied by the recycling factor ϕ. The PL decay function at very high injection ($\rho \gg N$) is calculated by substituting Eq. (178) into Eq. (55) and neglecting the $N_\rho(t)$ terms. These solutions apply to times $t < \tau_R$. Therefore an approximate solution can be written for the restricted time domain:

$$I_{PL}(t) = \frac{B\phi\rho_0^2}{(\phi + B\rho_0 t)^2} \quad (\tau_D < t < \tau_R). \tag{179}$$

If Eq. (178) is substituted into Eq. (55), the effective high-injection lifetime at $t = 0$ is

$$\tau_{eff} = \frac{\phi}{2B\rho_0}. \tag{180}$$

Here the initial, high-injection radiative lifetime $1/2B\rho_0$ is increased by the photon recycling factor ϕ. Examination of Eq. (178) indicates that the high-injection lifetime is increased by a factor of approximately ϕ from that obtained by neglecting photon recycling.

Implicit in this derivation is that $\phi(d)$ is independent of time. However, from the previous discussion, one sees that ϕ is time-independent only for $t > \tau_D$. When the excess e–h pairs are in the initial Beer's law distribution, one would expect ϕ to vary from the "flat band" case. In addition, the localized values of $\rho(x)$ are much larger. The total recombination rate, R, is a spatially varying function of x during the period $t < \tau_D$. At $t = 0$, one can write the rate $R(x)$ by substituting the Beer's law distribution into the recombination rate given by Eq. (52):

$$R(x) = B(\alpha I_0 N \exp(-\alpha x) + \alpha^2 I_0^2 \exp(-2\alpha x)). \tag{181}$$

The time-dependent solution to the diffusion/recombination problem requires a numerical calculation. Here I will take a simple model that assumes that the initial charge is uniformly distributed in a slab of thickness $1/\alpha$ at $t=0$. Again assuming unit cross-sectional area, the average charge density at the surface is obtained by integrating the Beer's law distribution from $x=0$ to $x=d$ and dividing by $1/\alpha$:

$$\rho_i = \alpha I_0 (1 - \exp(-\alpha d)). \quad (182)$$

Neglecting the loss due to recombination, the excess charge, after diffusion ($t > \tau_D$), uniformly fills the potential well with charge density

$$\rho_f = \frac{I_0(1 - \exp(-\alpha d))}{d}. \quad (183)$$

The quantity ρ_f is simply the total deposited charge, which is for $t > \tau_D$ uniformly distributed throughout the active volume. Therefore, the charge density ratio is simply

$$\frac{\rho_i}{\rho_f} \approx \alpha d. \quad (184)$$

An initial factor ϕ_i is defined to describe photon recycling while the charge is distributed in the Beer's law distribution. Using the high-injection case ($\rho > N$), one can estimate τ_i as the initial radiative lifetime. Here one substitutes ρ_i from Eq. (182) into Eq. (180):

$$\tau_i = \frac{\phi_i}{2B\rho_i}. \quad (185)$$

The ratio of τ_i to τ_{eff} (Eq. (180)) in high injection is approximately

$$\frac{\tau_i}{\tau_{\text{eff}}} \approx \frac{\phi_i}{\phi} \frac{\rho_f}{\rho_i} = \frac{1}{\alpha d} \frac{\phi_i}{\phi}. \quad (186)$$

This is the ratio of initial ($t=0$) lifetime to the lifetime at $t = \tau_D$. The ratio then depends on the absorption coefficient of the incident light. For weakly absorbed light, the ratio is unity. If $\alpha = 5 \times 10^4 \text{ cm}^{-1}$ and $d = 10 \ \mu\text{m}$, the ratio is $0.05 \ \phi_i/\phi$ and the initial lifetime is small compared to the "flat band" lifetime assuming that $\phi_i \cong \phi$.

2. MINORITY-CARRIER LIFETIME IN III–V SEMICONDUCTORS

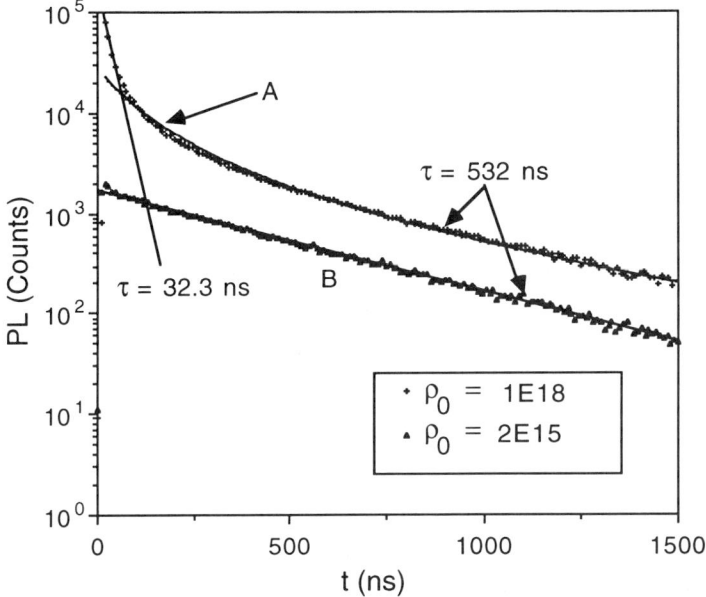

FIG. 20. Measured PL decay of a DH device ($N \sim 1 \times 10^{17}$ cm^{-3}) and $d = 10\,\mu$m at the indicated injection levels (Ahrenkiel and Levi, 1992).

Experimental observations (Ahrenkiel and Levi, 1992) have been made that the high-injection lifetimes are considerably longer than predicted by Eq. (66). Experimental data are shown in Fig. 20 for a GaAs DH device with $d = 10\,\mu$m and $N \sim 1 \times 10^{17}$ cm^{-3}. The measurement system response, using the microchannel plate detector, is less than 50 ps. The 600 nm incident laser radiation is measured with a calibrated power meter, and the energy per pulse is calculated from the known repetition rate. The energy density in photons per square centimetre is calculated by measuring the illuminated sample area with a calibrated microscope. The calculated injection level for curve A of the figure is $p_0 = 1.0 \times 10^{18}$ cm^{-3} and is accurate to within 50%. Using Eq. (66), the calculated lifetime at $t = 0$ is 2.38 ns. The maximum slope of the PL decay curve near $t = 0$ gives $\tau_{\text{eff}} \sim 32$ ns, which is much longer than the calculated value. The longer lifetime is a high-injection recycling effect, and the measured ratio is about 13.5. Taking the hole diffusivity as 5 cm^2/s, the diffusion transit time τ_D is about 20 ns. Therefore, the "flat band" conditions leading to Eq. (180) are met for most of the observed PL decay, and the steady state value of ϕ is approximately correct.

At times longer than about 1.5 μs, the decay becomes exponential with a lifetime of 532 ns. The predicted low-injection lifetime is 38.5 ns, giving a ϕ value of about 13.8. This is the steady-state or "flat band" value of ϕ for this device. The solid line associated with curve A is a fit of Eq. (178) substituted into Eq. (55) with $\phi = 13.8$. The fit is quite good except for $t < 50$ ns, as expected. Curve B is data for excitation at very low injection level (1×10^{15} cm^{-3}), and a simple exponential decay results. The slope gives a lifetime of 532 ns, in agreement with curve A.

In summary, the TRPL data must include the photon recycling effect at high injection levels in order to properly interpret the data. Failure to do so will lead to inaccurate calculations of the injection levels.

18. INTENSITY DEPENDENCE OF THE SHOCKLEY–READ–HALL LIFETIME

In Section II, the Shockley–Read–Hall (SRH) recombination mechanism was analyzed in the low-injection limit. An analysis of high injection on the SRH recombination rate in semiconductors will be reviewed here. The variation of minority-carrier lifetime with injection level has been observed in silicon for many years. The effects of high injection on SRH recombination were first described in a classic paper by Hall (1952). Hall showed that the effective lifetime becomes the sum of majority- and minority-carrier lifetimes during high injection. The enhancement of the high-injection silicon lifetime is reported (Yablonovitch and Gmitter, 1986) from measurements on low to moderately doped silicon. Blakemore (1962) has derived the steady-state SRH effective lifetime for an arbitrary injection level. He also shows that the transient decay at high injection is nonexponential and not fitted by a single lifetime. Here these early models will be extended to TRPL measurements on III–V double heterostructures. This analysis will be similar to recent treatments in the literature (Ahrenkiel et al., 1991b, 1991c; Marvin and Halle, 1991, 1992).

a. Bulk Recombination Effects

This analysis will again use the "flat band" condition that occurs when $t > \tau_D$. Recombination analysis of the diffusion-controlled period ($0 < t < \tau_D$) requires numerical solutions. The initial derivation will include only the SRH component of recombination. The radiative component will be added later.

Using Eq. (30) with $E_t = E_i$, the recombination rate can be written in terms of an injection density $\rho = \Delta n$, Δp. Here, N is the doping density

2. MINORITY-CARRIER LIFETIME IN III–V SEMICONDUCTORS

as before. The rate equation for recombination at a single defect level is

$$\frac{d\rho}{dt} = -\frac{\rho N + \rho^2}{\tau_{min}(N + \rho) + \tau_{maj}\rho}. \tag{187}$$

Here, τ_{min} and τ_{maj} are the SRH minority and majority-carrier lifetimes, respectively:

$$\tau_{min} = \frac{1}{\sigma_{min} v_{th} N_t},$$

$$\tau_{maj} = \frac{1}{\sigma_{maj} v_{th} N_t}. \tag{188}$$

As before, the instantaneous injection level $I(t)$ is

$$I(t) \equiv \frac{\rho(t)}{N}. \tag{189}$$

Equation (187) becomes

$$\frac{dI}{dt} = -\frac{I + I^2}{\tau_{min}[1 + I] + \tau_{maj}I}. \tag{190}$$

An analytical solution to Eq. (190) may not exist. A numerical solution shows nonexponential behavior, and thus a single value of lifetime cannot be defined. One can easily derive asymptotic solutions for low injection levels ($I \ll 1$) and for high injection levels ($I \gg 1$). In these regimes, the solution becomes exponential, and a lifetime can be defined for the specific range of I. These solutions of Eq. (190) are

$$I(t) = \exp(-t/\tau_{min}) \quad (I \ll 1),$$

$$I(t) = \exp(-t/[\tau_{min} + \tau_{maj}]) \quad (I \gg 1). \tag{191}$$

In high-injection, however, the effective lifetime becomes the sum of majority- and minority-carrier lifetimes, as was first noted by Hall (1952). At intermediate injection levels, Eq. (190) is solved numerically and the solution is nonexponential. To find solutions, one integrates Eq. (190) by the Runge–Kutta numerical technique (Press et al., 1988).

b. *Saturation of SRH recombination centers*

In order to understand saturation effects in the recombination process, one must look at the deep level capture rates for a single-center system. A similar analysis has been published for steady-state generation conditions (Blakemore, 1962). Taking an n-type material (N_D), assume that a steady-state density of excess carriers, Δp, is maintained by steady-state injection. One writes the capture rates for holes and electrons at a midgap center with capture cross-sections σ_p and σ_n as

$$R_p = \sigma_p v_{th} N^- \Delta p, \tag{192}$$

$$R_n = \sigma_n v_{th} N^0 (N_D^0 + \Delta n). \tag{193}$$

Here, N^0 and N^- are the densities of neutral and negatively charged levels for which the total density of midgap levels is N:

$$N = N^0 + N^-. \tag{194}$$

I have assumed that the deep level is an acceptor and may be either neutral or negatively charged. Assuming that steady-state or quasi-equilibrium conditions exist, one solves for the relative population of neutral and charged centers. While in low injection ($\Delta n < N_D$), quasi-equilibrium requires that $R_p = R_n$. Combining Eqs. (192), (193), and (194), one solves for the populations N^0 and N^-:

$$N^0 = \frac{N}{1 + \dfrac{\sigma_n}{\sigma_p}\dfrac{N_D}{\Delta p}} = \frac{N}{1 + \dfrac{\sigma_n}{\sigma_p}\dfrac{1}{I}} \tag{195}$$

and

$$N^- = \frac{N}{1 + \dfrac{\sigma_p}{\sigma_n}\dfrac{\Delta p}{N_D}} = \frac{N}{1 + I\dfrac{\sigma_p}{\sigma_n}}. \tag{196}$$

In low injection, $\sigma_n N_D \gg \sigma_p \Delta p$ (or $I\sigma_p/\sigma_n \ll 1$). Therefore, quasi-equilibrium conditions give $N^- \gg N^0$, and the deep acceptors stay filled with electrons. In low injection, the minority carrier capture rate determines the minority-carrier lifetime and is $1/(\sigma_p v_{th} N)$ by Eq. (188). At higher injection, one may find that $I\sigma_p/\sigma_n > 1$. This situation occurs either if I is sufficiently large and/or if $\sigma_n \ll \sigma_p$. In this example, one expects that $\sigma_n \ll \sigma_p$, as the neutral-center cross-section for electron

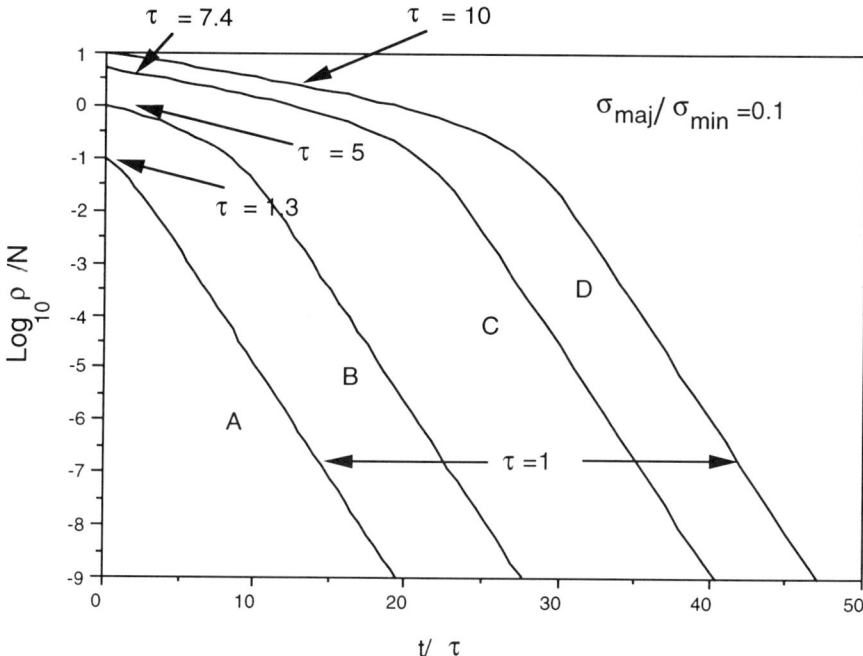

FIG. 21. Calculated minority-carrier decay versus time (in units of bulk lifetime) for SRH recombination with the indicated cross-section ratio. Curves A, B, C, and D correspond to injection ratios of 0.01, 0.10, 1.0 and 10.0, respectively (Ahrenkiel et al., 1991c, reprinted with permission).

capture is much smaller than the hole-capture cross-section of the negatively charged center. Here, majority-carrier capture controls the recombination lifetime even in low injection. The effective lifetime then is approximately given by $1/(\sigma_n v_{th} N)$ as $N^0 \sim N$.

In the case for which $I\sigma p/\sigma_n > 1$, the recombination center becomes saturated, and majority-carrier capture limits recombination. Center saturation decreases the recombination rate and increases the effective lifetime.

A hypothetical case was calculated, using Eq. (190) with some appropriate parameters, and the results are shown in Fig. 21. Here, the minority-carrier lifetime is 1.0 time units and the minority/majority carrier cross-section ratio is 10. Using $t = 0$ injection values of 0.01, 0.1, 1.0, and 10.0, curves A, B, C, and D, respectively, were calculated. The calculations show that saturation becomes observable at about $I = 0.1$, corresponding to $I\sigma_p/\sigma_n = 1.0$. For $I < 0.01$, the excess-carrier decay rate becomes 1.0 (equal to the minority-carrier lifetime). For $I > 0.01$, the slope of the curve produces decay times from 1.3 units to 10.0 units near $I = 10$. These calculations show that a single lifetime is not an appropriate model in this injection range.

c. Intensity Dependence of S

The single level interface model of Section II will be extended to the cases of intermediate and high injection. As before, one assumes that single-level midgap defects lie in two planes formed by the interfaces between the DH active layer and the window layers. The defect density is still N_{st} defects per square centimeter, and the capture cross-sections are σ_p and σ_n for holes and electrons, respectively. The recombination rate from deep levels at the surface can be derived by combining Eqs. (32) and (39):

$$\frac{d\rho_s}{dt} = -R_{vs} = -\frac{2}{d} \frac{[\rho_s N + \rho_s^2]}{\frac{1}{S_{min}}[N + \rho_s] + \frac{1}{S_{maj}}\rho_s}, \quad (197)$$

where S_{min} and S_{maj} are the minority- and majority-carrier surface recombination velocities, and ρ_s is the surface excess carrier density. Dividing by N, one gets

$$\frac{dI}{dt} = -\frac{2}{d} \frac{I + I^2}{(1 + I)/S_{min} + I/S_{maj}}. \quad (198)$$

Here, I is ρ_s/N and

$$S_{min} = N_{st} v_{th} \sigma_{min}, \quad S_{maj} = N_{st} v_{th} \sigma_{maj}. \quad (199)$$

Multiplying the numerator and denominator of Eq. (198) by $d/2$, one writes the recombination rate in terms of surface lifetimes:

$$\frac{dI}{dt} = -\frac{I + I^2}{\tau_{s,min}[1 + I] + \tau_{s,maj}I}, \quad (200)$$

where $\tau_{s,min}$ and $\tau_{s,maj}$ are the minority- and majority-carrier surface lifetimes, respectively:

$$\tau_{s,min} = \frac{d}{2S_{min}}; \quad \tau_{s,maj} = \frac{d}{2S_{maj}}. \quad (201)$$

From Eq. (200), the recombination velocities at low and high injection are the well-known values (Schroder, 1990, p. 365)

$$S_{\text{low}} = S_{\text{min}}, \quad S_{\text{high}} = \frac{S_{\text{min}} S_{\text{maj}}}{S_{\text{min}} + S_{\text{maj}}}. \quad (202)$$

To analyze data from DH structures with bulk SRH defects, interface recombination, and radiative recombination with photon recycling, the total rate equation must include all of these contributions. The radiative recombination is added to the bulk and interface recombination effect. Assuming that there are several species of bulk midgap defects,

$$\frac{d\rho}{dt} = -\left(\frac{B}{\phi(d)} + \frac{1}{\tau_{s,\text{min}}(N+\rho) + \tau_{s,\text{maj}}\rho} \right.$$
$$\left. + \sum_{i=1}^{k} \frac{1}{\tau_{\text{min}}^{i}(N+\rho) + \tau_{\text{maj}}^{i}\rho} \right)(\rho N + \rho^2). \quad (203)$$

Here, $\tau_{s,\text{min}}$ and $\tau_{s,\text{maj}}$ are the surface lifetimes for minority and majority carriers, respectively, and k is the number of different types of defect. Dividing through by N, the equation is written in terms of the injection ratio:

$$\frac{dI}{dt} = -\left(\frac{1}{\phi(d)\tau_R} + \frac{1}{\tau_{s,\text{min}}(1+I) + \tau_{s,\text{maj}}I} + \sum_{i=1}^{k} \frac{1}{\tau_{\text{min}}^{i}(1+I) + \tau_{\text{maj}}^{i}I} \right)(I + I^2). \quad (204)$$

Simulations by Marvin and Halle (1992) show that the radiative lifetime dominates the decay at high-injection levels, and that the minority/majority sum is not experimentally observed. However, that work neglects the photon recycling effect on high-injection radiative decay and therefore overestimates the bimolecular recombination rate.

d. Experimental Results

Minority-carrier lifetime measurements on the ternary alloy $Al_xGa_{1-x}As$ have indicated that SRH recombination is usually the dominant mechanism ('t Hooft et al., 1981; Timmons et al., 1990; Marvin and Halle, 1991). The TRPL measurements have consistently indicated intensity dependent lifetimes. Figure 22 shows the TRPL data from a p-type DH structure grown by metalorganic chemical vapor deposition (MOCVD) with the composition $Al_{0.9}Ga_{0.1}As/Al_{0.08}Ga_{0.92}As/Al_{0.9}Ga_{0.1}As$.

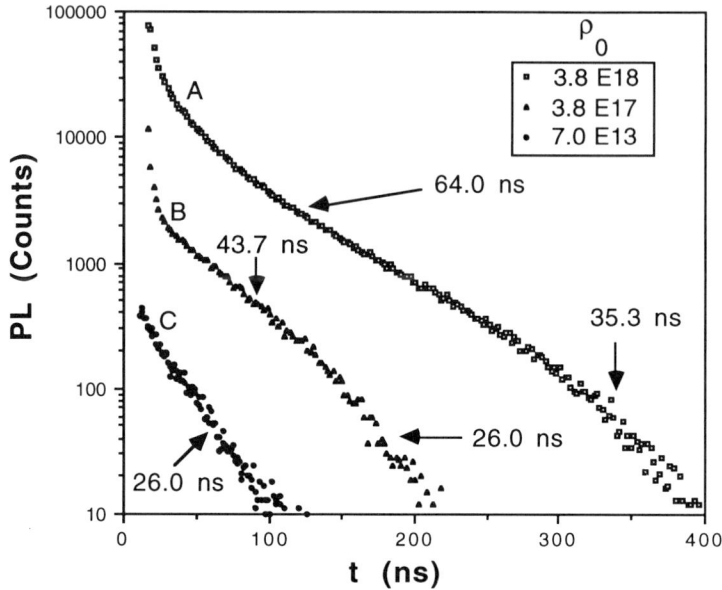

FIG. 22. TRPL data taken from a DH structure with the composition $AL_{0.9}Ga_{0.1}As/Al_{0.08}Ga_{0.92}As/Al_{0.9}Ga_{0.1}As$ a function of injection level. The injection levels (ρ_0) are indicated in the figure for curves A, B, and C (Ahrenkiel et al., 1991c).

The active layer doping density is $3 \times 10^{17}\,\text{cm}^{-3}$, and the active layer thickness is 5.0 μm. The three curves A, B, and C correspond to three different excitation intensities. The B-coefficients of the $Al_xGa_{1-x}As$ alloy series have apparently not been published. The value used in this analysis is $B = 1 \times 10^{-10}\,\text{cm}^3/\text{s}$ and $\phi = 9$ (Asbeck, 1977). Curve A corresponds to an initial e–h density of $3.8 \times 10^{18}\,\text{cm}^{-3}$, corresponding to an injection level (I) of 12.7. Curve B corresponds to about $3.8 \times 10^{17}\,\text{cm}^{-3}$ ($I = 1.0$), and curve C corresponds to about $7 \times 10^{13}\,\text{cm}^{-3}$ ($I = 2 \times 10^{-4}$). One sees two exponential regions of the PL decay with lifetimes of 64 ns and 26 ns. Curve C obviously corresponds to a low-injection, minority-carrier lifetime and has a simple exponential decay. Curve B displays two slopes with effective lifetimes varying between 26 ns and 43.7 ns. In addition, the initial pulse is indicative of a diffusion transient or bimolecular decay, or both. The calculated, high-injection radiative lifetime for Curve B is 132 ns at $t = 0$. Thus, one can assume that the 43.7 ns decay is primarily dominated by SRH recombination in Curve B. The partial saturation of a recombination level is indicated by the two-component decay. Finally, Curve A has an initial slope indicative of bimolecular decay. The initial decay time, using Eq. (181), is about 22 ns. The recombination rate declines to a value corresponding to a decay time of 64 ns between $t = 100$ and 300 ns. At that point in Curve A, the slope steepens and the decay time decreases to about 26 ns. The best fit to the data is with a single level with the lifetime ratio

$$\frac{\tau_{\text{maj}}}{\tau_{\text{min}}} = 2.5. \tag{205}$$

There are three regimes to describe the minority-carrier lifetime for these data. At $I > 10$, the recombination may be limited by bimolecular radiative recombination. At intermediate injection levels, the recombination becomes SRH-dominated with a PL decay time equal to the sum of majority- and minority-carrier lifetimes. At very low injection, one sees the SRH minority-carrier lifetime.

In summary, when SRH recombination dominates, the PL lifetime is intensity-dependent. The high-injection SRH lifetime increases relative to the low-injection lifetime. This behavior is contrasted with the radiative lifetime, which decreases at high injection. The intensity dependence may be used as a diagnostic tool for determining recombination mechanisms.

VI. GaAs Minority-Carrier Lifetime

19. GaAs Wafers

The literature on minority-carrier properties of GaAs wafers is sparse, because wafers are seldom used as the active elements in devices. Because of the development of epitaxy, with the great variety of structures that can be made, GaAs wafers are mainly of interest only as substrate material.

The most extensive lifetime work on bulk crystals is that of Hwang (1971), using the phase-shift technique. Hwang measured minority-carrier lifetimes in Te-doped (n-type) GaAs wafers with N_D ranging from 2×10^{16} cm^{-3} to 6.5×10^{18} cm^{-3}. The analysis of the experiments included the effects of surface recombination and self-absorption. Values of S were estimated from the combination of PL intensity and optical absorption. Figure 23 shows the dependence of the hole lifetime on the

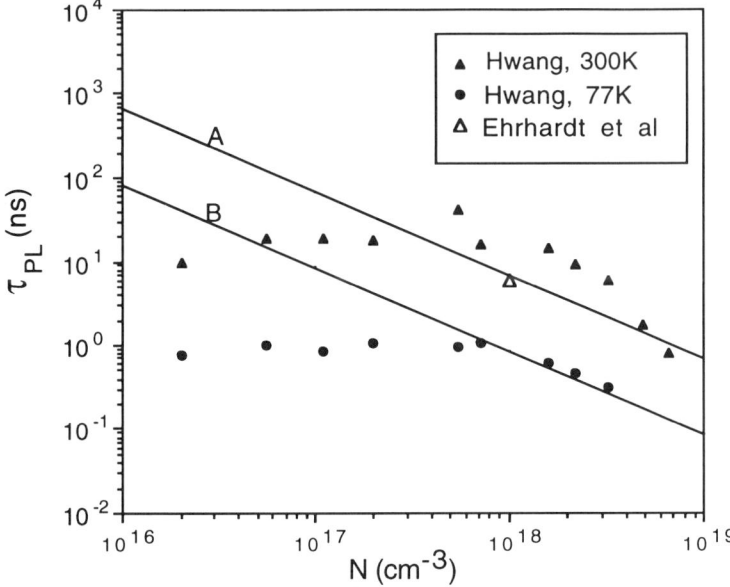

FIG. 23. The minority-carrier lifetime measured in GaAs wafer material. The solid points are from Hwang (1971) measured on n-type wafers at 300 K (▲) and 77 K (●). The point (△) is from Ehrhardt et al. (1991) for an AlGaAs passivated p-type wafer. The lines A and B are the calculated radiative lifetimes at 300 K and 77 K, respectively.

electron concentration at both 77 K and 300 K. The radiative lifetime is also plotted at both temperatures with B (300 K) equal to 2×10^{-10} cm^3/s. The room temperature data are doping-independent from about 1×10^{16} to 1×10^{18} cm^{-3}. The wafer lifetime is controlled by bulk SRH recombination over this concentration range. At about 2×10^{18} cm^{-3}, the lifetime decreases approximately as $1/N$ as τ_R becomes less than τ_{nR}. These data do indicate that some PL lifetimes are greater than the radiative lifetimes at higher doping levels. However, it will be shown later in this section that epitaxial materials are markedly superior to any reported wafer materials.

The measurements of Hwang were made on unpassivated wafers, and the surface recombination velocity was accounted for in the data analysis. Ehrhardt and co-workers (1991) passivated a 1×10^{18} cm^{-3} GaAs wafer with a thin LPE Al$_x$Ga$_{1-x}$As layer. These data are plotted in Fig. 24 for the wafer response before and after AlGaAs passivation. With no Al$_x$Ga$_{1-x}$As passivating layer, the lifetime is about 0.16 ns. The lifetime increased to 6 ns after the LPE layer was grown. From these data, Ehrhardt and co-workers calculated a free surface recombination velocity

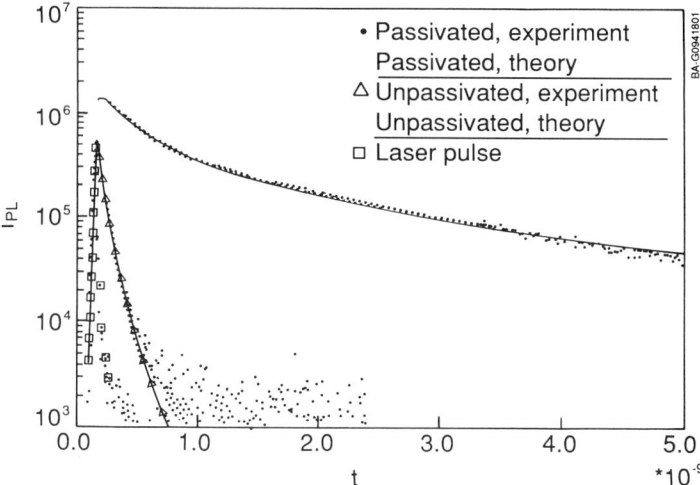

FIG. 24. The PL decay of a p-type ($\sim 1 \times 10^{18}\,\text{cm}^{-3}$) wafer with and without AlGaAs passivation (Ehrhardt et al., 1991, reprinted with permission).

of $S = 3 \times 10^7$ cm/s. This data point for the passivated wafer is plotted in Fig. 23 and agrees with the earlier wafer data.

Wong and co-workers (1990) markedly improved the diffusion length in n-type wafers by proximity annealing in sealed silica ampoules. They found that a hole trap designated HCX ($E_v + 0.57$ eV) was the dominant recombination center in their n-type wafers. The HCX defect is associated with excess arsenic. The hole lifetime appears to be inversely related to the HCX concentration, which is determined by deep-level transient spectroscopy (DLTS). They reported a post-annealing lifetime of 7 ns for a 1×10^{17} cm^{-3} wafer, which is still well below the radiative lifetime of about 50 ns.

The available wafer data then indicate lifetimes that are SRH-limited at doping concentrations below about 1×10^{18} cm^{-3}. The remaining sections (II, III, and IV) show that much larger lifetimes are measured in epitaxial GaAs and are primarily determined by the intrinsic or radiative lifetime.

20. UNDOPED GaAs DIAGNOSTIC STRUCTURES

a. Recombination Theory

Undoped devices have few applications in device technology, as Chapter 4 of this volume shows. However, for the evaluation of epitaxial quality, an undoped GaAs DH structure is a useful device. For epitaxial growth evaluation, a standard test is to grow an undoped $Al_xGa_{1-x}As/GaAs$ DH structure. The epitaxial GaAs background doping is commonly less than 10^{15} cm^{-3} and is itself a useful diagnostic. As the radiative recombination rate is BN (N is the background or unintentional doping), the radiative lifetime is about 5 μs at 10^{15} cm^{-3}. That value is increased by the appropriate photon recycling factor, and therefore nonradiative processes become more pronounced. Of course, if the SRH component dominates recombination, the measurement is difficult because of very weak light emission. Very useful diagnostics come from a relatively thin (1 μm or less), undoped DH devices. Usually, in undoped DHs, the PL lifetime is dominated by bulk SRH and interface recombination, with $\phi\tau_R \gg \tau_{SRH}$ and $\phi\tau_R \gg \tau_S$. From Eq. (119), the low-injection DH lifetime is

$$\frac{1}{\tau_{PL}} = \frac{1}{\tau_{nR}} + 2S. \tag{206}$$

If two devices are fabricated with different active-layer thicknesses, one can solve Eq. (206) for τ_{nR} and for S. Except for exceptionally defect-free material, Eq. (206) accurately describes the PL lifetime in undoped GaAs. The temperature dependence of the lifetime can be used to diagnose the active recombination mechanisms. If the active layer thickness d is sufficiently small, the surface lifetime ($2d/S \ll \tau_{SRH}$) will dominate the recombination. In this case,

$$\tau_{PL} = \frac{d}{2S}. \tag{207}$$

A figure of merit (FM) for thin samples may be defined as τ_{PL}/d:

$$FM = \frac{\tau_{PL}}{d} = \frac{1}{2S}. \tag{208}$$

Here, FM equals $1/2S$ and is indicative of the interface recombination velocity.

It is also very informative to sort out the background radiative lifetime. One means is to measure the temperature dependence of τ_{PL} over some range, typically 77 K to 300 K. As shown earlier, the radiative recombination lifetime varies with temperature approximately as $T^{1.5}$ (Karamon et al., 1985). Figure 4 ('t Hooft et al., 1983) shows that the radiative lifetime in GaAs varies with temperature as $T^{1.54}$ to $T^{1.84}$. Therefore, the radiative component may be uniquely identified by temperature measurements.

The temperature dependence ('t Hooft and van Opdorp, 1983) of interface recombination is weakly thermally activated. When the interface recombination controls the lifetime, τ_{PL} increases as the temperature is lowered. The interface recombination mechanism involves multiphonon coupling to the recombination center. For the $Al_xGa_{1-x}As/GaAs$ interface, the interface lifetime has been shown to be

$$\frac{1}{\tau_s} = \frac{2A \exp(-E_a/kT)}{d}. \tag{209}$$

Here, τ_s is the interface lifetime $(d/2S)$, and A is a constant proportional to S. Figure 25 shows the data of 't Hooft and van Opdorp (1983), plotting S versus temperature for an $Al_{0.12}Ga_{0.88}As/GaAs$ interface as calculated from temperature-dependent TRPL measurements. Here, S decreases from about $1{,}200$ cm/s at about 300 K to about 200 cm/s at about 110 K. The temperature dependence of S is described by Eq. (209), with E_a equal to 27 meV. This functional dependence is a signature of dominant interface recombination.

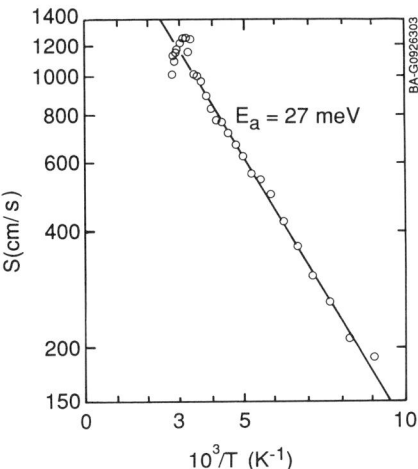

FIG. 25. The temperature dependence of interface recombination velocity (S) for a $GaAs/Al_{0.12}Ga_{0.88}As$ interface ('t Hooft and van Opdorp, 1983, reprinted with permission).

b. *Experimental Studies of Undoped GaAs*

One of the first reported undoped confinement structures (Dapkus *et al.*, 1970, 1979) was from the Illinois group that fabricated GaAs $n^+/n^-/n^+$ homostructures. The epitaxial isotype DH was reported as a diagnostic device for GaAs epitaxial growth about 20 years ago (Keune *et al.*, 1971; Nelson, 1978; Nelson and Sobers, 1978a, 1978b). Nelson and Sobers reported a 1.3-μs PL lifetime for an undoped ($p = 1.9 \times 10^{15}$ cm^{-3}) DH structure with $d = 16\,\mu$m (Nelson and Sobers, 1978a, 1978b). Their DH devices were grown by liquid phase epitaxy (LPE), and the lifetime was far larger than any previously measured. Nelson and Sobers calculated the photon recycling factors and found that $\phi\tau_R$ was greater than 50 μs for the undoped device. By measuring the total external quantum efficiency, they could calculate the total nonradiative lifetime (bulk and interface). Using Eq. (207), they calculated that S is 615 cm/s.

Since this early work, numerous researchers have found lifetimes greater than 1 μs in undoped GaAs DH structures. The work of Yablonovitch and co-workers (1987) also indicated that similar (approximately 1 μs) lifetimes were found in Al$_x$Ga$_{1-x}$As/GaAs DH structures grown by LPE, MOCVD, and molecular beam epitaxy (MBE) of unspecified thickness. Smith and co-workers (1990) reported lifetimes in the ~1 μs range for d in the range of 5 to 10 μm. The Philips group ('t Hooft *et al.*, 1985a) reported a 1.63-μs lifetime in a 9.6-μm, undoped Al$_{0.35}$Ga$_{0.65}$As/GaAs DH. By making a 0.2-μm-thick DH, these researchers calculated that S was less than 53 cm/s. Wolford and co-workers (1991) recently reported a 2.5 μs PL lifetime in an undoped 10-μm-thick, Al$_{0.30}$Ga$_{0.70}$As/GaAs DH. Hummel and co-workers (1990) reported a 0.4-μs lifetime in a 0.5-μm-thick, undoped Al$_{0.30}$Ga$_{0.70}$As/GaAs device grown by MOCVD. They used the arsine alternative tertiarybutylarsine dihydride (TBA) for the epitaxial growth. Their background doping was in the low 10^{14} cm^{-3} range, and the estimated S is calculated to be less than 62 cm/s.

Experiments by Molenkamp and co-workers (1989) measured the PL lifetime in undoped, 4.0-μm Al$_x$Ga$_{1-x}$As/GaAs DH structures grown by MOCVD. The lifetime was measured as a function of arsine partial pressure, and their data for these structures are shown in Fig. 26. This group reported a steep lifetime maximum of 4.9 μs in a device grown at 15×10^{-4} atmospheres of arsine pressure. Using Eq. (207), S is calculated as less than 41 cm/s and is probably the best value so far reported for the undoped Al$_x$Ga$_{1-x}$As/GaAs DH structure. However, very recently Lush and co-workers (1992b) reported values of S of 12 cm/s at higher doping

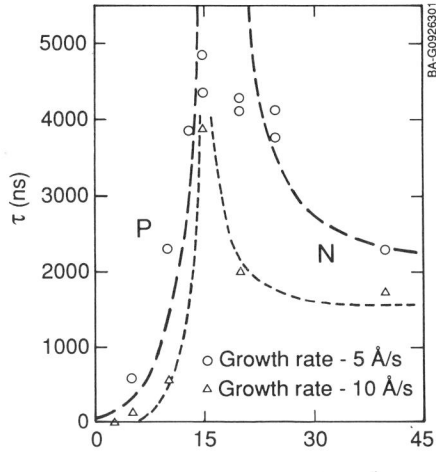

FIG. 26. The PL lifetime of a 4.0-μm-thick DH versus arsine partial pressure during MOCVD growth (Molenkamp et al., 1989. Reprinted with permission.) The devices are $Al_xGa_{1-x}As/GaAs$ DHs with active layer thicknesses of 4.0 μm.

levels ($n = 1.3 \times 10^{17}$ cm^{-3}). This work is described in detail in Section VI.22.

Dawson and Woodbridge (1984) found that buffer or prelayers grown between the substrate and the DH device reduced S below 100 cm/s for their DH structures grown by MBE. For devices without prelayers, S was found to be 4×10^3 to 6×10^3 cm/s. They measured a PL lifetime of 125 ns in a structure with $d = 0.2$ μm. This corresponds to $S \sim 80$ cm/s using Eq. (207). The PL lifetime of a typical, undoped MOCVD $Al_xGa_{1-x}As/GaAs$ device grown by A. Nozik and co-workers at NREL is shown in Fig. 27 (Ahrenkiel, 1992a). The active layer thickness d is 1.8 μm, and the PL lifetime is 1.1 μs. Using the interface lifetime approximation, S is less than 82 cm/s.

c. Undoped GaAs/GaInP Diagnostic Structures

The ternary $Ga_{0.5}In_{0.5}P$ is lattice-matched to GaAs and has a bandgap that varies from 1.8 to 1.9 eV depending upon growth conditions. $Ga_{0.5}In_{0.5}P$ has been found to be an effective window/passivating layer for GaAs (Olson et al., 1989). Figure 28 shows the PL decay data for undoped GaInP/GaAs DHs. The background electron concentrations for these structures are typically 2 to 5×10^{14} cm^{-3}. Because of the very low

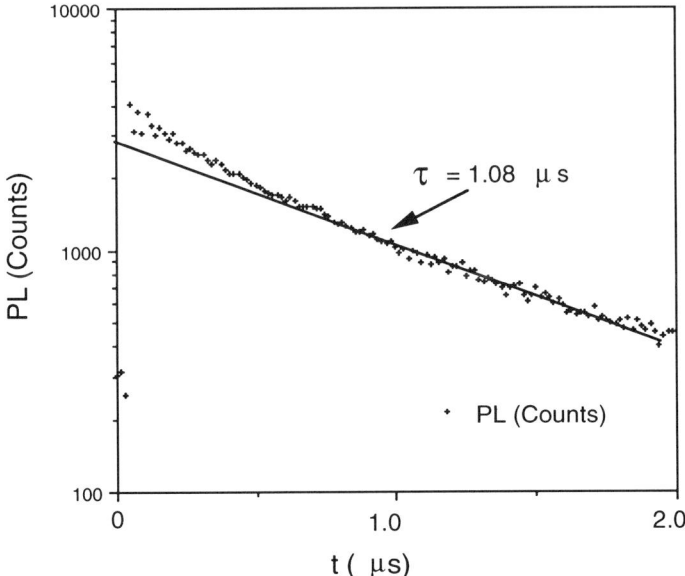

FIG. 27. The PL decay of an $AL_xGa_{1-x}As/GaAs$ DH grown at NREL by Nozik and co-workers. The GaAs is undoped, and the active layer thickness is 1.8 μm (Reprinted with permission from Ahrenkiel, 1992a, © 1992, Pergamon Press plc).

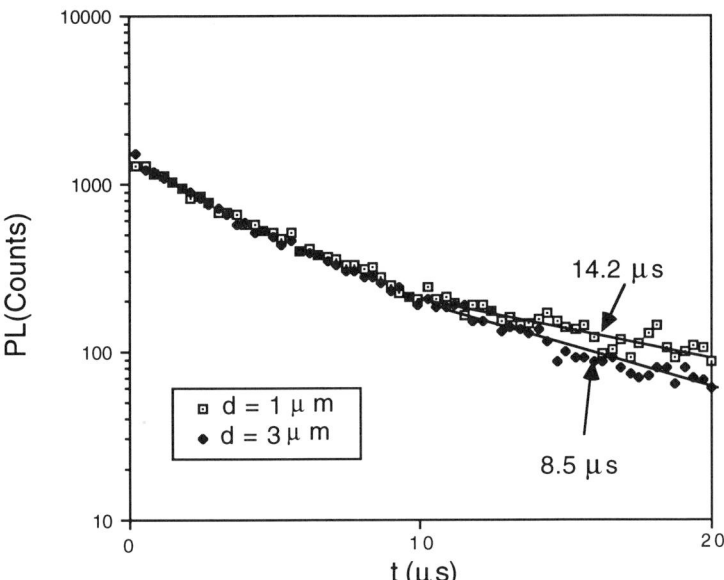

FIG. 28. The PL decay of two GaInP/GaAs DH devices with the indicated active layer thicknesses (Olson et al., 1989; reprinted with permission). The GaAs was undoped, and these are the longest reported GaAs lifetimes to date.

background doping, maintaining low-injection conditions becomes difficult. In these experiments, the laser intensity was lowered until low injection was approached. The injection level shown here is about 1×10^{15} cm^{-3}, and the devices clearly are initially in high injection. As the initial PL decay is bimolecular, the slope is measured at times greater than 10 μs. The 1.0-μm device has a PL lifetime of 14.2 μs; that is the largest lifetime so far reported for GaAs. The PL decay time of the 3 μm device is about 9 μs. The lack of thickness dependence here is indicative of dominant bulk recombination mechanisms. In other words, the $2S/d$ term of Eq. (206) is not greatly affecting the lifetime. The temperature dependence will show that radiative recombination dominates the lifetime even at room temperature. A determination of the S-values will be discussed in the next section.

d. Temperature Dependence of Undoped DHs

Figure 29 shows the figure of merit, τ_{PL}/d, versus temperature of a number of undoped GaAs DH structures that have been reported in the

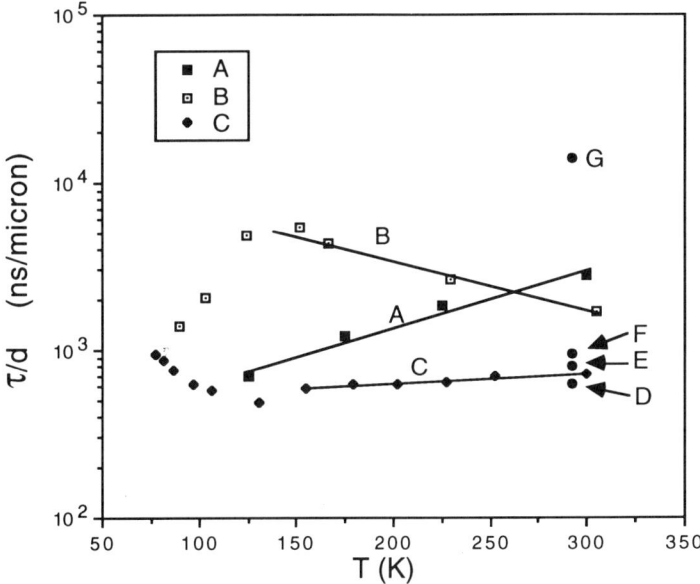

FIG. 29. Figure of merit defined as τ_{PL}/d versus temperature for a number of undoped GaAS DH structures reported in the literature. Curve A is a GaInP/GaAs DH (Olson et al., 1989; Ahrenkiel et al., 1990b). Curve B (Molenkamp and van't Bilk, 1988) and Curve C (Wolford et al., 1991) are for Al$_x$Ga$_{1-x}$As/GaAs devices. The other room-temperature points are: D (Dawson and Woodbridge, 1984); E (Hummel et al., 1990); F ('t Hooft et al., 1985a) and G (Olson et al., 1989).

literature. Curve A is the data for the 3 μm thick GaInP/GaAs DH that was recently reported (Olson et al., 1989; Ahrenkiel et al., 1990b, 1992b). The PL lifetime of this device varied as $T^{1.59}$ between 77 K to 300 K. The data of Curve B are for a 0.12 μm, undoped $Al_{0.3}Ga_{0.7}As$/GaAs structure and show the exponential dependence of Eq. (209) with E_a equal to 27 meV (Molenkamp and van't Bilk, 1988). In this device, the lifetime is dominated by the interface recombination. Applying Eq. (206), the authors calculate a room-temperature S of 18 cm/s. Curve C is the data of Wolford and co-workers (1991) from an undoped MOCVD-grown $Al_{0.30}Ga_{0.70}As$/GaAs DH with a thickness of 0.30 μm. The lifetime is weakly dependent on temperature from about 150 K to 300 K, indicating that the dominant recombination mechanism is nonradiative. The temperature dependence of the lifetime is not indicative of dominant radiative recombination in any temperature range. Between 50 K and 100 K, the lifetime is thermally activated with a slope similar to Curve B. This behavior is indicative of dominant interface recombination.

Other room-temperature measurements referred to in the preceding text are plotted as single points in Fig. 29. The datum of Dawson and Woodbridge (1984) is shown as point D. Other data shown are point E (Hummel et al., 1990), point F ('t Hooft et al., 1985a), and point G (Olson et al., 1989).

A strong radiative component is obvious from Curve A. Therefore, the calculation of surface recombination velocity S is not described by Eq. (206) and becomes more complicated. Because of the approximately $T^{1.5}$ lifetime dependence, one can write the temperature dependence of the total PL lifetime as

$$\frac{1}{\tau_{PL}(T)} = \frac{B(T)N}{\phi} + \frac{1}{\tau_{nR}^T} \tag{210}$$

Here I am combining bulk SRH and surface recombination into one term that is here labeled τ_{nR}^T. As the temperature is lowered, $B(T)$ increases, and the radiative component begins to dominate the total lifetime. Therefore, as τ_{Pl} decreases with lowered temperature, radiative recombination is at least partially dominating the total lifetime.

The calculation of S can be refined by knowing the radiative component of the total lifetime. It was shown earlier that $\phi \sim 5$ for the 3.0-μm device. In Fig. 30, the assumption is made that τ_{PL} equals $\phi\tau_R$ at 125 K. Extrapolated to a higher temperature, Eq. (211) can be used to calculate the nonradiative component. Here I used the temperature dependence $T^{1.7}$ suggested by earlier work ('t Hooft and van Opdorp,

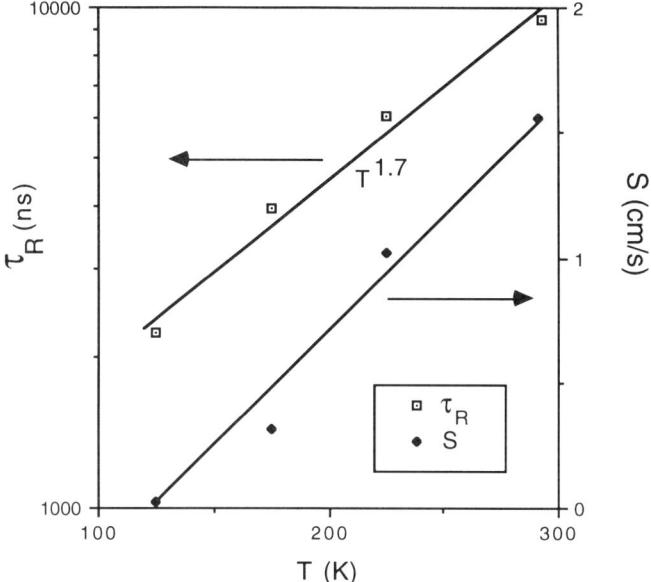

FIG. 30. The radiative lifetime and the surface recombination velocity are calculated as a function of temperature from the data for the GaInP/GaAs device (Olson *et al.*, 1989; Ahrenkiel *et al.*, 1990b).

1983), as τ_{PL} increases more rapidly than $T^{1.5}$. Therefore, S is given by

$$S = \frac{d}{2}\left(\frac{1}{\tau_{PL}(t)} - \frac{B(T)N}{\phi}\right). \tag{211}$$

The S-value computed from Eq. (211) is plotted in Fig. 30. The room-temperature value is about 1.5 cm/s and is in agreement with earlier estimates (Olson *et al.*, 1989).

These undoped GaAs DH devices are useful diagnostic structures. The best GaInP confinement layer appears to have lower S-values than the best $Al_xGa_{1-x}As$ confinement layers. Future work will find if that advantage continues.

21. LIFETIME OF ELECTRONS IN p-TYPE EPITAXIAL GaAs

The minority-carrier lifetime in GaAs has been more thoroughly studied than that in any other semiconductor except silicon. There have been more measurements of minority-carrier lifetime on p-type than

n-type GaAs, for several reasons. First, the very high mobility of minority-carrier electrons in p-GaAs is attractive for many devices. Second, early work showed that the light emission from p-type GaAs is usually stronger than that from n-type GaAs.

There were several reported lifetime measurements prior to the systematic work of Nelson and Sobers. Keune and co-workers (1971) made a DH device with the structure n^+-AlGaAs/n-GaAs/n^+-GaAs. They measured hole lifetimes at 77 K of 4.1 ns in a confined n-layer with $N_D = 1.5 \times 10^{15}$ cm^{-3}. Without the AlGaAs confinement layer, the measured hole lifetimes dropped to less than 1.0 ns. Acket and co-workers (1974) used the phase shift technique to study lifetimes in heavily Ge-doped, 2.0-μm thick GaAs films grown by LPE. They did not attempt to deduce surface effects, but developed a technique for including the effects of a surface recombination velocity.

Nelson (1978) and Nelson and Sobers (1978a, 1978b) made systematic studies of p-type, isotype heterostructures as a function of carrier concentration. These LPE DHs had the structure p-Al$_{0.5}$Ga$_{0.5}$As/p-GaAs/p-Al$_{0.5}$Ga$_{0.5}$As. They used time-correlated single photon counting and calculated the photon recycling factor for their devices as a function of d. In addition, they measured the external quantum efficiency and calculated the internal quantum efficiency from the data. Finally, Nelson (1978) systematically measured the PL lifetime as a function of d for p-type DHs. The GaAs layers have doping levels of 5×10^{15}, 2.9×10^{16}, and 1.7×10^{17} cm^{-3}. The slope of $1/\tau_{PL}$ versus $1/d$ indicated that S increased with doping level. The least squares fits to the data indicated that $S = 300$ cm/s, 350 cm/s, and 500 cm/s, respectively, for the three concentrations.

The internal quantum efficiency of a device can be related to radiative and nonradiative lifetimes as indicated by Eq. (60). Nelson and Sobers (1978a) showed that the external quantum efficiency can be expressed as

$$\eta_{ext} = G' \frac{\tau_{PL}}{\tau_R}. \qquad (212)$$

Here, G' is the fraction of photons that escape from the sample. G' was calculated by methods similar to that in calculating the photon recycling factor ϕ. By measuring τ_{PL} and the external quantum efficiency η_{ext}, one can calculate the radiative lifetime from Eq. (212). The experimental data agreed well with the B-coefficient calculations (Stern, 1976; Casey and Stern, 1976). Figure 31 shows the Nelson and Sobers PL decay data (Curve A) and the corrected bulk lifetime (Curve B). The latter

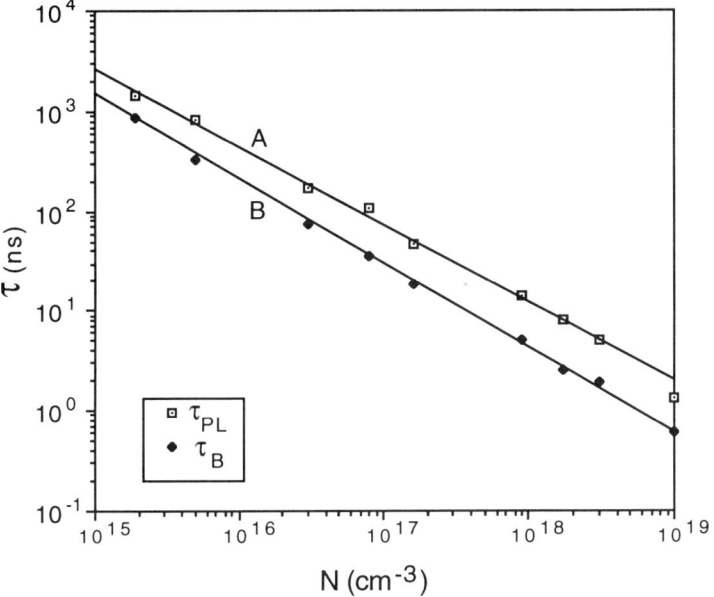

FIG. 31. The PL lifetimes (Curve A) and the bulk electron lifetimes (Curve B) for p-type GaAs as a function of hole density (Nelson and Sobers, 1978a).

was calculated by using the values of S (Nelson, 1978) in Eq. (119). The data show that

(1) the PL lifetime is dominated by radiative recombination and varies approximately inversely with N, and
(2) the measured lifetime $\tau_{Pl} > \tau_R$ indicates that photon recycling is an important factor in measurements of this type.

These data appear to be the first in the literature that show that the lifetime in epitaxial GaAs is controlled by radiative recombination. Also, the data showed, for the first time, that minority-carrier lifetimes greater than 1 μs could be obtained in undoped GaAs.

G. W. 't Hooft and co-workers (1981) grew MOCVD DHs with the structure p-$Al_{0.6}Ga_{0.4}As/p$-$GaAs/p$-$Al_{0.6}Ga_{0.4}As$. Lifetime measurements on these structures produced values in relative agreement with the earlier LPE data of the group, which corrected their data by subtracting a surface lifetime and taking S as 400 cm/s. Recent work (Ahrenkiel et al., 1990a) studied the bulk lifetime in p-type 4×10^{18} cm^{-3} DH structures with AlGaAs confinement layers of both 30% and 80% aluminum content. These data indicated lower S-values with the 30% AlGaAs confinement layer.

Ahrenkiel and co-workers (1988d) measured lifetimes in free-standing MOCVD p-$Al_{0.90}Ga_{0.10}As/GaAs$ DHs (5×10^{16} cm^{-3}) that were etched from the host substrate after growth. The liftoff process decreased the bulk lifetime from about 64 ns to 32 ns.

22. Lifetime of Holes in Epitaxial, n-Type GaAs

Until recently, there has been much less known of hole lifetimes in n-GaAs than of electron lifetimes in p-GaAs. The early measurements in n-type wafers were described in Section I. One of the first reported lifetime estimates on epitaxial material was by Casey and co-workers (1973). Diffusion length measurements were made on LPE-grown n-GaAs over a large concentration range. The Phillips group ('t Hooft and van Opdorp, 1983) measured lifetimes in n-type MOCVD DHs over a concentration range from low 10^{15} cm^{-3} to 1×10^{19} cm^{-3}. Recent work by Puhlmann and co-workers (1991) examined lifetimes in LPE-grown GaAs. The author and co-workers (Ahrenkiel et al., 1989) measured the hole lifetime in n-type (2×10^{17} cm^{-3}) $Al_{0.30}Ga_{0.70}As/GaAs$ DHs as a function of growth temperature. Other work (Ahrenkiel et al., 1990b) described measurements of the hole lifetime in GaInP/GaAs DHs over the range of 3 to 5×10^{17} cm^{-3}.

Very recently, an extensive study was undertaken by Lush and co-workers (1991, 1992a, 1992b) on MOCVD DHs over the concentration range of 1.3×10^{17} cm^{-3} to 3.8×10^{18} cm^{-3}. For each doping concentration, five DH devices were grown with thicknesses ranging from 0.25 μm to 10 μm. These devices were grown by MOCVD at 740°C by H. F. MacMillan at the Varian Research Center. The data from this study enabled the group of workers to resolve bulk lifetimes, interface recombination velocities, and photon recycling factors. This work will be discussed in detail here.

Some of the TRPL data from that study were shown in Fig. 20 for the 10 μm thick, n-type DH doped to 1.3×10^{17} cm^{-3}. The low-injection lifetime varied from 480 ns to 532 ns across the film and is comparable to that of many undoped devices. Using the B-coefficient as 2×10^{-10} cm^{-3} and neglecting the surface lifetime and the bulk τ_{nR}, ϕ (10 μm) = 14.0. By analyzing the very thin (1.25 μm) devices in this concentration series, one calculates that $S \sim 12$ to 109 cm/s. That translates to a surface lifetime of between 4.6 and 41.7 μs in the 10 μm device. Thus, surface recombination appears to be negligible in these devices and truly "surface-free' behavior is observed.

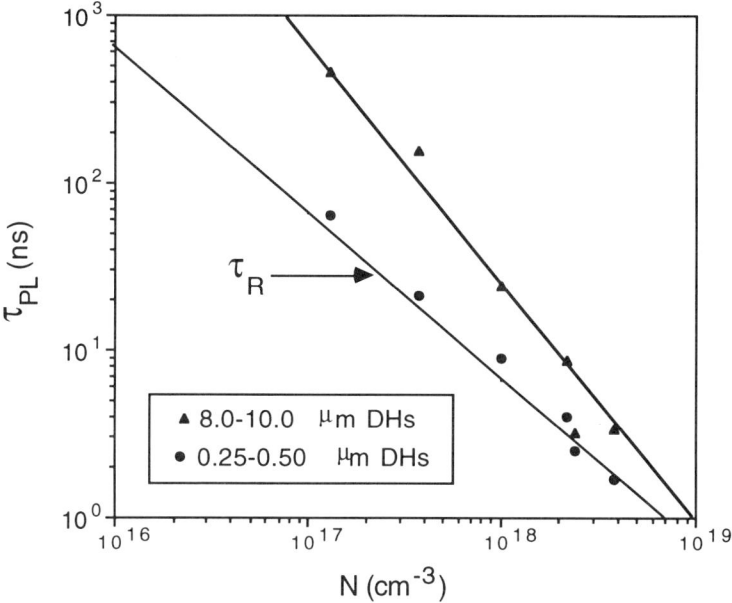

FIG. 32. The PL lifetimes for thick (▲) and thin (●) n-type DHs grown by MOCVD as a function of electron concentration (Lush *et al.*, 1992a, reprinted by permission). The radiative lifetime is indicated by the solid line.

Figure 32 plots the PL lifetime of the thinnest and thickest DH structures from each doping concentration. The solid line is a plot of the radiative lifetime, again taking $B = 2 \times 10^{-10}$ cm^3/s. All of the data points fall on or above the radiative lifetime. The circular data points are from devices either 0.25 μm or 0.50 μm thick, and the points generally lie slightly higher than τ_R. The thick devices (triangular points) are either 8.0-μm or 10.0-μm thick. The PL lifetime of the 10-μm device from the 1.3×10^{17} cm^{-3} series is 12.2 times larger than the radiative lifetime. As the doping density increases, τ_{PL}/τ_R decreases. This decrease is a result of the drop in $\phi(d)$ with increased doping and will be discussed later. The ▲ data points are representative of the radiative lifetime and agree quite well with the accepted B-value.

Detailed studies (Lush *et al.*, 1992a) show that there is a negligible component of bulk SRH recombination in this series of devices except at doping levels of 1×10^{18} cm^{-3} and higher. The higher doping levels showed an increase in lifetime at higher injection levels that is characteristic of SRH-dominated recombination (see Fig. 21). Thus, similarly to the case of p-type DH structures, the minority-carrier lifetimes in n-type epitaxial devices are controlled by radiative recombination. As verified by the data plotted in Fig. 32, the effective radiative lifetime may increase up to an order of magnitude because of photon recycling.

The radiative lifetimes are consistent with the data of Casey and Stern (1976) and Stern (1976). The lifetime minimum predicted by van Cong (1981) is certainly not consistent with the available data. The data also show no indication of the onset of Auger recombination at higher doping levels.

23. INTERFACE RECOMBINATION IN GaAs STRUCTURES

a. Calculating S from TRPL Data

The measurement of interface recombination in $Al_xGa_{1-x}As/GaAs$ DHs was first shown in the measurements of Nelson (1978). Equation (119) for the DH lifetime indicates that $1/\tau_{PL}$ is linear in $1/d$ except for the thickness dependence inherent in the photon recycling factor. Therefore, measuring the slope of $1/\tau_{PL}$ versus $2/d$ allows S to be calculated. Figure 33 shows TRPL data by Nelson (1978) on 15

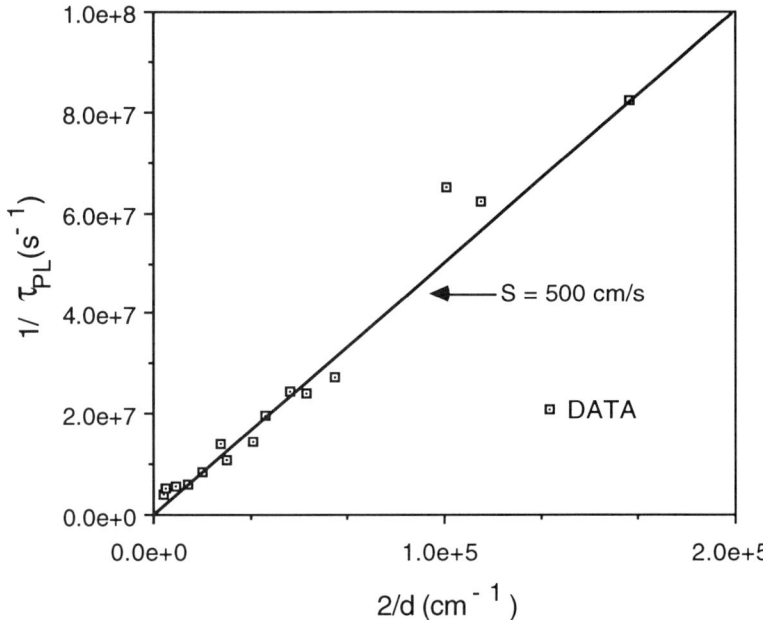

FIG. 33. The inverse of the PL lifetime versus $2/d$ (from the data of Nelson, 1978).

Al$_x$Ga$_{1-x}$As/GaAs DH devices with a range of active layer thicknesses. This study was done very carefully, and many devices were fabricated with $d < 1$ μm. The figure shows a least squares fit to the data over the range $2/d > 0.25 \times 10^5$ cm^{-1}. A straight line fits the data fairly well, although there is appreciable scatter.

It is not necessary to make numerous diagnostic devices to obtain useful estimates of S. A value may be calculated by measuring the PL lifetime of as few as two devices of different thicknesses. The photon recycling must be estimated if radiative recombination is a significant component. The analysis is simplified if the two d values are less than 1 μm so that $\phi(d)$ changes very little (see Fig. 15). Good choices are $d = 0.25$ and 0.50 μm, or $d = 0.50$ and 1.0 μm.

b. Other Experimental Data

Nelson (1978) noted that for Al$_x$Ga$_{1-x}$As/GaAs DHs, S increases with the GaAs doping level. Nelson finds that S is about 350 cm/s at 3×10^{16} cm^{-3} and increases to 550 cm/s at 2×10^{17} cm^{-3}. The S-values increase with doping to between 450 and 650 cm/s in the low 10^{18} cm^{-3} range. Many larger values are reported in the literature.

Recent studies measured the recombination velocity between GaAs and lattice-matched GaInP. The previous section indicated that for undoped DHs, S is less than 2 cm/s. For GaAs doped in the 1 to 5×10^{17} cm^{-3} range, S was measured as 196 cm/s using GaInP as the window layer (Ahrenkiel et al., 1990b).

As noted earlier, very recent work by Lush and co-workers (1992b) found S less than 11 cm/s. These DH devices were grown by MOCVD at 740°C using a Al$_{0.30}$Ga$_{0.70}$As window layer and active layer doping of 1.3×10^{17} cm^{-3}. This is the lowest reported value of S in moderately doped Al$_x$Ga$_{1-x}$As/GaAs DH devices.

c. Dependence of S on Growth Temperature

Studies found that the recombination velocity at the MOCVD n-Al$_{0.30}$Ga$_{0.80}$As/GaAs interface changed by orders of magnitude, depending on growth temperature. These systematic studies (Ahrenkiel et al., 1989, 1990a) had the purpose of improving the efficiency of MOCVD GaAs solar cells. The Al$_{0.30}$Ga$_{0.80}$As/GaAs interface was used as a back surface minority-carrier reflector for the 1×10^{17} cm^{-3} n-type base. Double heterostructures, 4 μm thick and doped n-type to 1–2×10^{17} cm^{-3}, were made in a series of thicknesses. At each growth

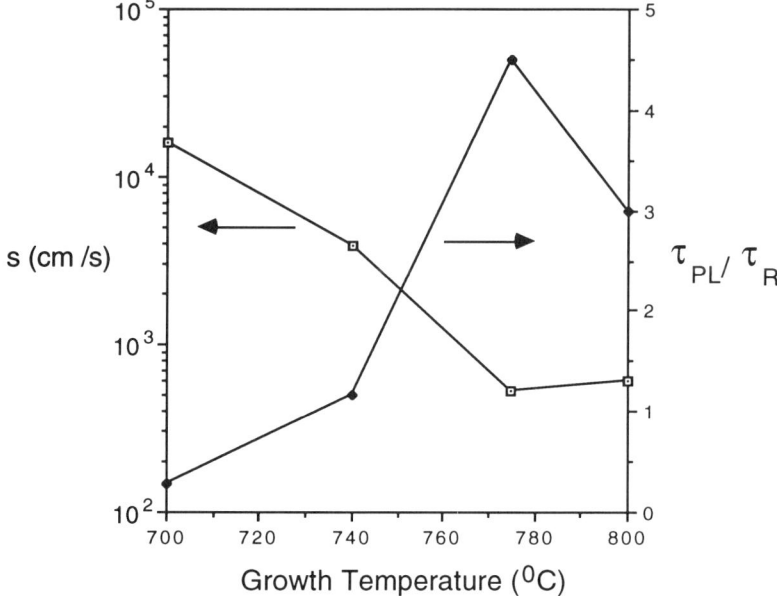

FIG. 34. The interface recombination velocity (S) versus MOCVD growth temperature for n-type, $Al_{0.30}Ga_{0.70}As/GaAs$ DHs (Ahrenkiel et al., 1989, reprinted by permission). The ratio τ_{PL}/τ_R is plotted as the right-hand ordinate.

temperature, DH diagnostic devices were grown with the GaAs layer thickness (d) being about 1, 2, 4 and 8 μm. Figure 34 shows the variation of the recombination velocity S and the ratio of τ_{PL}/τ_R with growth temperature. The active layer thickness is 4 μm for this series. The experimental analysis included the photon recycling effect and ϕ, which was calculated (Miller, 1989) specifically for these devices as a function of thickness. The data show that S drops many orders of magnitude for growth temperatures greater than 740°C. The interface recombination velocity was found to vary from a minimum of about 500 cm/s at a growth temperature of 775°C to 20,000 cm/s at 700°C. The devices grown at temperatures above 740°C were very much improved, with $\tau_{Pl} \gg \tau_R$. The 775°C growth temperatures appears to be optimum for this particular MOCVD system. As noted earlier, the very low values of S reported by Lush and co-workers (1992a) were obtained from MOCVD devices grown at 740°C. In general, the interface recombination velocity of MOCVD-grown $Al_xGa_{1-x}As/GaAs$ minimizes with growth in the range 740 to 780°C.

24. THE PHOTON RECYCLING EFFECT IN GaAs

The photon recycling effect has been observed in the laboratory for a number of years. The early work by Ettenberg and Kressel (1976) showed lifetimes that were five times larger than the predicted radiative lifetime (Casey and Stern, 1976). TRPL data by Nelson and Sobers (1978a, 1978b) found PL lifetimes in p-type DH devices that were significantly larger than the calculated radiative lifetimes. These authors analyzed the data by calculating a photon recycling factor as a function of active layer thickness.

Recent TRPL measurements (Ahrenkiel et al., 1989) on n-type $Al_{0.30}Ga_{0.70}As/GaAs$ DHs found large τ_{PL}/τ_R ratios in 4- to 8-μm-thick DH devices. The active layers were doped to about 1×10^{17} cm^{-3}, and the measured lifetimes exceeded the radiative lifetimes by factors of 3 to 8 for active layers of 4- to 8-μm thickness, respectively. Bensaid and co-workers (1989) measured PL decay in angle-lapped $Al_xGa_{1-x}As/GaAs$ DH devices. They found a monotonic lifetime increase as the beam was scanned from the thin to the thick regions of the active layer. Partain and co-workers (1990) measured the spectral response and diffusion length in heavily doped GaAs homojunctions. The data indicated diffusion lengths that were longer than a "theoretical" maximum using $(D\tau_R)^{0.5}$ in which D is the majority-carrier diffusivity at a given doping level. The interpretation offered is that photon recycling increases the lifetime above the theoretical limit, producing an enhanced diffusion length. Recent calculations (Renaud et al., 1992) also indicate that the minority-carrier diffusivity in GaAs is increased by the photon recycling effect. Thus, both D and τ may have been increased by recycling in these experiments.

A definitive test for photon recycling is to measure PL lifetime in a series of DH devices with a range of active layer thicknesses. The DH devices must be identically grown (constant S) with the same carrier concentration. If photon recycling is not important, then Eq. (111) predicts that $1/\tau_{PL}$ versus $2/d$ will be a linear plot. Most of the available data indicate that such plots are nonlinear.

The work by Lush and co-workers (Lush et al., 1991, 1992a) on n-type $Al_{0.30}Ga_{0.70}As/GaAs$ DH devices was described earlier. Five devices, ranging in thickness from 0.25 to 10 μm, were made at each doping level. Low-injection lifetime data for three of these doping series are shown in Fig. 35 (Ahrenkiel et al., 1992a). The electron concentrations for the three series are A: 1.3×10^{17} cm^{-3}; B: 3.7×10^{17} cm^{-3}; and C: 1×10^{18} cm^{-3}. The data are plotted as $1/\tau_{PL}$ versus $2/d$ for each doping level. It is quite obvious that the plots are not linear, which indicates that the

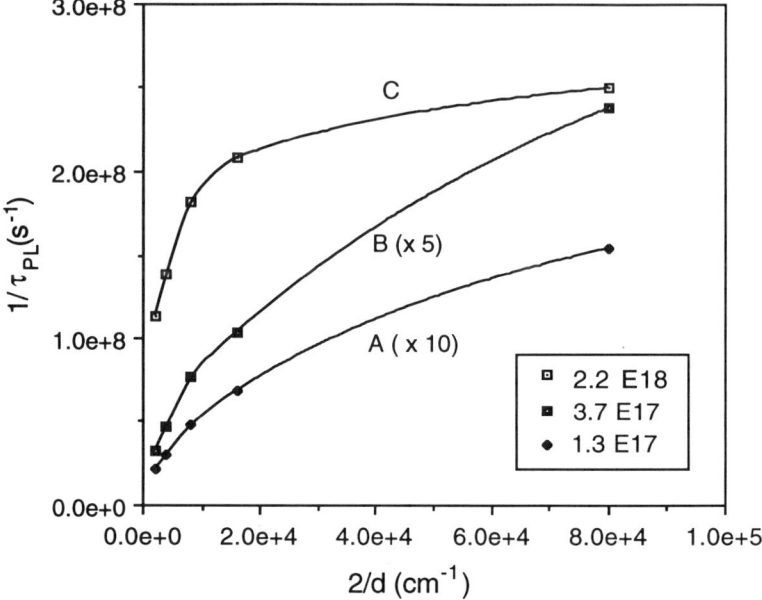

FIG. 35. The plots of $1/\tau_{PL}$ versus $2/d$ for n-type DHs (Lush et al., 1992a). The electron concentrations are A (1.3×10^{17} cm^{-3}); B (3.7×10^{17} cm^{-3}); and C (2.2×10^{18} cm^{-3}). To allow plots to be compared on the same graph, the data ($1/\tau_{PL}$) are multiplied by factors of 10 and 5 for curves A and B, respectively.

$\phi(d)$ variation is significant. Equation (119) can be solved for $\phi(d)$ as

$$\phi(d) = \frac{BN}{\dfrac{1}{\tau_{PL}} - \dfrac{1}{\tau_{nR}} - \dfrac{2S}{d}}. \tag{213}$$

Assuming that S is constant for the entire concentration series, $\phi(d)$ can be calculated from the PL measurement. Here B was taken as the Stern value (1976) of 2×10^{-10} cm^3/s over this range of electron concentration. The best measurements of S are obtained from devices with a submicron active-layer thickness. In these particular samples, the bulk nonradiative lifetime also appeared to be an insignificant factor. This conclusion was reached by data fits and also by the lack of the typical SRH intensity dependence for the PL decay.

The S value was deduced from the devices with the smallest values of d

in the series. Solving for S, one finds

$$S = \frac{d}{2}\left[\frac{1}{\tau_{PL}} - \frac{BN}{\phi_0}\right]. \tag{214}$$

In this analysis, ϕ_0 is defined as the ϕ-value for $d < 1\,\mu$m and was taken as 1.95 according to the calculations of Miller (1989). The recycling factor, at small values of d, is most affected by total internal reflection. This phenomenon is shown in the calculations of Asbeck and others. The strategy for doing this is as follows. The PL lifetime of the 0.25-μm, 1.3×10^{17} cm^{-3} device is 70.0 ns. At τ_R is 38.5 ns, the τ_{PL}/τ_R ratio is 1.82. Asbeck's calculations produced values of ϕ of 1.6 ($x = 0.25$) to 1.9 ($x = 0.50$) for d less than 1.0 μm. Equation (213) could only fit the $d = 0.25$-μm data if ϕ ($d = 0.25\,\mu$m) is greater than 1.82. Thus, the 1.95 value of ϕ_0 is most consistent with the experimental data and overlaps the calculations. Therefore, this analysis assumes that ϕ ($d < 1\,\mu m$) *is independent of doping* and equal to 1.95 for all doping levels. This approximation neglects the weak linear independence given by Eq. (137).

Figure 36 is a plot of the photon recycling factor ϕ using Eq. (213) with

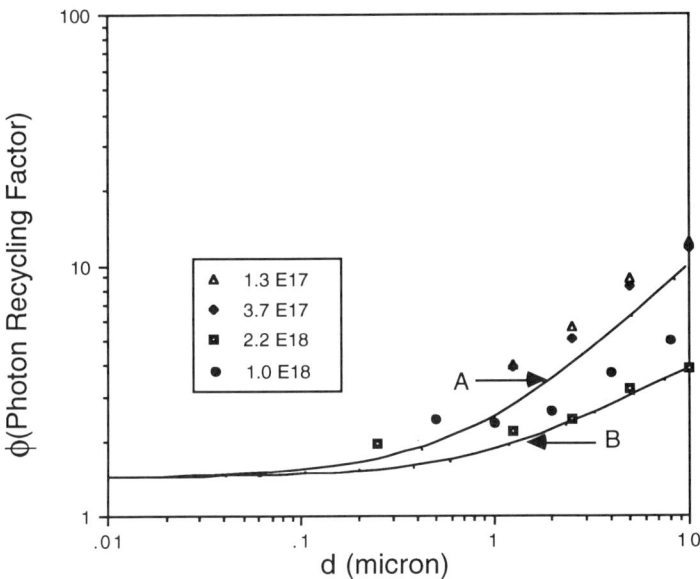

FIG. 36. The photon recycling factor ϕ versus active layer thickness as deduced from PL lifetime measurements on n-type Al$_{0.30}$Ga$_{0.70}$As/GaAs DHs (Ahrenkiel et al., 1992a, reprinted by permission). Curves A and B are the calculated values of ϕ by Asbeck (1977) for undoped and $n = 2 \times 10^{18}$ cm^{-3} structures, respectively.

S determined as described. Here, the bulk SRH term was dropped and was thought to be insignificant. This procedure was carried out using three electron concentrations with the results shown in the figure. There is a significant drop in ϕ at $d = 5$ to $10\,\mu$m at the higher concentrations relative to the lowest concentration $(1.3 \times 10^{17}\,\text{cm}^{-3})$. The values are compared with the Asbeck (1977) calculated values at the same n-type doping level. The solid lines are Asbeck's values of $\phi(d)$ for A (undoped GaAs) and B $(n = 2 \times 10^{18}\,\text{cm}^{-3})$. These data are remarkably consistent with the calculations. The data show the decrease in ϕ ($d > 5\,\mu$m) with doping concentration as predicted by the calculations. The overlap between the internal emission spectra and the strong absorption region lessens remarkably at high doping and reduces the photon recycling effect.

An enhancement of the photon recycling effect was reported by Lush and co-workers (1992b). A hole was etched through the substrate of the DH structures described earlier. The TRPL measurements were made in these regions with the substrate etched away and compared with the

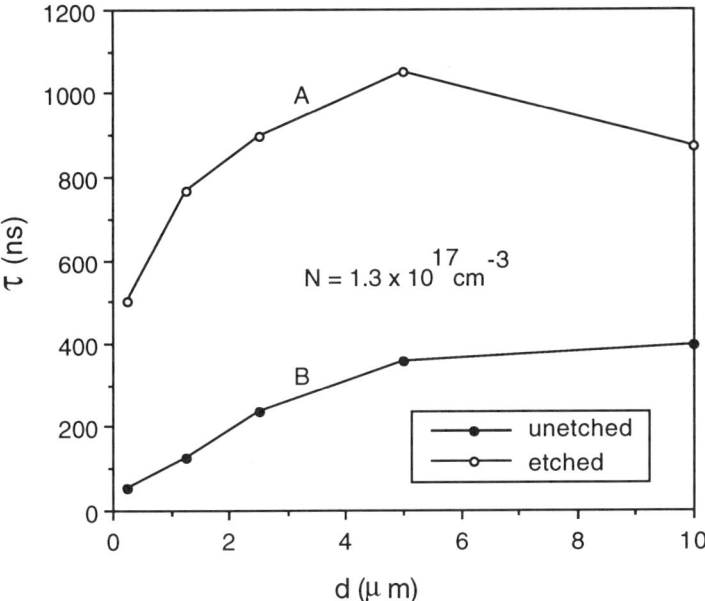

FIG. 37. The PL lifetime versus active layer thickness d in microns for $Al_{0.30}Ga_{0.70}As/GaAs$ with $n = 1.3 \times 10^{17}\,\text{cm}^{-3}$ (Lush et al., 1992b). Curve A represents measurements taken at a region where the GaAs substrate has been etched away. Curve B represents measurements where no etching has occurred. The radiative lifetime for these devices is 38.5 ns.

unetched regions. The data for the thickness series with $N = 1.3 \times 10^{17}$ cm^{-3} are shown in Fig. 37. The PL lifetime for the $d = 5.0$ μm device is 1.07 μs and is 28 times the radiative lifetime. This result verifies calculations that photon recycling is greatly enhanced by substrate removal (Garbuzov et al., 1977). As shown by Fig. 37, the PL lifetime varies from about 500 ns for thin DHs to over 1 μs for thicker DHs. These data suggest that very long lifetimes are feasible in more heavily doped, free-standing thin-film devices because of the photon recycling effect. This finding has applications for the photovoltaic technology (Lush and Lundstrom, 1991), as well as other devices.

VII. Al$_x$Ga$_{1-x}$As Lifetimes

A review of lifetime measurements in Al$_x$Ga$_{1-x}$As can be found in some recent literature (Ahrenkiel and Dunlavy, 1989; Ahrenkiel, 1992a, 1992b). Among the first reported lifetime measurements on Al$_x$Ga$_{1-x}$As was that of van Opdorp and 't Hooft (1981). They grew DH structures with Al$_y$Ga$_{1-y}$As confinement layers and Al$_x$Ga$_{1-x}$As ($y > x$) active layers. The composition range grown was $0.4 < y < 0.62$ and $0.10 < x < 0.17$. These structures were grown both by LPE and by MOCVD. The active layer doping levels $|N_D - N_A|$ varied from 1.7×10^{16} to 1.5×10^{17} cm^{-3}. The data were analyzed assuming an interface recombination velocity of 400 cm/s. Their composite published data are shown in Fig. 38 ('t Hooft, 1981a). These data indicate that the Al$_x$Ga$_{1-x}$As lifetime drops sharply with increasing aluminum concentration. This was the first work to report that the PL lifetime in Al$_x$Ga$_{1-x}$As is dependent upon incident light intensity. The increase in lifetime at high injection levels is indicative of SRH-dominated recombination as described in Section V.

By measuring the efficiency of DH lasers in the high-injection regime, van Opdorp and 't Hooft (1981) were able to measure the nonradiative component in their LPE structures. The measurements revealed that nonradiative centers exist both at the interface and in the bulk of the active layers.

By the zero-field time-of-flight technique, Ahrenkiel and co-workers (1986) measured the hole lifetime and mobility in MOCVD n-Al$_{0.25}$Ga$_{0.75}$As. Data analysis produced a lifetime of 10.0 ns at $|N_D - N_A| = 1 \times 10^{17}$ cm^{-3}.

The alloy changes from a direct- to an indirect-gap semiconductor near $x = 0.37$ with $E_g = 1.92$ eV (Casey and Panish, 1969). Measurements using both PL decay and the zero-field time-of-flight (ZTOF) technique

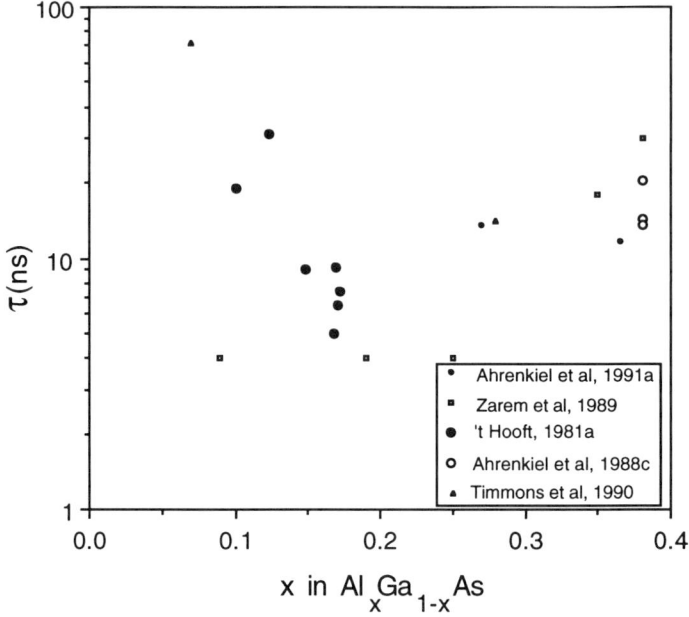

FIG. 38. A composite of reported minority-carrier lifetime measurements in $Al_xGa_{1-x}As$ as a function of the aluminum concentration x. As noted in the text, the $Al_xGa_{1-x}As$ lifetime is generally a function of the excess carrier density.

provided minority-carrier lifetimes in MOCVD-grown n- and p-type $Al_{0.37}Ga_{0.63}As$ (Ahrenkiel et al., 1987). The lifetimes in p-type active layers varied from 2.5 to 7.0 ns for $|N_A - N_D|$ equal to 2×10^{18} and $1-3 \times 10^{17}$ cm^{-3}, respectively. Similar values were deduced for n-type active layers of comparable doping density. These radiative contribution to the total lifetime was discernible in these measurements.

PL decay measurements were made on LPE-grown DHs (Ahrenkiel et al., 1988b) of n-$Al_{0.38}Ga_{0.62}As$. The window layers were composed of 0.5 μm thick layers with n-$Al_{0.8}Ga_{0.2}As$ active layer composition. These devices typically had hole lifetimes greater than 10 ns for active layers between 3.0 and 6.5 μm thick. Again a weak dependence of the lifetime on majority-carrier concentration was observed, showing that $\tau_{nR} < \tau_R$ in $Al_xGa_{1-x}As$. A record low-injection lifetime of 20.2 ns was measured for a 5.0-μm device doped to 4×10^{15} cm^{-3} (Ahrenkiel and Dunlavy, 1989).

A study of the MOCVD $Al_xGa_{1-x}As$ minority-carrier lifetime versus growth temperature indicated that the best lifetimes were obtained at high (800°C) growth temperatures (Timmons et al., 1989; Ahrenkiel et al., 1988a). Other work involved MOCVD-grown devices with an

Al$_{0.08}$Ga$_{0.92}$As active layer (Timmons et al., 1990). A PL lifetime of 67.5 ns was measured in a 5 μm thick DH structure. Several thicknesses were done, and S-values of about 4×10^4 cm/s were found. A lifetime of 11.0 ns was measured in an undoped DH, with an active region width of 4.5 μm and a composition of Al$_{0.28}$Ga$_{0.72}$As (Ahrenkiel et al., 1988a).

Zarem and co-workers (1989) measured the diffusion length and PL lifetime in MBE-grown, undoped Al$_x$Ga$_{1-x}$As DHs over the range $0 < x < 0.38$ ($x = 0.0$, 0.9, 0.19, 0.25, 0.35, and 0.38). High-injection densities were used for the TRPL measurements with $\rho \sim 3 \times 10^{18}$ cm^{-3}. The lifetimes were constant and about 4 ns for $x < 0.35$. Using Eq. (66) and the B-coefficient of GaAs, the high-injection radiative lifetime is about 1 ns at $\rho \sim 3 \times 10^{18}$ cm^{-3}. Thus, radiative recombination can account for the short lifetime at this injection level. With an expected smaller B-value for increasing aluminum content, combined with photon recycling, these data are compatible with theory. At $x = 0.35$, τ_{PL} lifetime was about 20 ns, and at $x = 0.38$, the τ_{PL} lifetime was about 30 ns. The lifetime increase is seen in those devices that have compositions near the direct–indirect crossover. The authors described these effects using a model of population of the indirect L and X valleys. The model was based on thermal population of the indirect valleys according to a Boltzmann factor:

$$\frac{n_x}{n_\Gamma} = \left(\frac{m_x}{m_\Gamma}\right)^{1.5} \exp(-\Delta E/kT). \tag{215}$$

Here n_x and n_Γ are the electron population of the X and Γ valleys, respectively, and ΔE is the energy separation between them. Also, m_x and m_Γ are the density-of-states effective masses in the respective valleys. For larger x-values, the lifetime reduction should increase sharply, according to the theory.

A recent paper (Ahrenkiel et al., 1991a) presents data on MBE-grown Al$_x$Ga$_{1-x}$As with $x = 0.27$ (p-type) and $x = 0.365$ (n-type). A thin AlAs confinement window layer was grown adjacent to the active layer. An outer window layer of Al$_{0.6}$Ga$_{0.4}$As was used to protect the AlAs. The low-injection lifetimes in 4.0 μm devices were 13.6 ns for the p-type structure and 11.6 ns for the n-type structure. By doing a thickness series, researchers found the S-value to be about 9×10^3 cm/s for the n-type DHs. Reproducible growth (in terms of lifetime) was reported for these structures.

Also shown in Fig. 38 is a composite of Al$_x$Ga$_{1-x}$As lifetime data as a function of aluminum content. These data are a mixture of low- and high-injection studies by necessity. The lifetime does not vary monotonically with $1/N$ as for GaAs. The scatter in the data and the intensity dependence of the lifetime are indicative of SRH-dominated recombination. The radiative lifetime of $Al_xGa_{1-x}As$ is not measurable because of the high concentration of SRH centers in the material. With improved growth techniques, the intrinsic recombination properties of $Al_xGa_{1-x}As$ will become measurable.

VII. Summary

A recombination theory appropriate to the III–V semiconductors was developed. TRPL is the common and very powerful measurement technique for measuring minority-carrier lifetime in the light-emitting materials. The isotype double heterostructure or DH is an ideal diagnostic device for lifetime analysis. From appropriately planned experiments, one can obtain both bulk lifetime and interface recombination velocities from the TRPL data analysis. High-injection effects are important in the analysis of both radiative and SRH-dominated recombination. The minority-carrier lifetime in bulk GaAs wafers is SRH-limited from surveys of the limited available data. The lifetime in epitaxial GaAs is limited by the radiative lifetime and is inversely proportional to majority-carrier density in low injection. $Al_xGa_{1-x}As$/GaAs interface recombination velocities (S) less than 50 cm/s are common for undoped DH devices. For undoped GaInP/GaAs DHs, interface recombination velocities less than 2 cm/s have been reported. The interface recombination velocity increases with majority-carrier concentration. Recently, S-values less than 100 cm/s have been reported for GaAs devices at the 1×10^{17} cm^{-3} majority-carrier concentration level. Photon recycling is clearly an important phenomenon found in the direct-bandgap semiconductors. Recent work indicates PL lifetimes that are 10 to 38 times the calculated radiative lifetime for GaAs DH devices. The application of photon recycling to improved device performance appears to be a promising technology in the future.

The lifetime in epitaxial $Al_xGa_{1-x}As$ is defect-limited, with wide scatter in the reported data. Because the lifetime is SRH-controlled, the lifetime is very injection-dependent.

Acknowledgments

The author wishes to thank his co-workers D. J. Dunlavy, B. M. Keyes, and D. Levi for their inestimable contributions to this work. D. J.

Dunlavy was primarily responsible for designing and building the time-resolved photoluminescence as well as the photoluminescence spectroscopy facilities. B. M. Keyes and D. Levi have measured and analyzed hundreds of samples using the TRPL technique. I wish to thank M. L. Timmons, S. M. Vernon, A. Nozik, and J. Olson for providing interesting samples or devices for advancing our fundamental studies of minority-carrier lifetime in III–V semiconductors. The author is grateful to Mr. G. Lush of Purdue University for use of his very important data prior to its publication. I also wish to thank my wife Cindy for her support during the preparation of this volume.

The author wishes to thank a number of workers in this field for taking the time to do a critical reading of this manuscript prior to publication. Besides doing a critical technical edit of the manuscript, those named below made important comments and suggested additions to the manuscript. I am deeply grateful for their taking the time to read and make the suggestions which greatly improved the quality of this manuscript.

The technical reviewers include Dr. G. W. 't Hooft of the Phillips Research Laboratories (Eindhoven), Dr. A. Ehrhardt and Professor W. Wettling of the Fraunhofer Institute (Stuttgart), and Professor Peter Landsberg of the University of South Hampton. Also, making important contributions to the editing were co-workers B. M. Keyes (NREL), M. S. Lundstrom (Purdue), and G. Lush (Purdue).

I wish to thank Penny Sadler for her contributions to the artwork preparation. The NREL research reported here was sponsored by the U.S. Department of Energy under Contract No. DE-AC02-83Ch10093.

References

Acket, G. A., Nijman, W. and 't Lam H. (1974). *J. Appl. Phys.* **45**, 3033.
Acket *et al.* (1974) p. 133.
Ahrenkiel, R. K. (1987). *J. Appl. Phys.* **62**, 2937.
Ahrenkiel, R. K. (1988). In *Current Topics in Photovoltaics,* Vol. 3. Academic Press, London, Edited by T. J. Cootts and J. D. Meakin.
Ahrenkiel (1990b) p. 135.
Ahrenkiel, R. K. (1991). In Properties of Indium Phosphide,' EMIS. Datareviews Series No. 6, p. 77. INSPEC, IEE, London and New York.
Ahrenkiel, R. K. (1992a). *Solid St. Electron.* **35**, 239.
Ahrenkiel, R. K. (1992b). In "Properties of AlGaAs," EMIS Datareview. INSPEC, IEE, in press.
Ahrenkiel, R. K., and Dunlavy, D. J. (1989). *J. Vac. Sci. Technol.* **A7**, 822.
Ahrenkiel, R. K., and Levi, D. (1992), unpublished data.
Ahrenkiel, R. K., Dunlavy, D. J., Hamaker, H. C., Green, R. T., Lewis, C. R., Hayes, R. E., and Fardi, H. (1986). *Appl. Phys. Lett.* **49**, 725.

Ahrenkiel, R. K., Greenberg, D., Dunlavy, D. J., Hanak, T., Schlupmann, J., Hamaker, H. C., and Lewis, C. R. (1987). *Proceedings of the 19th IEEE Photovoltaic Specialists Conference — 1987,* p. 1180.
Ahrenkiel, R. K., Dunlavy, D. J., and Timmons, M. L. (1988a). *Proceedings of the Twentieth IEEE Photovoltaics Specialists Conference — 1988, Las Vegas, Nevada,* pp. 611–615.
Ahrenkiel, R. K., Dunlavy, D. J., Loo, R. Y., and Kamath, G. S. (1988b). *J. Appl. Phys.* **63,** 5174.
Ahrenkiel, R. K., Dunlavy, D. J., and Hanak, T. (1988c). *Solar Cells* **24,** 339.
Ahrenkiel, R. K., Dunlavy, D. J., Benner, J., Gale, R. P., McCleland, R. W., Gormely, M. M., and King, B. D. (1988d). *Appl. Phys. Lett.* **53,** 598.
Ahrenkiel, R. K., Dunlavy, D. J., Keyes, B. M., Vernon, S. M., Dixon, T. M., Tobin, S. P., Miller, K. L., and Hayes, R. E. (1989). *Appl. Phys. Lett.* **55,** 1088.
Ahrenkiel, R. K., Dunlavy, D. J., Keyes, B. M., Vernon, S. M., Tobin, S. P., and Dixon, T. M. (1990a). *Proceedings of the Twentieth IEEE Photovoltaics Specialists Conference — 1990,* p. 432.
Ahrenkiel, R. K., Olson, J. M., Dunlavy, D. J., Keyes, B. M., and Kibbler, A. E. (1990b). *J. Vac. Sci. Technol.* **A8,** 3002.
Ahrenkiel, R. K., Keyes, B. M., Shen, T. C., Chyi, J. I., and Morkoc, H. (1991a). *J. Appl. Phys.* **69,** 3094.
Ahrenkiel, R. K., Keyes, B. M., and Dunlavy, D. J. (1991b). *Solar Cells* **30,** 163.
Ahrenkiel, R. K., Keyes, B. M., and Dunlavy, D. J. (1991c). *J. Appl. Phys.* **70,** 225.
Ahrenkiel, R. K., Keyes, B. M., Lush, G. B., Melloch, M. R., Lundstrom, M. S., and MacMillan, H. F. (1992a). *J. Vac. Sci. Technol.* **A10,** 990.
Ahrenkiel, R. K., Keyes, B. M., Dunlavy, D. J., and Kazmerski, L. L. (1992b). In *Proceedings of the 10th EC Photovoltaic Energy Conference,* pp. 533–536. Kluwer Academic Publishers, The Netherlands.
Asbeck, P. (1977). *J. Appl. Phys.* **48,** 820.
Bachrach, R. Z. (1972). *Rev. Scientific Instruments* **43,** 734.
Bardeen, J. (1947). *Phys. Rev.* **71,** 727.
Bensaid, B., RAymond, F., Lerous, M., Verie, C., and Fofana, B. (1989). *J. Appl. Phys.* **66,** 5542.
Blakemore, J. S. (1987). In *Semiconductor Statistics,* p. 263. Pergamon Press, New York.
Boulou, M., and Bois, D. (1977). *J. Appl. Phys.* **48,** 4713.
Braunstein, R. (1955). *Phys. Rev.* **99,** 1892.
Carslaw, H. S., and Jaeger, J. C. (1978). In *Conduction of Heat in Solids,* 2nd Ed. Oxford University Press, Oxford.
Casey, H. C., Jr., Miller, B. I., and Pinkas, E. (1973). *J. Appl. Phys.* **44,** 1281.
Casey, H. C., Jr., and Panish, M. B. (1969). *J. Appl. Phys.* **40,** 4910.
Casey, H. C., Jr., and Buehler, E. (1977). *Appl. Phys. Lett.* **30,** 2550.
Casey, H. C., Jr., and Panish, M. B. (1978). In *Heterostructure Lasers.* Academic Press, Orlando, edited by P. F. Liao and P. Kelley.
Casey, H. C., Jr., and Stern, F. (1976). *J. Appl. Phys.* **47,** 631.
Casey, H. C., Jr., Sell, D. D., and Wecht, K. W. (1975). *J. Appl. Phys.* **46,** 250.
Crank, J. (1975). In *The Mathematics of Diffusion,* 2nd Ed. Clarendon, Oxford.
Dapkus, P. D., Holonyak, N., Jr., Burnham, R. D., and Keune, D. L. (1970). *Appl. Phys. Lett.* **16,** 93.
Dapkus, P. D., Holonyak, N., Jr., Burnham, R. D., Keune, D. L., Burd, J. W., Lawley, K. L., and Walline, R. E. (1979). *J. Appl. Phys.* **41,** 4194.

Dawson, P., and Woodbridge, K. (1984). *Appl. Phys. Lett.* **45**, 1227.
del Alamo, J. A. (1988). In "Properties of Silicon," EMIS Datareviews Series No. 4, Ch. 5, pp. 148–149. INSPEC, IEE, London and New York.
Demas, J. N. (1983). In *Excited State Lifetime Measurements*. Academic Press, New York.
Dhariwal, S. R., and Mehrotra, D. R. (1988). *Solid-St. Electron.* **31**, 1355.
Dumke, W. P. (1957). *Phys. Rev.* **105**, 139.
Ehrhardt, A., Wettling, W., and Bett, A. (1991). *Appl. Phys.* **A53**, 123.
Ettenberg, M., and Kressel, H. (1976). *J. Appl. Phys.* **47**, 1538.
Garbuzov, D. Z. (1982). *J. Luminescence* **27**, 109.
Garbuzov, D. Z. (1986). In *Semiconductor Physics* (V. M. Tuchkevich and V. Y. Frenkel, eds.), p. 53. Consultants Bureau, New York.
Garbuzov, D. Z., Ermakova, A. N., Rumyantsev, V. D., Trukan, M. K., and Khalfin, V. B. (1977). *Sov. Phys. Semicond.* **11**, 419.
Garbuzov, D. Z., Khalfin, V. B., Trukan, M. K., Agafonov, V. G., and Abdullaev, A. (1978). *Sov. Phys. Semicond.* **12**, 809.
Gel'mont, B. L., Sokolova, Z. N., and Khalfin, V. B. (1983). *Sov. Phys. Semicond.* **17**, 180.
Grove, A. S. (1967). In *Physics and Technology of Semiconductor Devices*, p. 129. John Wiley and Sons, Inc., New York.
Hall, R. N. (1952). *Phys. Rev.* **87**, 387.
Hall, R. N. (1960). *Proc. Inst. Elect. Eng.* **106B**, Suppl. 17, 983.
Haug, A. (1983). *J. Phys. C* **16**, 4159.
Henry, C. H., Levine, B. F., Logan, R. A., and Bethea, C. G. (1982). *IEEE J. Quantum Electron.* **QE-19**, 905.
Henry, C. H., Logan, R. A., Merritt, R. R., and Bethea, C. G. (1984). *Electronics Letters* **20**, 359.
't Hooft, G. W., (1981). *Appl. Phys. Lett.* **39**, 389.
't Hooft, G. W. (1992). Private communication.
't Hooft, G. W., and van Opdorp, C. (1983). *Appl. Phys. Lett.* **42**, 813.
't Hooft, G. W., and van Opdorp, C. (1986). *J. Appl. Phys.* **60**, 1065.
't Hooft, G. W., van Opdorp, C., Veenliet, H., and Vink, A. T. (1981). *J. of Crystal Growth* **55**, 173.
't Hooft, G. W., Leys, M. R., and Roozeboom, F. (1985a). *Jpn. J. Appl. Phys.* **24**, L761.
't Hooft, G. W., Leys, M. R., and Thalen-van der Mheen, M. J. (1985b). *Superlattices and Microstructures* **1**, 307.
Hummel, S. G., Beyler, C. A., Zou, Y., Grodzinski, P., and Dapkus, P. D. (1990). *Appl. Phys. Lett.* **57**, 695.
Hwang, C. J. (1971). *J. Appl. Phys.* **42**, 4408.
Hwang, C. J., and Dyment, J. C. (1973). *J. Appl. Phys.* **44**, 3240.
Ioannou, D. E., and Gledhill, R. J. (1984). *J. Appl. Phys.* **56**, 1797.
Kameda, S., and Carr, W. N. (1973). *J. Appl. Phys.* **44**, 2910.
Kamiya, T., Hirose, S., and Yanai, H. (1979). *J. Luminescence* **18**, 910.
Karamon, H., Masumot, T., and Makino, Y. (1985). *J. Appl. Phys.* **57**, 3527.
Keune, D. L., Holonyak, N., Jr., Burnham, R. D., Scifres, D. R., Zwicker, H. W., Burd, J. W., Craford, M. G., Dickus, D. L., and Fox, M. J. J. (1971). *J. Appl. Phys.* **42**, 2048.
Knox, R. S. (1963). In *Theory of Excitons, Solid State Physics* (F. Seitz and D. Turnbull, eds.), Supplement 5. Academic Press, New York.
Kuriyama, T., Kamiya, T., and Yanai, H. (1977). *Jap. J. Appl. Phys.* **16**, 465.

Landsberg, P. T. (1987). *Appl. Phys. Lett.* **50**, 745.
Landsberg, P. T. (1991). In *Recombination in Semiconductors*. Cambridge University Press, Cambridge.
Landsberg, P. T., and Browne, D. C. (1988). *Semicond. Sci. Technol.* **3**, 193.
Landsberg, P. T., and Robbins, D. J. (1978). *Solid-St. Electron.* **21**, 1289.
Loebner, E. E., and Poor, E. W. (1959). *Phys. Rev. Letters* **3**, 23.
Louis, T. A., Ripamonti, G., and Lacaita, A. (1990). *Rev. Sci. Instrum.* **611**, 11.
Lundstrom, M. S., Klausmeier-Brown, M. E., Melloch, M. R., Ahrenkiel, R. K., and Keyes, B. M. (1990). *Solid-State Electronics* **33**, 693.
Lush, G. B., and Lundstrom, M. S. (1991). *Solar Cells* **30**, 337.
Lush, G. B., MacMillan, H. F., Keyes, B. M., Ahrenkiel, R. K., Melloch, M. R., and Lundstrom, M. S. (1991). In *Proceedings of the Twenty-Second IEEE Photovoltaic Specialists Conference, Las Vegas, Nevada*, p. 182. IEEE New York.
Lush, G. B., MacMillan, H. F., Keyes, B. M., Levi, D. H., Ahrenkiel, R. K., Melloch, M. R., and Lundstrom, M. S. (1992a), *J. Appl. Phys.* **72**, 1436.
Lush, G. B., Melloch, M. R., Lundstrom, M. S., Levi, D. H., Ahrenkiel, R. K., and MacMillan, H. F. (1992b). *Appl. Phys. Lett.* **61**, p. 2440.
Many, A., Goldstein, Y., and Grover, N. B. (1971). In *Semiconductor Surfaces*, p. 194. North-Holland, Amsterdam.
Marvin, D. C., and Halle, L. F. (1991). In *Proceedings of the Twenty-First IEEE Photovoltaics Specialists Conference — 1990*, p. 353. IEEE, New York.
Marvin, D. C., and Halle, L. F. (1992). In *Twenty-Second IEEE Photovoltaic Specialists Conference — 1991, Las Vegas, Nevada*, p. 198. IEEE, New York.
Mayburg, S. (1961). *Solid-St. Electron.* **2**, 195.
Miller, G. L., Robinson, D. A. H., and Ferris, S. D. (1978). *Proc. Electrochem. Soc.* **78**, 1.
Miller, K. L. (1989). "Analysis of Transient Photoluminescence from $Al_xGa_{1-x}As$/GaAs Double Heterostructure Samples," M.S.E.E. Thesis, University of Colorado at Boulder.
Molenkamp, L. W., and van't Bilk, H. F. J. (1988). *J. Appl. Phys.* **64**, 4253.
Molenkamp, L. W., Kampschoer, G. L. M., de Lange, W., Maes J. W. F. M., and Roksnoer, P. (1989). *Appl. Phys. Lett.* **54**, 1992.
Moss, T. S. (1957). *Proc. Phys. Soc. (London)* **B70**, 247.
Nathan, m. I., and Burns, G. (1963). *Phys. Rev.* **129**, 125.
Nelson, R. J. (1978). *J. Vac. Sci. Technol.* **15**, 1475.
Nelson, R. J., and Sobers, R. G. (1978a). *J. Appl. Phys.* **49**, 6103.
Nelson, R. J., and Sobers, R. G. (1978b). *Appl. Phys. Lett.* **32**, 761.
Nicollian, E. H., and Brews, J. R. (1982). In *MOS (Metal Oxide Semiconductor) Physics and Technology*, p. 645. John Wiley & Sons, New York.
Oberhettinger, F., and Baddi, L. (1973). In *Table of Laplace Transforms*. Springer-Verlag, Berlin.
Olson, J. M., Ahrenkiel, R. K., Dunlavy, D. J., Keyes, B. M., and Kibbler, A. E. (1989). *Appl. Phys. Lett.* **55**, 1208.
Orton, J. W., and Blood, P. (1990). In *The Electrical Characterization of Semiconductors: Measurement of Minority-Carrier Measurements Properties*, pp. 51–122. Academic Press, London.
Palik, E. D. (1985). In *Handbook of Optical Constants of Solids*. Academic Press, Orlando, Florida.
Pankove, J. I. (1971). In *Optical Processes in Semiconductors*. Dover Publications, New York.
Pankove, J. I., and Massoulie, M. J. (1962). *Bull. Am. Phys. Soc.* **7**, 88.

Partain, L. D., Liu, D. D., Kuryla, M. S., Ahrenkiel, R. K., and Asher, S. E. (1990). *Solar Cells* **28**, 223.
Press, W. H., Flannery, B. P., Teukolsky, S. A., and Vetterling, W. T. (1988). In *Numerical Recipes,* p. 248. Cambridge University Press, New York.
Puhlmann, N., Oelgart, G., Gottschalch, V., and Nemitz, R. (1991). *Semicond. Sci. Tecnol.* **6**, 181.
Renaud, P., Raymond, F., Bensaid, B., and Verie, C. (1992). *J. Appl. Phys.* **71**, 1907.
Roberts, G. E., and Kaufman, H. (1966). In *Table of Laplace Transforms.* W. B. Saunder Company, Philadelphia and London.
Rose, A. (1978). In *Concepts in Photoconductivity and Allied Problems,* p. 11. Robert Krieger Publishing Company, Huntington, New York.
Sah, C. T., Noyce, R. N., and Shockley, W. (1957). *Proc. IRE* **45**, 1228.
Schroder, D. K. (1990). In *Semiconductor Material and Device Characterization,* pp. 359–447. John Wiley & Sons Inc., New York.
Sell, D. D., and Casey, H. C., Jr. (1974). *J. Appl. Phys.* **45**, 800.
Shockley, W. (1932). *Phys. Rev.* **56**, 317.
Shockley, W., and Read, W. T. (1952). *Phys. Rev.* **87**, 335.
Smith, L. M., Wolford, D. J., Martinsen, J., Venkatasubramanian, R., and Ghandhi, S. K. (1990). *J. Vac. Sci. Technol.* **B8**, 787.
Spenke, E. (1958). In *Electronic Semiconductors,* p. 45. McGraw-Hill, New York.
Stern, F. (1976). *J. Appl. Phys.* **47**, 5382.
Stern, F., and Woodall, J. M. (1974). *J. Appl. Phys.* **45**, 3904.
Su, C. B., and Olshansky, R. (1980). *Appl. Phys. Lett.* **41**, 833.
Sugimura, A. (1983). *IEEE J. Quantum Electron.* **QE-19**, 930.
Sze, S. M. (1981). In *Physics of Semiconductor Devices.* Wiley-Interscience, New York.
Tamm, I. E. (1932). *Z. Phys.* **76**, 849.
Timmons, M. L., Hutchby, J. A., Ahrenkiel, R. K., and Dunlavy, D. J. (1989). In *Gallium Arsenide and Related Compounds, Inst. Phys. Conf. Ser. No. 96,* p. 289.
Timmons, M. L., Colpits, T. S., Venkatasubramanian, R., Keyes, B. M., Dunlavy, D. J., and Ahrenkiel, R. K. (1980). *Appl. Phys. Lett.* **56**, 1850.
Tyagi, M. S., Nus, J. F., and van Overstraeten, R. J. (1982). *Solid-St. Electron.* **25**, 411.
Vaitkus, J. (1976). *Phys. Status Solidi* **A34**, 769.
van Cong, H. J. (1981). *Phys. Chem. Solids* **42**, 95.
van Opdorp, C., and 't Hooft, G. W. (1981). *J. Appl. Phys.* **52**, 3827.
van Opdorp, C., Vink, A. T., and Werkhoven, C. (1977). Inst. Phys. Conf. Ser., No. 33b, p. 317.
van Roosbroeck, W. (1955). *J. Appl. Phys.* **26**, 380.
van Roosbroeck, W., and Shockley, W. (1954). *Phys. Rev.* **94**, 1558.
Willander, M., and Grivickas, V. (1988). [In Properties of Silicon, EMIS Datareviews Series No. 4, Ch. 8, pp. 195–197. INSPEC, IEE, London and New York.
Wolford, D. J., Gilliland, G. D., Keuch, T. F., Smith, L. M., Martinsen, J., Bradley, J. A., Tsang, C. F., Venkatasubramanian, R., Ghandi, S. K., And Hjalmarson, H. P. (1991). *J. Vac. Sci. Technol.* **B9**, 2369.
Wong, D., Schlesinger, T. E., and Milnes, A. G. (1990). *J. Appl. Phys.* **68**, 5588.
Yablonovitch, E., and Gmitter, T. (1986). *Appl. Phys. Lett.* **49**, 587.
Yablonovitch, E., and Gmitter, T. (1992). *Solid St. Electron.* **35**, 261.
Yablonovitch, E., Allara, D. L., Chang, C. C., Gmitter, T., and Bright, T. B. (1986). *Phys. Rev. Letts.* **57**, 249.
Yablonovitch, E., Bhat, R., Harbison, J. P., and Logan, R. A. (1987). *Appl. Phys. Lett.* **50**, 1197.

Zarem, H. A., Lebens, J. A., Nordstrom, k. B., Sercel, P. C., Sanders, S., Eng, L. E., Yariv, A., and Vahala, K. (1989). *J. Appl. Phys. Lett.* **55,** 2622.

Zwicker, H. R., Keune, D. L., Holonyak, N., Jr., and Burnham, R. D. (1971a). *Solid St. Electron.* **14,** 1023.

Zwicker, H. R., Scifres, D. R., Holonyak, N., Jr., Dupuis, R. D., and Burnham, R. D. (1971b). *Solid State Comm.* **9,** 587.

CHAPTER 3

High Field Minority Electron Transport in p-GaAs

Tomofumi Furuta

NTT LSI LABORATORIES
KANAGAWA, JAPAN

I. INTRODUCTION	.	151
II. DRIFT VELOCITY	.	153
1. *Introduction*	.	153
2. *Principle of Time-of-Flight Measurement*	.	155
3. *Experimental Results*	.	156
III. ENERGY TRANSFER PROCESS	.	162
4. *Electron Temperature*	.	162
5. *Energy Loss Rate*	.	167
6. *Energy and Momentum Relaxation Times*	.	168
VI. ULTRAFAST ENERGY RELAXATION PROCESS	.	170
V. MONTE CARLO SIMULATION RESULTS	.	177
7. *Scattering Models*	.	178
8. *Calculated Results*	.	180
VI. SUMMARY	.	189
ACKNOWLEDGMENT	.	190
REFERENCES	.	190

I. Introduction

Recent progress, combined with a strong need for handling high-speed information processing, has led to increased interest in high-speed devices. This has led to much attention being focused on III–V compound semiconductors, because these materials possess suitable material parameters for high-frequency device operation. Some of the most fundamental parameters are the carrier transport properties in a high electric field. This review here is primarily from recent experimental work on the high field minority electron transport in p-GaAs, which relates to the carrier dynamics in heterojunction bipolar transistors (HBTs) and solar cells.

Historically, the first experimental study of minority carrier drift mobility in semiconductors was performed by Haynes and Shockley (1949). In a follow-up study (Haynes and Shockley, 1951) and similar subsequent experiments (Vilims and Spicer, 1965; Ettenberg et al., 1973), minority carrier transport properties were thought to be determined by the scattering processes of impurities and phonons. In other words, the carrier mobility was assumed to be independent of whether the quasi-particle is a minority or a majority carrier. However, the scattering mechanisms of these two carrier types may be different when the important role of carrier–carrier scattering in carrier transport is considered (Debye and Conwell, 1954). For minority carriers, Paige and McLean first pointed out the important role of electron–hole interaction (Paige, 1960; McLean and Paige, 1960), which directly affects the total momentum of both types of carrier. They also predicted the negative absolute mobility of minority carriers due to the carrier drag effect, where the minority carriers drift in the same direction as the majority carriers. Since then, while a number of experiments have been performed, the effect of electron-hole interaction on drift mobility has not been clearly observed. This interaction has generally been described in tems similar to ionized impurity scattering (Prince, 1953; Ehrenreich, 1957). Thus, the overall concentration of Coulomb scattering centers has been treated as twice as the ionized impurity concentration. Walukiewicz et al. (1979) calculated the low field minority electron mobility in p-GaAs over wide ranges for hole concentration and temperature. Their calculated results indicated that the minority electron mobility is somewhat lower than the majority electron mobility with the same doping concentration. In the 1980s, several experimental techniques were applied to the investigation of low field minority electron drift mobility in p-GaAs. These include electron diffusion length in combination with electron lifetime (Nelson, 1979), spin relaxation time (Lagunova et al., 1985), diffusion transit time (Ahrenkiel et al., 1986, 1987; Lovejoy et al., 1990), HBT characteristics (Ito and Ishibashi, 1986; Nathan et al., 1988; Klausmeirer-Brown et al., 1990; Tiwari and Wright, 1990), and the photo-Hall method (Ito and Ishibashi, 1989). The experimental results revealed that minority electron drift mobilities were lower than those of majority electrons in p-GaAs, a finding that agrees very well with theoretical predictions (Walukiewicz et al., 1979). This reduction in mobility of minority electrons is generally attributed to the momentum transfer by electron–hole interaction where holes behave as ionized impurity scattering centers because of the heavy effective mass of holes.

However, the electron–hole interaction is essentially different from the ionized impurity scattering in a number of respects: (1) The effective masses of carriers are finite. (2) Electrons and holes are mobile along the electric field, but in opposite directions. (3) Both carriers are able to screen several interactions and each other. In addition, the electron–hole interaction also affects the energy distribution of carriers, in which energy transfer between electrons and holes may occur. Given these differences, one would expect minority electrons to have different transport properties from majority electrons in the high electric field region.

Here, it should be noted that the electron–hole interaction is particularly important in direct-gap semiconductors such as GaAs, because the conduction and valence band extrema in direct-gap semiconductors are not separated in wave-vector space. On the other hand, the strength of the electron–hole interaction in indirect-gap semiconductors such as Si is smaller than that in direct-gap semiconductors. Here, the conduction band minima in indirect-gap semiconductors are widely separated in wave-vector space from the valence band maxima. From this perspective, attention should focus on how the electron–hole interaction affects the dynamics of hot carriers and high field transport properties in III–V compound semiconductors and related materials.

This chapter is organized as followed: Sections II to IV will survey recent experimental work on minority electron transport properties assuming hot-electron conditions. This work uses optical measurements such as the time-of-flight method and time-integrated and time-resolved photoluminescence. This will be followed with a discussion of the data obtained on drift velocity, electron temperature, energy loss rate, and relaxation processes. Finally, in Section V, I will present the calculated results by a Monte Carlo simulation, including electron–hole interaction, and compare these results with the experimental ones.

II. Drift Velocity

1. INTRODUCTION

Generally, the switching speed of conventional electronic devices is governed by the drift velocity of carriers in the high electric field region. Information about high-field drift velocity is thus indispensable for predicting the ultimate frequency performance of the electronic devices. The investigation of drift velocity in high fields has been an important area of research, not only for practical device applications, but also to elucidate semiconductor physics.

So far a number of experimental studies have been conducted to investigate the drift velocity in semiconductors under high electric fields.

The current–voltage measurement method is one useful and convenient approach. This involves deducing the drift velocity from the current density through a region of the sample where the field can be related to the applied voltage. However, this technique provides only limited information about drift velocity. At higher fields, current instabilities occur because of transfer of electrons to the upper valleys and the carrier concentration cannot be accurately determined. This technique is thus only able to provide information on drift velocity up to the threshold field.

The time-of-flight method, on the other hand, is a powerful technique for measuring the drift velocity–electric field relationship because it provides direct information about the drift velocity. The basic time-of-flight technique was first proposed by Haynes and Shockley (1949). The most important experimentally measurable quantity is the transit time. This is the time for carriers to drift across the sample under a known electric field. Since then, many groups have adopted this method to study the carrier transport properties (for example, diffusion and recombination) in semiconductors.

The first application of this technique to the study of high-field drift velocity of carriers in semiconductors was performed by Ruch and Kino (1967). They measured the drift velocity–electric field relationship for electrons in intrinsic GaAs employing a Schottky diode structure. Evans and Robson (1974) improved this technique and applied it to the measurement of thin epitaxial films. Meanwhile, Smith et al. (1980) used this technique to measure electron drift velocity for fields up to 220 kV/cm in GaAs. In all of these studies, majority electrons were measured. For minority carrier transport properties, Neukermans and Kino (1973) applied this method to the measurement of the high-field t velocity of electrons in p-InSb.

More recently, because of dramatic progress in generating ultrafast laser pulses, short optical pulses combined with high time-resolution electronic instruments have provided extremely precise measurements using the time-of-flight method. This electro-optic technique has been widely adopted for recent time-of-flight measurements. Minority carrier transport properties have been investigated in p-Si (Morohashi et al., 1985; Tang et al., 1986), in p-GaAs (Furuta et al., 1990; Furuta and Tomizawa, 1990), in p-InGaAs (Degani et al., 1981), in p-AlGaAs/GaAs quantum-well structures (Höpfel et al., 1985, 1986a; Shigekawa et al., 1990), and in p-doped compositionally graded AlGaAs (Levine et al., 1982, 1983).

2. PRINCIPLE OF TIME-OF-FLIGHT MEASUREMENT

A schematic of the basic time-of-flight technique is shown in Fig. 1. Two electrodes are connected in series to a sample with length L, and a pulse generator applies voltage pulses to the sample. When an optical excitation source irradiates the sample surface, electrons and holes are generated with concentrations Δn and Δp. The uniform electric field through the sample forces carriers to drift across the sample. The drifting carriers induce a transient current, which is monitored in the external electronic circuit by high time-resolution electronics. For minority electrons in p-type semiconductors, the current intensity is given by

$$\Delta I = \frac{e \Delta n v_d}{L}, \qquad (1)$$

where v_d is the drift velocity of electrons. The duration of the transient current signal is the time required for electrons to drift across the sample. Hence, the drift velocity is obtained by dividing the drift length W by the duration of the photocurrent signal ΔT:

$$v_d = \frac{W}{\Delta T}. \qquad (2)$$

From Eq. (1), the intensity of curent also provides the drift velocity. Since it is difficult to determine the injected carrier density with a high degree of accuracy, however, the drift velocity is generally determined from the current duration. A typical transient current signal, generated by a picosecond pulsed laser, is shown in Fig. 2. When the sample is

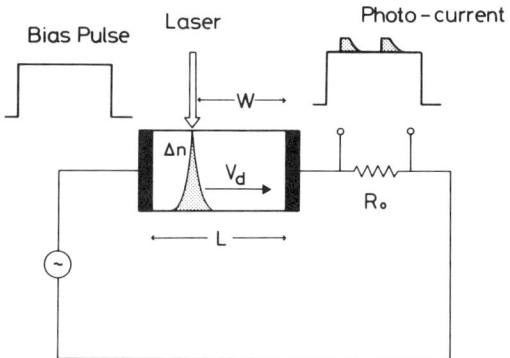

FIG. 1. Schematic illustration of time-of-flight measurement.

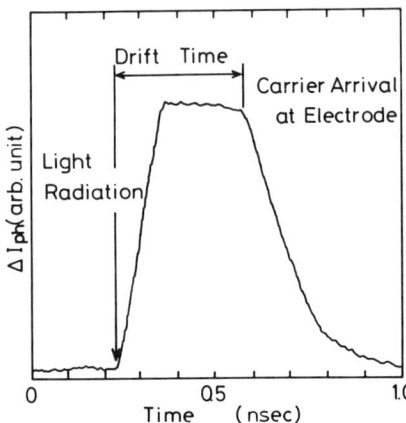

FIG. 2. Typical photocurrent signal for minority electrons in p-GaAs.

irradiated, the photocurrent rises quickly and remains almost constant. During this time, photoexcited electrons are drifting across the sample. When these electrons reach the positive electrode, the photocurrent falls off. Hence, the drift transit time is the time between the laser pulse and the photocurrent drop. Dividing the transit time by the drift length of electrons gives the drift velocity. A detailed discussion of experimental procedures and their application to time-of-flight measurement can be found in Canali et al. (1985). In the following, I present experimental results describing on the drift velocity–electric field relationships of minority electrons in GaAs.

3. Experimental Results

a. Drift Velocity of Minority Electrons in p-GaAs

First, I will consider the experimental time-of-flight measurements in bulk GaAs (Furuta et al., 1990; Furuta and Tomizawa, 1990). In these experiments, electron–hole pairs are optically injected by a dye laser that is pumped by a mode-locked YAG laser. The dye laser has a pulse width of 5 ps and an excitation wavelength of 750 nm. The samples were p-AlGaAs/p-GaAs/AlGaAs double heterostructures grown by molecular beam epitaxy. The p-GaAs layer, was 5,000 Å thick, and the AlAs mole fraction in the AlGaAs layers was 0.5, so that photoexcited electron–hole pairs were only generated in the GaAs layers. The hole concentrations in the p-GaAs range from 10^{17} to 10^{19} cm^{-3}.

3. HIGH FIELD MINORITY ELECTRON TRANSPORT IN p-GaAs

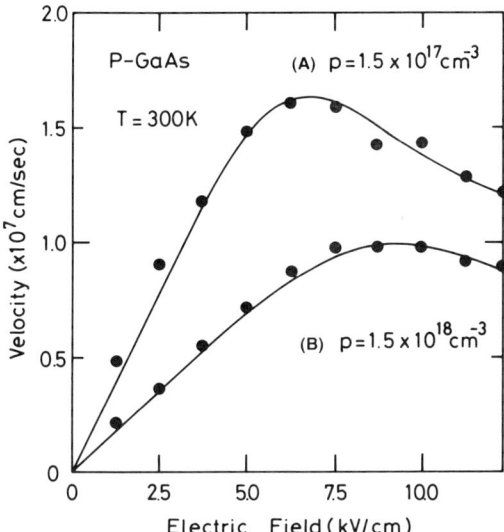

FIG. 3. Drift velocity versus electric field for minority electrons in p-GaAs with hole concentrations of 1.5×10^{17} and 1.5×10^{18} cm^{-3} (From Furuta et al., 1990).

The measured drift velocities of minority electrons in p-GaAs, with hole concentration of 1.5×10^{17} (A) and 1.5×10^{18} cm^{-3} (B), are plotted as a functions of electric field in Fig. 3. It can be seen that the velocity–electric field relationships are characterized by low field mobility μ_e, a peak velocity v_p, and a threshold electric field F_{th} and strongly depend on the hole concentration. In the low electric field region, the drift velocities for both samples increase linearly with electric field, and these slopes indicate a strong dependence on the hole concentration. From the slope of the low-field regions, calculated drift mobilities are $\mu_e = 3,200$ and $1,300$ cm^2/V · s for samples A and B, respectively. Drift mobilities of the majority electrons, corresponding to samples A and B, are $4,500$ and $2,500$ cm^2/V · s, respectively. The reduction in mobility is attributed to the effect of electron–hole momentum transfer. Here, holes behave like impurity scattering centers because of the heavy effective mass of the holes.

With increasing electric field, the velocity of sample A reaches a peak at an electric field around $F_{th} = 6.25$ kV/cm, and its value is 1.6×10^7 cm/s. As the electric field exceeds this value, the velocity falls gradually and shows a negative differential resistance up to $F = 12.5$ kV/cm. At $F = 12.5$ kV/cm, the velocity is 1.2×10^7 cm/s and continues to fall. For sample B, the velocity also reaches a maximum of

$v_p = 1.0 \times 10^7$ cm/s at $F = 7.5$ kV/cm, and the value stays almost constant up to 10.0 kV/cm. When the electric field exceeds this region, the velocity gradually begins to decrease, but the rate of decline is not so apparent ($v = 0.9 \times 10^7$ cm/s at $F = 12.5$ kV/cm). The observed negative differential resistance is thus not as strong in this case. Similar experiments were reported by Degani et al. (1981) in p-$In_{0.53}Ga_{0.47}As$ with a hole concentration of 5×10^{16} cm^{-3}. In their work, the drift velocity increased linearly at low field and gradually saturated at high field. They found no evidence of negative differential resistance in the drift velocity–electric field relationship. This result is similar to that in p-GaAs with a hole concentration of 1.5×10^{18} cm^{-3}.

As the hole concentration increases, the drift velocity values decline and the shape of the velocity–electric field relationship tends to increase monotonically and to saturate gradually with increasing electric field. In this case, it is difficult for electrons to transfer from the Γ valley to the upper valleys. This suggests that the hot-electron energy relaxation process is enhanced by scattering between electrons and hole plasma. The influence of the electron–hole interaction on the energy transfer process is discussed in the next section.

Next to be considered in the drift velocity–electric field relationship for samples with a hole concentration around 10^{19} cm^{-3} (Furuta and Tomizawa, 1990). In practical device structures, such as HBTs, the doping concentrations exceed 10^{19} cm^{-3}. The transport properties, under such high doping conditions, are very important for characterizing the device performance and designing the device structures. These samples had similar structures and parameters to those described earlier. Doping concentrations in p-GaAs were 5×10^{18} (C), 1×10^{19} (D), 2×10^{19} (E), and 4×10^{19} cm^{-3} (F).

The minority electron drift velocities for samples C, D and E are shown in Fig. 4. The drift velocities increase monotonically with electric field and gradually saturate for all samples. When the hole concentration increases from 5×10^{18} to 1×10^{19} cm^{-3}, the drift velocity decreases, as expected from the large electron–hole interaction. In the sample with a hole concentration of 2×10^{19} cm^{-3}, the drift velocity increases slightly as shown in the figure, although the hole concentration continues to increase. An increase in velocity is also observed for the sample with a hole concentration of 4×10^{19} cm^{-3}.

In the highly doped samples, the hole Fermi energy (E_F^h) becomes large with increasing hole concentration. For example, E_F^h is about 7.5 meV for $p = 1.0 \times 10^{18}$ cm^{-3}, whereas E_F^h is 60 meV for $p = 2.0 \times 10^{19}$ cm^{-3}. In these cases, E_F^h lies in the valence band, and the degeneracy is expected to play an important role because many hole states have

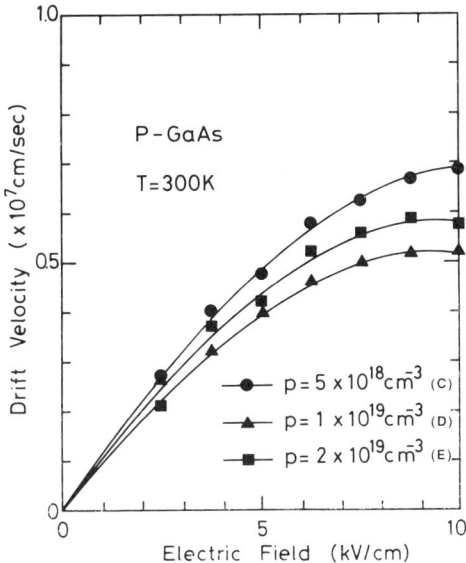

FIG. 4. Drift-velocity versus electric field for minority-electrons measured at 300 K in the samples with hole concentrations of 5.0×10^{18}, 1.0×10^{19}, and $2.0 \times 10^{19}\,\text{cm}^{-3}$ (From Furuta and Tomizawa, 1990).

already been occupied. As a result, the degeneracy effectively reduces the intensity of the electron–hole interaction according to Pauli's exclusion principle, and the drift velocity does not show any notable decrease.

These results qualitatively agree with the theoretical calculation by M. Combescot and R. Combescot (1987). These authors observed that the scattering rate reaches a maximum at large majority carrier concentrations. As the hole concentration increases beyond the maximum point, the scattering rate decreases because the momentum transfer rate decreases with increasing Fermi velocity. In Section V, these experimental data are compared with the results calculated by Monte Carlo simulation and discussed in some detail.

b. *Drift Velocity of Minority Electrons in Quantum-Well Structure*

Modulation-doped quantum-well structures are very attractive systems for studying minority electron transport mechanisms because they allow the strength of the electron–hole interaction to be modified by varying

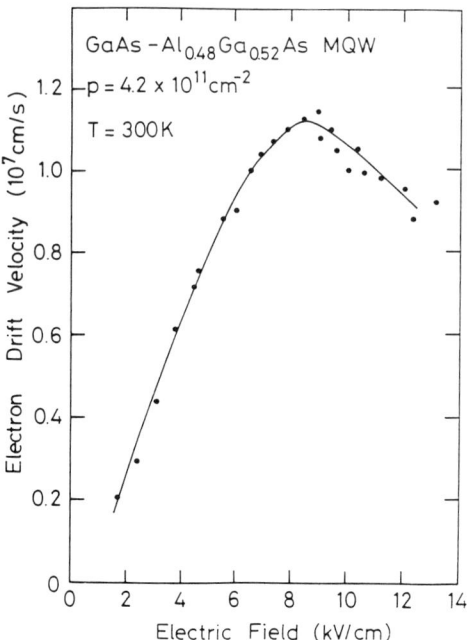

FIG. 5. Drift velocities of minority electrons in p-AlGaAs/GaAs modulation-doped quantum-well structures with a well width of 90 Å (from Höpfel et al., 1986a).

such structural parameters as the well layer thicknesses and the doping concentrations. The first observation of minority electron drift velocity characteristics in these systems was performed in modulation-doped p-AlGaAs/GaAs quantum-well structures by Höpfel et al. (1985, 1986a).

Figure 5 shows the drift velocity of minority electrons in a 90-Å-wide p-AlGaAs/GaAs quantum well with hole concentration of 4.2×10^{11} cm^{-2} at room temperature. The drift velocity increases linearly with electric field. At fields around 8 kV/cm, the drift velocity reaches a maximum of 1.2×10^7 cm/s. For even higher electric fields, the velocity decreases and shows negative differential resistance. From the slope of low electric field regions, the drift mobility is about 1,500 cm^2/V · s, which is a much lower value than the intrinsic bulk GaAs mobility at room temperature (8,000 cm^2/V · s). This reduction in mobility is due to electron–hole momentum scattering. The effective doping concentration in well region P_I (the sheet hole concentration divided by the well thickness) is about 5×10^{17} cm^{-3} in this case. The electron mobility in bulk GaAs with this impurity concentration is 4,000 cm^2/V · s, which is higher than that in a two-dimensional minority electron system. Höpfel et al. (1986a, 1988) pointed out that the reduction of screening, the two-dimensional density of states, and the shape of wavefunctions are crucial factors in explaining the minority electrons drift velocity in quantum-well structures.

The spatial overlap of wavefunctions for minority electrons and majority holes is one of the most important factors explaining the two-dimensional minority electron transport. Recently, systematic measurements have been carried out in modulation-doped p-AlGaAs/GaAs quantum-well structures with well layer thicknesses of 200, 600, and 2,000 Å as parameters (Shigekawa et al., 1990). All the samples had nearly equal sheet hole concentrations of 1×10^{12} cm^{-3}. The drift velocity–electric field relationships, for sample G (200 Å), sample H (600 Å), and sample I (2,000 Å), are shown in Fig. 6.

The drift velocity of sample G is suppressed in comparison with those for samples H and I. The drift velocities for samples H and I are identical. Values of drift mobility, estimated from the slopes at low electric fields, are approximately 3,500, 5,500, and 5,500 cm^2/V · s for samples G, H, and I, respectively. The peak velocity and the threshold field are 1.3×10^7 cm/s and 6.5 kV/cm for sample G, and 1.5×10^7 cm/s

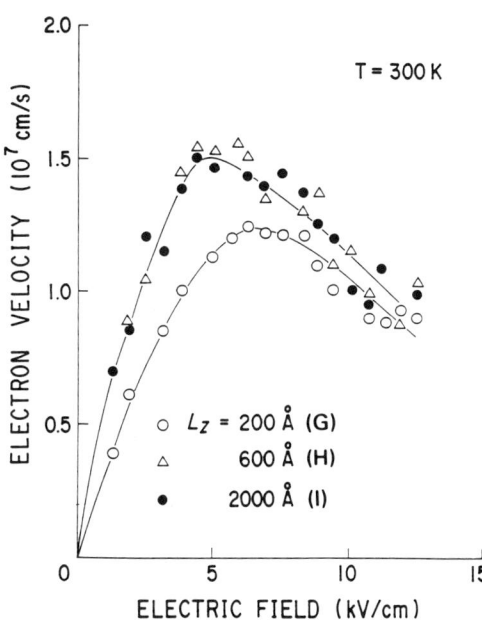

FIG. 6. Drift velocities of minority electrons in p-type single quantum-well structures with well widths of 200, 600, and 2,000 Å (from Shigekawa et al., 1990).

and 4.5 kV/cm for samples H and I. The suppression of drift velocity for sample G is attributed to enhanced electron–hole interaction due to an overlap of the wavefunction of electrons and holes. For this sample, the spatial distribution of majority holes is located near the AlGaAs/GaAs heterointerfaces, while electrons are distributed around the center of the well. Although the wavefunctions of electrons and holes are spatially separated from each other, the effect of the electron–hole interaction due to spatial overlap still influences the drift velocity in this sample. The distribution of two-dimensional carriers is on the order of a few hundred angstroms, which is comparable to the well layer thickness of sample G. This interpretation is consistent with the data for a 90-Å-wide well sample shown in Fig. 5. In this case, both wavefunctions for electrons and holes are located at the center of the well, which enhances the electron–hole interaction. Thus, the drift velocities have markedly smaller values.

III. Energy Transfer Process

When an external electric field is applied, the carriers in semiconductors are easily heated up, and the thermal equilibrium between the carrier and the phonon bath collapses. This heating of carriers can be characterized in terms of a carrier temperature. For defining the carrier temperature, the following condition must be satisfied: The rate of input power at which carriers receive energy from an electric field is equal to the rate of energy loss at which carriers transfer their energy to the phonon bath. This energy transfer process is carried out by inelastic scattering such as that produced by carrier–phonon and carrier–carrier interactions. These interactions are thus very important in hot carrier physics; they determine the behavior of carriers under high electric field and therefore the characteristics of ultrasmall devices. Regarding the minority electron transport, the electron–hole Coulomb interaction is a very important energy loss process as is electron–phonon interaction.

In the following, I first discuss the dependence of electron temperature on electric field. After that, I discuss the energy transfer mechanism for minority electrons in p-GaAs using the energy loss rate and the relaxation times.

4. ELECTRON TEMPERATURE

So far, several experimental techniques have been applied to determine the carrier temperature in semiconductors including the

Schubnikov–de Haas oscillation, far-infrared emission, and so on. Photoluminescence measurement is most sensitive over a wide carrier temperature range, and is thus one of the most widely and well-established techniques for studying the electron temperature (Shah, 1978, 1989). When electrons and holes are thermalized and have the same temperature T_{eff}, the photoluminescence intensity as a function of photon energy $h\nu$ is given by

$$I(h\nu) \propto (h\nu)^2 \exp\left(-\frac{h\nu}{k_B T_{\text{eff}}}\right) \quad (3)$$

for direct electron–hole recombination. Therefore, the effective temperature can be obtained from the slope of the semilog plot of photoluminescence intensity for energies greatly exceeding the quasi-Fermi energy.

Figures 7a and 7b show the room-temperature photoluminescence spectra as a function of electric field for p-GaAs samples with hole concentrations of $p = 1.5 \times 10^{17}$ (a) and 1.5×10^{18} cm^{-3} (b), respectively (Furuta et al., 1990). The photoluminescence spectra near the GaAs bandgap (1.42 eV) are attributed to the recombination of photoexcited minority electrons with majority holes. The spectra have high energy tails that decrease exponentially with the photon energy. This suggests that the carrier distributions are well thermalized and characterized by a Fermi–Dirac distribution with an effective carrier temperature. As the electric field increases, the slopes of the tails decrease, which is indicative of carrier heating. The decrease in the slope is not so apparent for the sample with high hole concentration. For quantitative study of the electron heating process and its dependence on hole concentration, the carrier temperatures are derived from the slope of the exponential high-energy tails of the photoluminescence spectra. The photoluminescence spectra are the product of the electron and hole distribution functions. It should be noted that the derived temperature is an effective T_{eff} that includes both electron temperature, T_e, and hole temperature, T_h. Here, assuming that electrons and holes are described by Fermi–Dirac distribution functions with different temperatures T_e and T_h, the band structures of both carriers are parabolic and characterized the effective mass of the electron, m_e^*, and the hole, m_h^*. One finds that the experimentally measured effective temperature can be expressed by

$$\frac{1}{T_{\text{eff}}} = \frac{1}{(m_e^* + m_h^*)} \times \left(\frac{m_h^*}{T_e} + \frac{m_e^*}{T_h}\right). \quad (4)$$

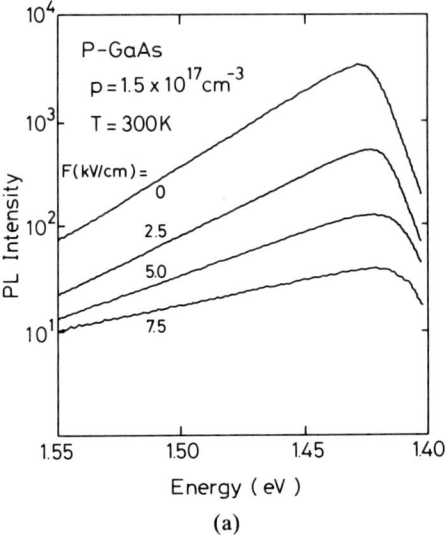

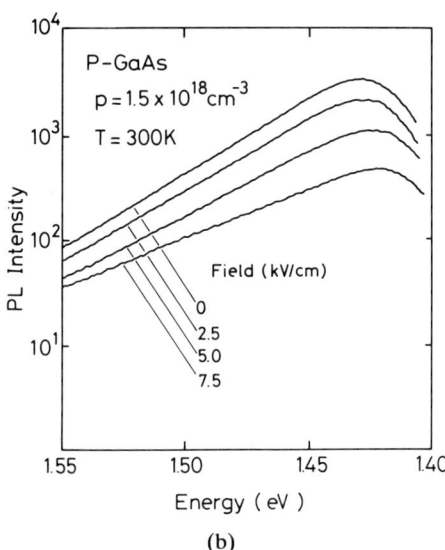

FIG. 7. Photoluminescence spectra for samples with hole concentrations of 1.5×10^{17} (a) and 1.5×10^{18} cm^{-3} (b).

3. HIGH FIELD MINORITY ELECTRON TRANSPORT IN p-GaAs

The unknown parameter, T_h, is obtained from the hole current–bias voltage relationship. At room temperature, the mobility of holes is mainly limited by nonpolar optical phonon scattering. By numerical analysis of scattering theory (Conwell, 1967), the hole mobility varies with hole temperature as

$$\mu_h \propto T_h^{-0.9} \quad (5)$$

in the 300 K lattice temperature range (Höpfel et al., 1986b, 1988). At this stage, the hole concentration is set to be constant over the entire applied bias voltage range. This procedure enables one to derive the drift velocity of holes. The hole velocities of both samples A and B increase almost linearly with electric field, which suggests that the hole temperature stays cold even in the high electric field region. The hole temperatures at various electric field strengths are deduced from hole-current bias–voltage relationships by fitting the hole drift-velocity electric field relationships to Eq. (5). Using Eq. (4) with an effective temperature T_{eff} and hole temperature T_h, the minority electron temperature T_e can be obtained.

Temperatures of minority electrons and majority holes versus the electric field strength are shown in Figs. 8a and 8b, respectively (Furuta et al., 1990). A temperature difference between electrons and holes is clearly observed. Next, I will consider the implications of these results. Electrons gain much more power per carrier from the electric field than do holes because the electron velocity is higher than that of holes. In addition, holes are coupled more strongly to the lattice phonon because of interaction of nonpolar optical as well as polar optical phonons. Therefore, it is difficult for holes to gain additional energy from the electric field. This situation implies the existence of thermally nonequilibrium energy states between minority electrons and majority holes. The hole temperatures in all samples do not notably change with increasing electric field (from 300 K to 400 K). The electron temperature, on the other hand, drastically changes with the electric field. The electron temperature also shows a strong dependence on the hole concentration. In samples with a low hole concentration, the electron temperature T_e varies from 300 K to 1600 K in fields of up to 8 kV/cm. On the other hand, T_e varies from 300 K to 800 K in fields of up to 10 kV/cm at high hole concentrations.

This reduction in electron temperature, with increasing hole concentration, and the thermal-nonequilibrium state between electrons and holes are attributed to the electron–hole interaction. The latter transfers energy from hot minority electrons to cold majority holes. In other

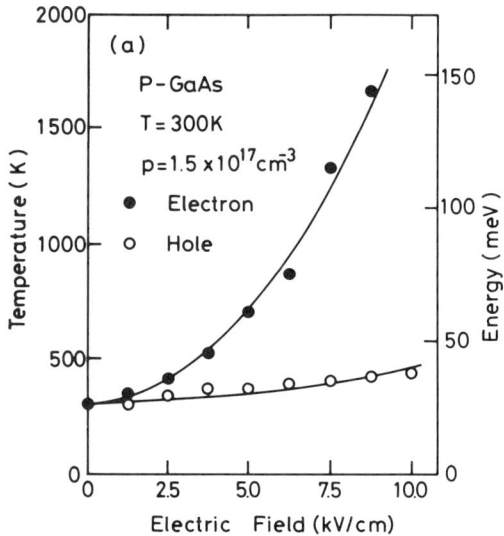

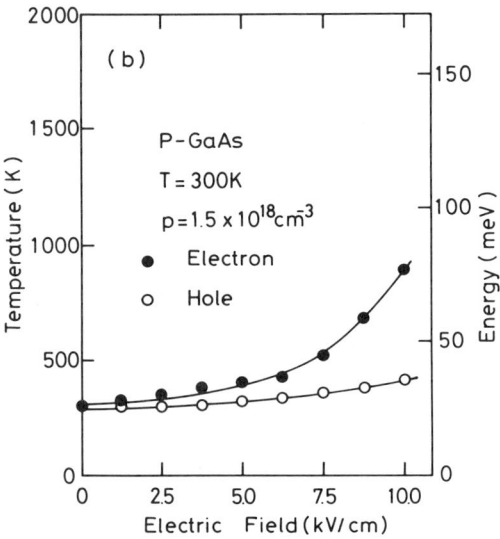

FIG. 8. Dependence of electron and hole temperatures on electric field for samples with hole concentrations of 1.5×10^7 (a) and 1.5×10^{18} cm^{-3} (b) (from Furuta et al., 1990).

words, this interaction makes it difficult for minority electrons to gain energy and transfer from the Γ valley to the upper valleys. Hence, the absence of negative differential resistance in the samples, with higher hole concentration, can be qualitatively explained by the electron–hole interaction.

5. Energy Loss Rate

To illustrate the influence of the electron–hole interaction on energy transfer, I will consider the energy loss rate of minority electrons in p-GaAs. The energy loss rates provide fundamental insight into the carrier–phonon and the carrier–carrier interactions because they are directly related to inelastic interactions. From the experimentally measured drift velocity, v_d, and electron temperature, T_e, as a function of electric field, the input power per electron, ev_dF, can be obtained. For the steady-state situation, the energy loss rate per electron must equal the input power per electron. This relation provides the unique means for directly measuring the energy loss rate as a function of electron temperature.

Using the data of drift velocity (Fig. 3) and electron temperature (Fig. 8) for samples with hole concentrations of 1.5×10^{17} and 1.5×10^{18} cm^{-3}, an electron temperature, T_e, is plotted against the input power per electron. The latter is equal to the energy loss rate per electron, as shown in Fig. 9 (Furuta et al., 1990). This plot allows one to compare the experimentally measured total energy-loss-rate of minority electrons with that of majority electrons. For the latter, only energy loss by polar optical phonon emission is possible. This is possible because the total energy loss rate per electron can be written as the sum of loss rate due to the LO phonon cascade process and the electron–hole interaction:

$$\left\langle \frac{dE}{dt} \right\rangle_{\text{total}} = \left\langle \frac{dE}{dt} \right\rangle_{\text{el-ph}} + \left\langle \frac{dE}{dt} \right\rangle_{\text{el-h}}, \qquad (6)$$

In a first-order approximation, these two scattering processes are independent of each other. Over the entire electron temperature range, the energy loss rate is larger at higher hole concentrations. In other words, the energy relaxation occurs more quickly as the hole concentration increases. Since the polar optical-phonon interaction depends weakly on hole concentration, the difference in the energy loss rate per electron between the samples is attributed to the electron–hole interaction.

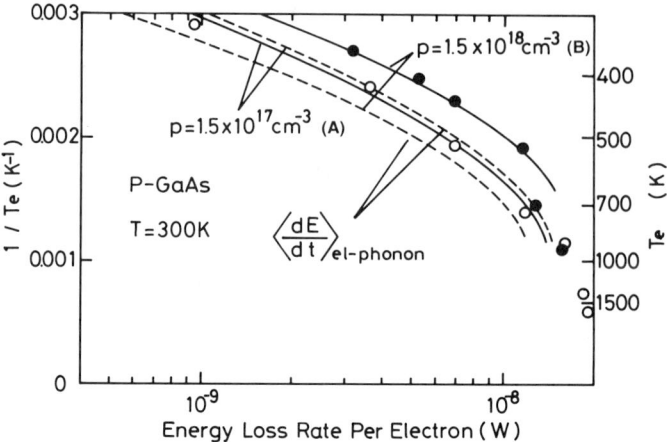

FIG. 9. Energy loss rates measured at 300 K plotted as a function of minority-electron temperature. Also included are results for majority electrons with the same doping concentrations (from Furuta et al., 1990).

To investigate the energy loss mechanism involved in the electron–hole interaction, the measured total energy-loss rate per electron is compared with the calculated rate for which only the screened electron–longitudinal polar optical phonon interaction is taken into account. The calculated results are also shown in Fig. 9. It can be seen that the calculated curve for the low hole concentration is in good agreement with the measured values, while the calculated curve for the high hole concentration is lower than the experimental one. This result shows that the emission of longitudinal polar optical phonons is the dominant energy relaxation mechanism at a hole concentration of 1.5×10^{17} cm^{-3}. The energy transfer from hot electrons to cold holes by the electron–hole interaction becomes effective when the hole concentration reaches 1.5×10^{18} cm^{-3}.

Other experimental studies, on the energy loss mechanism, have been reported for p-GaAs (Shah and Leite, 1964), p-InGaAs (Shah et al., 1982), and quantum-well structures (Höpfel et al., 1986b, 1988; Shah et al., 1983, 1984, 1985).

6. Energy and Momentum Relaxation Times

From the hydrodynamic approximation, the momentum relaxation time, τ_m, and energy relaxation time, τ_e, are

$$\tau_m = \frac{m_e^* v_d}{eF}, \qquad (7)$$

$$\tau_e = \frac{(E - E_0)}{ev_d F}, \qquad (8)$$

where E ($=3/2kT_e$) is average electron energy and E_0 ($=3/2kT_L$) is the thermal energy of an electron at the lattice temperature T_L.

I will consider the energy relaxation time τ_e. First, Fig. 10 shows the energy relaxation time as a function of average electron energy (Furuta et al., 1990). The hole concentration dependence is obvious. The energy relaxation time, for a hole concentration of 1.5×10^{17} cm^{-3} (A), increases monotonically with average electron energy and varies from 500 femtoseconds (fs) to 750 fs. The energy relaxation time for a hole concentration of 1.5×10^{18} cm^{-3} (B), on the other hand, is somewhat lower than that at low hole concentrations and ranges from 200 to 500 fs. A small value of τ_e, for high hole concentration, comes from the energy relaxation via the electron–hole interaction.

Next, I will discuss the momentum relaxation time. Using Eq. (4), the momentum relaxation time τ_m is derived from the experimental data. The results for samples A and B are shown as solid lines in Fig. 11 (Furuta et al., 1990), where $0.067m_0$ is assumed for the effective mass of electrons. From the figure, it can be seen that the momentum relaxation times decrease monotonically as the average electron energy increases. The relaxation time for sample B varies from 50 to 35 fs and is somewhat lower than that for sample A, which ranges from 130 to 55 fs. This means that that the electron–hole interaction reduces the momentum relaxation

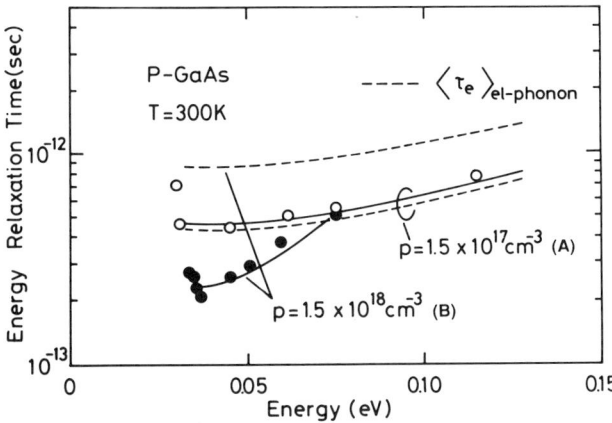

FIG. 10. Energy relaxation times plotted as a function of minority electron energy. Also included are the calculated results with screened electron–LO phonon interaction for the same doping concentrations (dotted lines) (from Furuta et al., 1990).

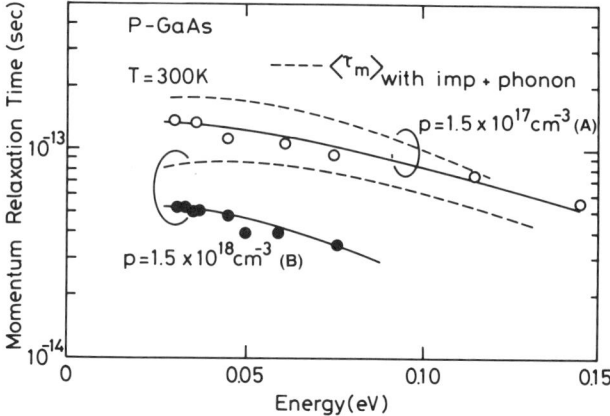

FIG. 11. Momentum relaxation times plotted as a function of minority-electron energy. The calculated results, including impurity and LO phonon scattering, are also included in the figure (dotted lines) (from Furuta et al., 1990).

time. To confirm this, the momentum relaxation time produced by impurity and phonon scattering, at constant doping concentrations, was calculated. The calculated results are also shown in Fig. 11 as dotted curves. The calculated values are larger than the experimental values for the entire electron-energy region. To investigate the effect of electron–hole interaction on momentum relaxation time, the ratio between the experimental and the calculated momentum relaxation times was calculated for low electron energies. Values of about 0.8 for sample A and 0.6 for sample B are obtained. These values are consistent with the low electric field mobility ratio between minority and majority electrons.

Höpfel et al. (1988) also studied momentum relaxation times of minority carriers in modulation-doped AlGaAs/GaAs quantum-well structures. They found that the momentum relaxation time of minority electrons is less than 100 fs, whereas that of minority holes is almost two orders of magnitude larger. They attributed the large difference in momentum relaxation times between minority electrons and holes to three factors: differences in effective mass, degeneracy, and different screening behaviour, which strongly depends on dimensionality. The effect of degeneracy will be discussed later.

IV. Ultrafast Energy Relaxation Process

The energy relaxation process of carriers in semiconductors is of fundamental interest and of great importance in device physics. The energy relaxation process is, in general, classified into two time stages. The first time stage is the initial thermalization of photoexcited carriers by carrier–carrier and carrier-phonon interactions. This process occurs on a femtosecond time scale. In contrast, the second stage, which is the cooling of the thermalized distribution function, is characterized by carrier temperature on a time scale of a few picoseconds. Time-resolved spectroscopy with a femtosecond time resolution provides important insight into the carrier dynamics determined by carrier–carrier and carrier–phonon interactions. Derivation of time-resolved photoluminescence and absorption saturation with an excite-and-probe technique satisfy these high time-resolution criteria and have been extensively used for the study of ultrafast phenomena. So far, several researchers have used these techniques to study the energy relaxation process of photoexcited carriers in semiconductors (Shah, 1989). Here, it should be noted that the results obtained from these techniques are compatible because the intensity of photoluminescence is proportional to the product of the electron and hole distribution functions, whereas small changes in transmission are proportional to the sum of these distribution functions. In this section, we will discuss the time-resolved studies that yield the thermalized hot carriers distribution cooling data. For p-type samples, energy relaxation of photoexcited electrons is strongly influenced by electron–hole interaction.

Furuta and Yoshii (1991) have performed time-resolved photoluminescence measurements employing the up-conversion technique on p-GaAs as a function of hole concentration. This method provides very high time resolution measurements. An experimental setup for time-resolved

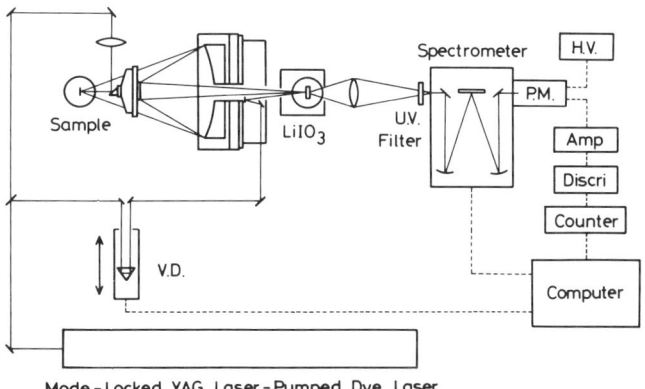

FIG. 12. Schematic of the time-resolved photoluminescence measurement system using the up-conversion technique.

photoluminescence measurements is depicted in Fig. 12. Femtosecond optical pulses of a dye laser, which is pumped by a mode-locked YAG laser, were used to generate electron–hole pairs in the samples. The dye laser had a pulse width of 130 fs and an excitation energy of 1.90 eV. The photoluminescence from the sample was collected by dispersion-free optics and focused on a 0.5-mm-thick $LiIO_3$ nonlinear optical crystal. The delayed laser beam was also focused noncollinearly at the same point in the nonlinear crystal. The up-converted signal was dispersed by a monochromator, and then detected by a photon-counting technique. From the cross-correlation trace, the time resolution of the total experimental system was 130 fs. The detailed experimental technique for the up-conversion method has been reviewed by Shah (1988). The samples were p-AlGaAs/p-GaAs/AlGaAs double heterostructures grown by molecular beam epitaxy. The thickness of the Be-doped GaAs was 100 nm. Doping concentrations were 1×10^{16}, 1×10^{17}, 1×10^{18}, and 1×10^{19} cm^{-3}. The AlAs mole fraction was 0.5, so photoexcited electron–hole pairs were only generated in the GaAs layers.

Figure 13 shows the time evolution of photoluminescence intensity for

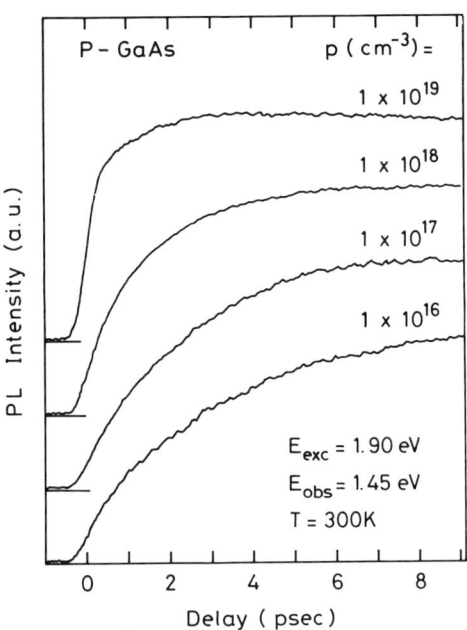

FIG. 13. Time-resolved near–band-edge photoluminescence intensity in p-GaAs with hole concentrations of 1.0×10^{16}, 1.0×10^{17}, 1.0×10^{18}, and 1.0×10^{19} cm^{-3} at 300 K. The excitation energy is 1.9 eV, the pulse width is 130 fs, and the injected carrier density is 10^{16} cm^{-3} (from Furuta and Yoshii, 1991).

band-to-band transition at 1.45 eV. The data reveal two features of special interest. The first observation is that the rise of photoluminescence intensity, which corresponds to the time for photoexcited carriers to relax from excited energy states to the bottom of the conduction band edge, becomes shorter with increasing hole concentration. The second point of interest is that the rise of photoluminescence for the most highly doped sample consists of two time constants. The samples with lower doping have a single time constant. A fast time constant (less than 1 ps) appears immediately after excitation. A slower time constant is observed near saturation.

The initial energy relaxation of photoexcited minority electrons is produced by competition between electron–phonon and electron–hole interaction. The energy relaxation effect due to electron–hole interaction becomes stronger as the hole concentration increases, and thus the rise time becomes shorter. While the electron–hole interaction is expected to dominate for the highest hole concentrations, the combination of phonon cascade and electron–hole interactions may be important in the two samples with moderate hole concentrations. The two rise-time components of the photoluminescence intensity, which are observed for the highest hole concentration, can be understood as follows. At $p = 1.0 \times 10^{19}$ cm^{-3}, the hole Fermi energy is approximately 40 meV from the top of the valence band at 300 K. The chemical potential of the photoexcited electron system is much higher than that of the holes immediately after excitation. Hence, the excess energy of the electrons is transferred to the holes through electron–hole interactions because of the large chemical potential difference. This energy transfer finally ceases when the chemical potentials of the electrons and holes coincide. When the chemical potentials are equal, the electrons weakly interact with the holes. As a consequence, the electrons further relax by emitting optical phonons and approach a distribution function appropriate to the lattice temperature. This is a slower process, as was explained before.

To examine in detail the energy relaxation process due to electron–hole interaction, Furuta and Yoshii (1991) investigated the dependence of photoluminescence intensity rise-time on photon energy for the sample doped p-type to 1.0×10^{19} cm^{-3}. Figure 14 shows the time-resolved photoluminescence data at five different photon energies. The initial rise time becomes shorter with increasing photon energy. Above 1.54 eV, the photoluminescence intensity exhibits different characteristics: specifically, a rapid fall toward the saturation point after a steep rise in intensity. As the photon energy increases, the rise time does not change, but the decay time is shorter. At 1.64 eV, more than 180 meV above the band gap, the decay time is comparable to the rise time. The observed rapid rise and fall in photoluminescence intensity, at higher photon energies, imply

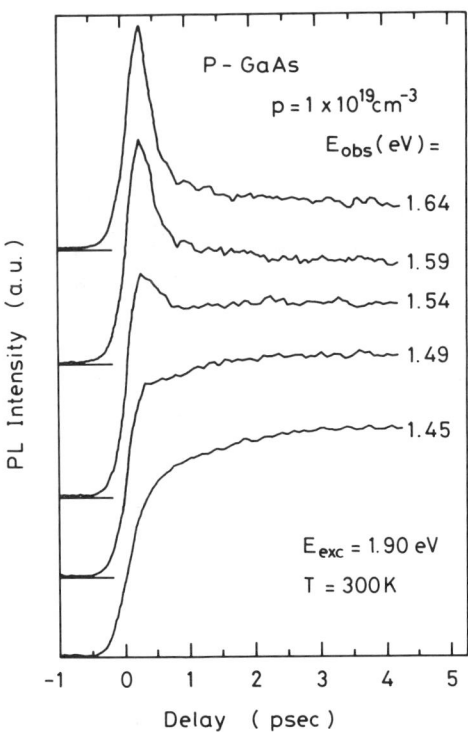

FIG. 14. Time-resolved photoluminescence intensity in p-GaAs with hole concentration of 1.0×10^{19} cm^{-3} monitored at five different photon energies within the band-to-band spectra (from Furuta and Yoshii, 1991).

rapid changes in the distribution of electrons. The fast rise suggests that electrons are scattered from higher-energy states to the measured energy states. The fast intensity decay indicates that the electrons are scattered from the observed energy states to lower-energy states. At photon energies of 1.45 and 1.49 eV, the fall in photoluminescence intensity is not observed because these energy states are very near the bottom of the conduction band. There are very few energy levels available for these electrons to scatter into. Similar rise and decay phenomena are observed in samples with other hole concentrations. The rise and fall times are not nearly as short as they are for the 1.0×10^{19} cm^{-3} sample, although they become shorter with increasing hole concentration.

To elucidate the thermalization effects, a time-resolved photoluminescence spectrum, which is proportional to the distribution function, is plotted. Figure 15 shows the results for the sample doped p-type to

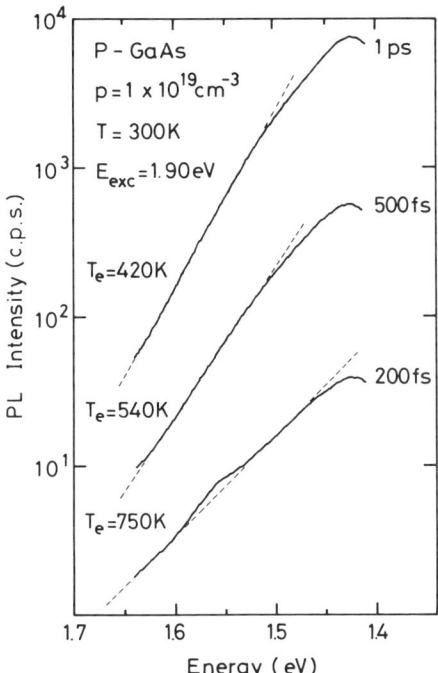

FIG. 15. Time-resolved photoluminescence spectra of p-GaAs with hole concentration of 1×10^{19} cm^{-3} for three different time delays after excitation (from Furuta and Yoshii, 1991).

1.0×10^{19} cm^{-3}. The spectra in the high-energy region are characterized by the Fermi–Dirac distribution function because of the slope of their semilog plot. The spectrum at 200 fs is approximated by a straight line from the slope of the dashed line drawn in the figure. A small deviation from the line, however, is observed for the energy between 1.5 and 1.6 eV. The deviation disappears for other curves that correspond to 500 fs and 1 ps. This implies that before 200 fs, electrons still remain in the nonequilibrium state. Therefore, it can be found that electrons are thermalized as rapidly as 200 fs from the nonequilibrium state created by excitation.

The time evolution of electron temperatures can be deduced from the high-energy tails of time-resolved photoluminescence spectra in Fig. 15 (dashed line). However, it should be noted that the relaxation processes of electrons and holes are different, which accounts for the temperature difference between the carriers. Recent experiments have shown that the temperatures of electrons and holes during the cooling process remain

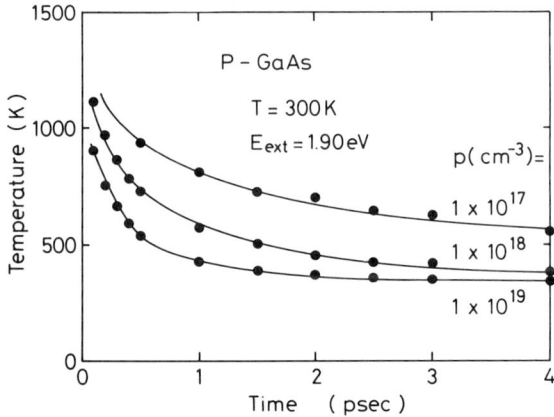

FIG. 16. Cooling curves of photoexcited minority electrons in p-GaAs with hole concentrations of 1.0×10^{17}, 1.0×10^{18}, and 1.0×10^{19} cm^{-3}.

different on a time scale of picoseconds (Bradley et al., 1989; Zhou et al., 1989). Here, I set the hole temperature at 300 K because of the low excitation condition ($n_{ext} \approx 10^{16}$ cm^{-3}). Figure 16 shows the time evolution of electron temperature for three hole concentrations. The reduction of the electron temperature has a strong dependence on the hole concentrations. For the most highly doped sample, the cooling proceeds faster than for the less highly doped samples because of the stronger electron–hole interaction on the cooling rate. While two cooling rates are clearly observed for the sample with a doping level of 1.0×10^{19} cm^{-3}, the other samples have a single exponential cooling rate. For the time delays changing from 200 to 500 fs, the electron temperature decreases from 750 to 540 K. This time period corresponds to the time when the fast decay of photoluminescence intensity can be observed (shown in Fig. 14). Therefore, this cooling process is based on the electron–hole interaction, as was explained before. After 500 fs, the electron temperature decreases slowly. At 1 ps, the temperature is at 450 K and after 3 ps, it reaches 350 K. The slower cooling rate is controlled by the electron–phonon interaction.

Very recently, the influence of electron–hole interaction on photoexcited electron–hole plasma thermalization in doped GaAs was calculated by Leo and Collet (1991). They presented calculations of cooling for different electron and hole temperatures that includes electron–hole scattering and nonequilibrium optical phonons. They pointed out that for p-GaAs, the cooling is mainly determined by the energy transfer between electrons and holes due to electron–hole Coulomb scattering, rather than by the emission of optical phonons. These results are consistent with the experimental results presented above.

Further experimental evidence of the electron–hole interaction on thermalization of nonthermal electrons distribution function is given by Knox et al. (1986, 1988). They performed an absorption saturation measurement with the excite-and-probe technique for modulation-doped GaAs quantum-well structures. They measured an electron thermalization time of 60 fs in a p-type quantum-well structure with a hole concentration of 3×10^{11} cm^{-2}. This time is a little longer than that in an n-type quantum well with the same carrier concentration.

V. Monte Carlo Simulation Results

The transport theory that describes high-field transport in semiconductors is based on the solution of the time-dependent Boltzmann transport equation. So far several methods have been proposed to solve this equation (Nag, 1980). Recently, the Monte Carlo method has been well applied to the study of carrier transport in semiconductors. This method involves computer simulation of a carrier motion in k-space under the influence of both applied electric fields and collisions. Random numbers are applied to set both the scattering rate and the direction of carrier motion after each scattering event. Statistical convergence is obtained after the number of events exceeds 10^5. The advantages of this technique are that it provides a very accurate microscopic description of the physical processes, and that it does not require any assumption regarding the electron distribution function. In fact, this method explains well all qualitative features of macroscopic quantities such as drift velocity and carrier energy. The detailed procedure for carrying out a Monte Carlo calculation and its application to semiconductor physics have been reviewed by several authors (Fawcett et al., 1970; Littlejohn et al., 1977; Jacoboni and Reggiani, 1983). For minority carriers, the steady state transport has been studied by several authors in p-GaAs (Sadra et al., 1988; Saito et al., 1989; Taniyama et al., 1990) and in p-InGaAs (Osman and Grubin, 1987). The Monte Carlo method has been also applied to the study of electron–hole interaction in nonequilibrium transport (Osman and Ferry, 1987a; Joshi and Grondin, 1989) and ultrafast energy relaxation (Osman and Ferry, 1987b; Goodnick and Lugli, 1988).

The aim of this section is to present the calculated results for minority-electron transport in p-GaAs by Monte Carlo simulation and to compare the calculated results with the experimental data. First, I briefly introduce the scattering models.

7. Scattering Models

The dominant scattering mechanisms that affect carrier motion in semiconductors are:

(1) ionized impurity;
(2) optical phonon;
(3) acoustic phonon;
(4) intervalley phonon;
(5) carrier–carrier; and
(6) plasmon scattering.

Modeling of all the scattering mechanisms is carried out within a first-order, time-dependent perturbation theory (golden rule). Consequently, only the two-body interaction is analyzed. The standard scattering models ((1) to (4)) are well understood, and several reviews have already been published (Conwell, 1967; Nag, 1980). Here, the physical models describing scattering specific for minority electron transport are discussed.

In polar semiconductors, the free carrier plasma frequency at degenerate carrier concentrations is comparable to the LO phonon frequency. In this case, the LO phonons are strongly coupled with the free carrier system (plasma), and a coupled-plasmon LO phonon mode results. Hence, the dispersion of LO phonons is modified by the carrier concentrations. It is well known that the coupled mode consists of two modes. One is the collective plasma mode and the other is the single-particle excitation mode. For brevity, the cutoff wave number k_c is introduced to separate these two modes. The cutoff wave number k_c is that point at which the plasmon dispersion curve intersects the LO phonon dispersion curve (Moglestue, 1984; Xiaoguang et al., 1985).

Below k_c, the LO phonon mode couples with the collective motion of carriers, and subsequently, the coupled plasmon–LO phonon mode is formed (Kim et al., 1978). For k greater than k_c, on the other hand, the collective mode decays into the single-particle mode and is Landau-damped. In this region, the electron–LO phonon interaction and the carrier–carrier interaction become the dominant scattering mechanisms. The scattering rate of coupled plasmon–LO phonon scattering is derived from a dielectric response function using the random phase approximation.

3. HIGH FIELD MINORITY ELECTRON TRANSPORT IN p-GaAs

At large wave numbers, the electron–hole scattering becomes important. The electrons and holes are assumed to interact through a screened Coulomb potential. The Coulomb potential is represented by two different models: the Brooks–Herring model and the Conwell–Weisskopf model (Ridley, 1982). In a former model, a screened Coulomb potential is expressed as

$$V(r) = \frac{e^2}{4\pi\varepsilon r} \exp\left(-\frac{r}{\lambda}\right), \tag{9}$$

where ε is the dielectric constant of the material and λ is the screening length. This model thus includes the effect of screening by free carrier on carrier–carrier scattering and on impurity scattering. In the case of the Conwell–Weisskopf model, on the other hand, the extent of the Coulomb potential is determined by the average distance between impurities. It is assumed that carriers do not shield ionized impurities. Several detailed comparisons of these models have been published (Seeger, 1985; Chattopadhyay and Queisser, 1981).

Two different models of screening length are presented, the Debye–Hückel and the Thomas–Fermi types. The former is applicable when carrier frequency is negligible, while the latter is applicable when the carrier degeneracy is important. Degeneracy occurs when the carrier density is high and the carrier temperature is low. Moreover, the screening length is closely related to the cutoff wave number k_c (Pines, 1956; Madelung, 1978), because the plasmon cutoff is attributed to the Landau damping. In order to accurately obtain the cutoff wavenumber, excitations of both heavy and light holes should be taken into consideration (Levi and Yafet, 1987).

From the Fermi golden rule, the scattering rate from the initial states (k_0, k) to the final states (k'_0, k') for electrons interacting with holes is

$$S(k_0, k; k'_0, k') = \left(\frac{2\pi}{\hbar}\right) |M|^2 f_e(k_0) f_h(k) \times [1 - f_e(k'_0)][1 - f_h(k')]$$

$$\times \delta(E_{k'_0} + E_{k'} - E_{k_0} - E_k),$$

$$M = \frac{e^2}{V\varepsilon} \frac{\delta_{k_0+k, k'_0+k'}}{|k'_0 - k_0|^2 + 1/\lambda^2}, \tag{9}$$

where M is the matrix element of the scattering, k_0 and k'_0 are the electron wave numbers, and k and k' are the hole wave numbers. In this expression, the Brooks–Herring model is adopted for the Coulomb potential, and $f_{e(h)}$ and $E_{k(k')}$ are the occupation probability for electrons (holes) and the carrier energy of wave number $k(k')$, respectively. Here, V is the volume of the crystal, and the initial and final states are represented by plane waves.

The hole distribution function is needed to obtain the screened electron–hole scattering rate. However, it is too complicated to simultaneously vary the electron and hole distribution functions by simulation. Therefore, only the electron distribution varies by simulation, and the hole distribution function is kept constant. For simplicity, the hole distribution function under the electric field is set to be a drifted Maxwellian distribution function f_h as follows:

$$f_h(k) = \left(\frac{\hbar^2}{2\pi m_h^* k_B T_h}\right)^{3/2} \exp\left(\frac{-\hbar^2 |k - k_h|^2}{2 m_h^* k_B T_h}\right), \tag{10}$$

$$k_h = \frac{m^* \mu_h E}{\hbar},$$

where μ_h^* is the hole mobility, m_h^* is the effective mass of the hole, E is the electric field, and T_h is the hole temperature.

Furthermore, the effect of degeneracy has a significant effect on carrier transport in the highly doped sample. In this case, the E_F lies in the valence band and the degeneracy is expected to play an important role because many hole states are occupied. Therefore, the effect of degeneracy effectively reduces the electron–hole scattering according to Pauli's exclusion principle. The procedure for implementing the effect of degeneracy in the Monte Carlo method is as follows (Lugli and Ferry, 1985). If the final states of holes lie at a lower energy level than the Fermi level, the scattering process is prohibited.

With respect to band structure, three conduction band valleys — Γ, L, and X — and one valence band for heavy holes are included. For the conduction band, the nonparabolicity is taken into account by introducing energy-independent nonparabolicity coefficients (Littlejohn et al., 1977).

8. Calculated Results

Minority electron transport in p-GaAs is calculated by the Monte Carlo method on the basis of the present modeling (Taniyama et al., 1990). The calculated results are then compared with the experimental results and with the calculated results for majority electron transport in n-GaAs at equivalent carrier concentrations. Three carrier concentrations are included in the calculation, $p = N_A = 1.5 \times 10^{17}$, 1.5×10^{18}, and 1.0×10^{19} cm^{-3}, where N_A is the acceptor concentration. The Thomas–Fermi model is adopted for screening length. Throughout the calculation, hole temperature is fixed at 300 K and hole mobility is taken from the Hall measurement results for the drifted Maxwellian distribution of holes. The material and scattering parameters adopted in the calculation are essentially the same as those used by Osman and Ferry (1987b). The number of simulated electrons is 10^5. All of the scattering events are calculated every 25 fs, and the total sampling time is typically set at 3 ps.

a. Scattering Rate

The energy dependent scattering rate at $T = 300$ K is shown in Fig. 17a for a hole concentration of $p = N_A = 1.5 \times 10^{17}$ cm^{-3}. This figure shows that, except for the low-energy region, the electron–LO phonon scattering rate is much larger than the electron–hole scattering rate. For a higher carrier concentration, $p = N_A = 1.5 \times 10^{18}$ cm^{-3}, the electron–hole scattering rate becomes larger than the electron–LO phonon scattering rate up to about 0.4 eV, as shown in Fig. 17b. Furthermore, at $p = N_A = 10^{19}$ cm^{-3}, the electron–hole scattering rate becomes dominant over a wide range of energies, as indicated in Fig. 17c. It can be seen from these figures that as the hole concentration increases, the electron–hole scattering rate becomes larger than the Γ–L intervalley scattering rate. The latter dominates the electron–phonon scattering in an energy region over $\Delta E_{\Gamma\text{-L}} = 0.29$ eV. The electron–phonon scattering rate is slightly reduced by the screening effect. The scattering rates and the angular distributions are quite similar to those seen for electron–impurity scattering, because the hole effective mass is much larger than the electron effective mass. This fact suggests that electron–hole scattering acts like impurity scattering when the energy transfer from the minority electron ensemble to the majority hole ensemble can be ignored. Therefore, one would expect the electron mobility to degrade and the negative differential resistance to reduce with increasing hole concentrations. On the other hand, coupled plasmon–LO phonon scattering is small in comparison with the other scattering mechanisms, because the plasmon cutoff wave number is estimated to be small because of the excitation of holes from the Fermi sphere (Landau damping).

b. Drift Velocity

Based on these scattering rates, the minority-electron drift velocity versus electric field is calculated by Monte Carlo simulation. First, I will

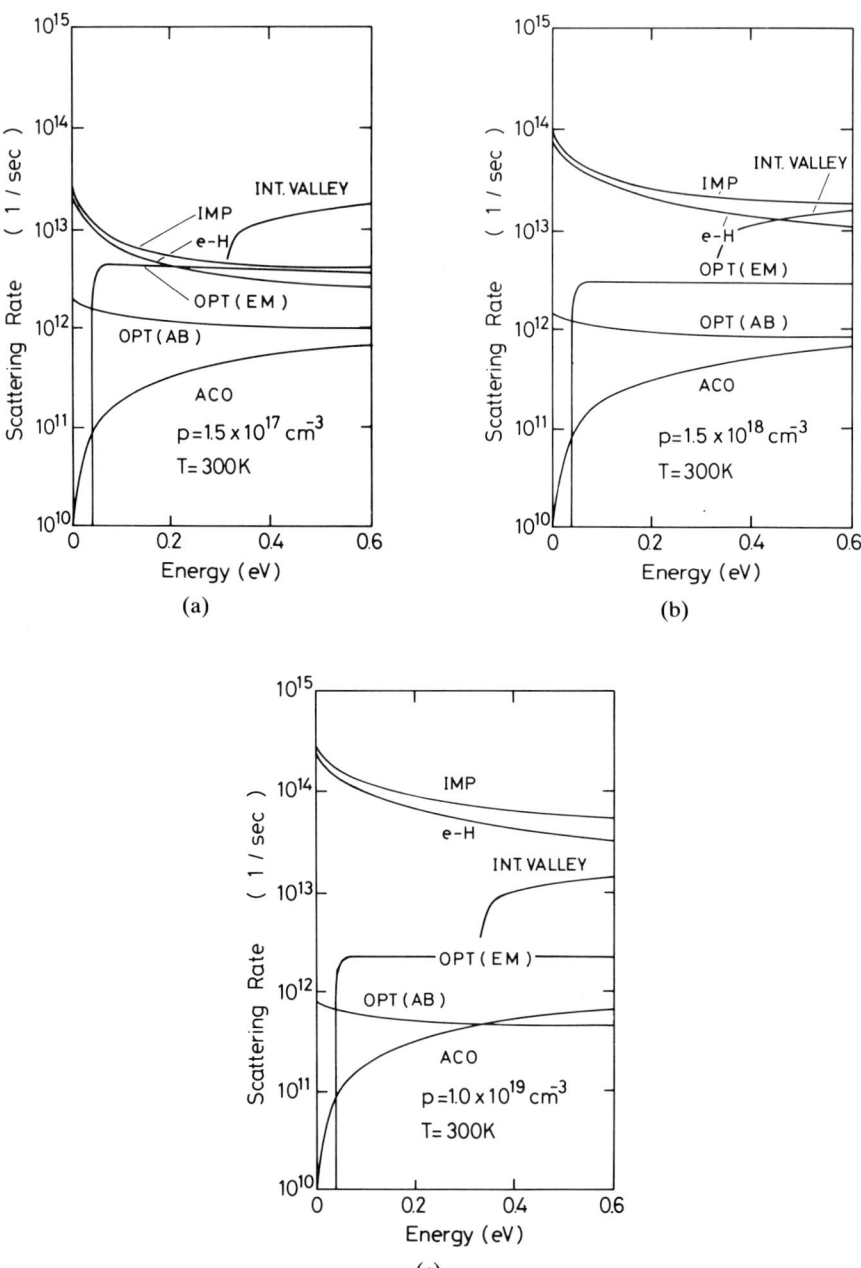

FIG. 17. Energy dependence of the scattering rates for electrons in p-GaAs with hole concentrations of 1.5×10^{17} (a), 1.5×10^{18} (b), and 1.0×10^{19} cm^{-3} (c).

discuss the calculated results for carrier concentrations of $p = N_A = 1.5 \times 10^{17}$ and $1.5 \times 10^{18}\,\mathrm{cm}^{-3}$, which are shown in Figs. 18a and 18b, respectively (Furuta *et al.*, 1990; Taniyama *et al.*, 1990). In these figures, the dashed line curves are the calculated results including electron–hole interaction. These curves describe minority-electron transport in p-GaAs. In order to clarify the effect of the electron–hole interactions on electron transport, the majority electron velocities in n-GaAs are also calculated. This calculation includes the electron–electron interactions, which are represented by chain-dotted line curves. Also included are the experimental results, which are represented by solid line curves. These figures show that the minority-electron velocity and the saturation velocity become lower with increasing hole concentration. These results are well supported by the experimental results. The velocity degradation for a low electric field can be ascribed to electron–hole and electron–impurity scattering. The majority electron velocity is not reduced as much compared with the minority-electron velocity. The difference between minority and majority electrons can be attributed to differences in carrier–carrier scattering. Electron–electron scattering does not affect the velocity, because the effective mass of electrons is much smaller than that of holes. The majority electron–impurity scattering dominates the low electric field mobility.

In high electric fields, the calculated minority-electron drift velocity for a hole concentration of $1.5 \times 10^{17}\,\mathrm{cm}^{-3}$ is almost the same as that calculated for majority electrons because the electron–phonon interaction is dominant. At higher hole concentrations, on the other hand, the carrier–carrier interaction becomes the dominant scattering mechanism on drift velocity. The electron–hole interaction, in particular, is dominant over electron–electron interaction because of the longer screening length. Therefore, the differences between minority- and majority-electron velocity become larger with increasing carrier concentration. Furthermore, the calculated results showed a slightly higher drift velocity than that of the experiments. This disagreement can be understood as follows. In the present calculation, the hole energy distribution is assumed to be drifted Maxwellian, with hole temperature and mobility set to be constant. However, this assumption is not always strictly true. For the higher fields, the increment of hole temperature is slightly higher than 300 K, as confirmed by the experiments shown in Fig. 8. The increase in temperature is not large, owing to the very low mobility caused by the heavy effective mass of holes. This implies that the energy transfer rate from hot electrons to cold holes decreases and the transition from Γ to upper valleys increases. Accordingly, the drift velocity is reduced and better agreement between the experiments and calculations can be expected for

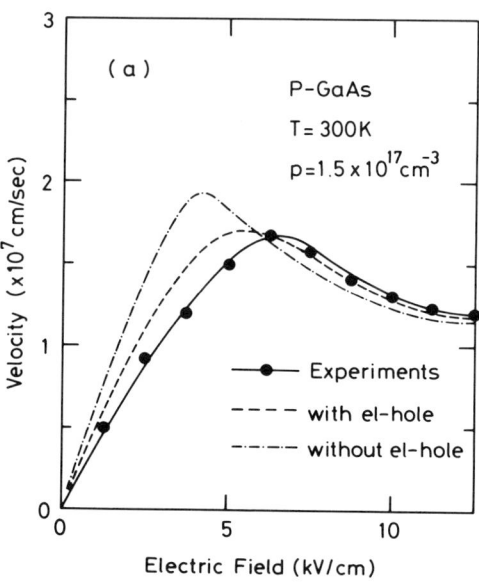
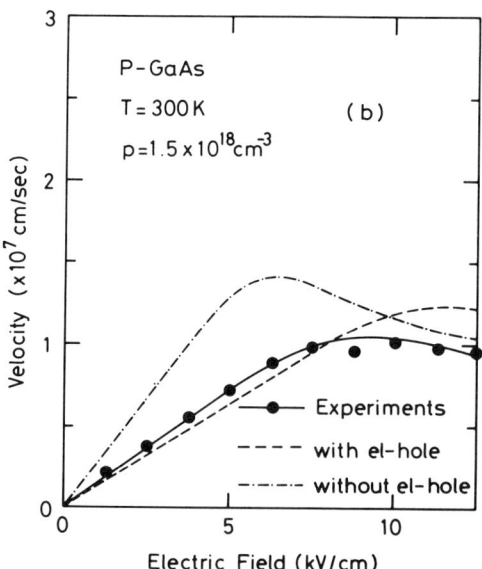

FIG. 18. Monte Carlo calculations for the velocity versus electric field of minority electrons in p-GaAs. The hole concentrations are 1.5×10^{17} (a) and 1.5×10^{18} cm^{-3} (b). Dotted and chain dotted lines indicate the calculated results with and without the electron–hole interaction, respectively (from Furuta et al., 1990).

the high-field region. The result, including the effect of the hole distribution function dependence on electric field, will be discussed later.

Next, I discuss the calculated results for the high doping. The calculated results for a hole concentration of $p = N_A = 1.0 \times 10^{19}$ cm^{-3} are shown in Fig. 19 (Furuta and Tomizawa, 1990; Taniyama et al., 1990). Using the just-mentioned models, the calculated result is shown by the chain-dotted line. This calculated result is much smaller than the experimental data over the entire range of applied electric fields. This discrepancy is mainly attributed to the fact that the degeneracy effect is not taken into account in the calculation. In the highly doped sample, the Fermi energy E_F lies in the valence band, and many hole states are occupied. This means that only the holes near the Fermi energy interact with electrons. The degeneracy effect thus reduces the strength of the electron–hole interaction according to Pauli's exclusion principle.

The dashed-line curve in Fig. 19 represents the calculations that include the degeneracy effect. The calculated velocity increases because of the reduction in electron–hole interaction. At low fields, it more closely agrees with the experimental data. However, the velocities at the

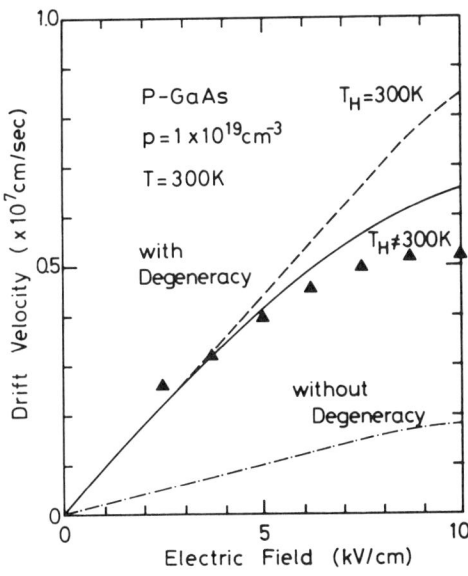

FIG. 19. Monte Carlo calculations of drift velocity of minority-electrons in p-GaAs with a hole concentration of 1.0×10^{19} cm^{-3}. Dashed and chain dotted lines indicate the calculated results with and without the effect of degeneracy, respectively. Solid line indicates the calculated result, including the effect of degeneracy, and hole mobility and temperature dependence on the electric field (from Furuta and Tomizawa, 1990).

high electric field region diverge and overestimate the experimental data. As mentioned above, the energy distribution of holes is assumed to be constant over the entire electric field range in this calculation. In reality, both hole temperature and mobility change with increasing electric field and affect the hole-energy distribution, so that the electron–hole scattering rate is modified. It is, therefore, necessary to include the hole temperature and mobility dependence on the electric field more precisely. As mentioned in Section III, these parameters are obtained from the hole current–bias voltage measurement. Using these data, the modified drifted Maxwellian can be determined.

The result, including the hole-distribution dependence on electric field, is also shown in Fig. 19 (the solid line). In low electric fields, the velocity does not indicate any notable deviation from the results shown by the dashed curve, because the changes in hole temperature and mobility with electric field are small. On the other hand, the velocity is smaller than that represented by the dashed curve and converges with the experimental curve at high electric fields. This velocity reduction in the high electric fields is caused by enhanced electron–hole scattering accompanied by a hole temperature increase.

The dependence of drift mobility of minority electrons on hole concentration is shown in Fig. 20 (Furuta and Tomizawa, 1990). In this figure, the dashed, chain-dotted, and solid lines have the same significance as in Fig. 19. The calculated results, including the effect of

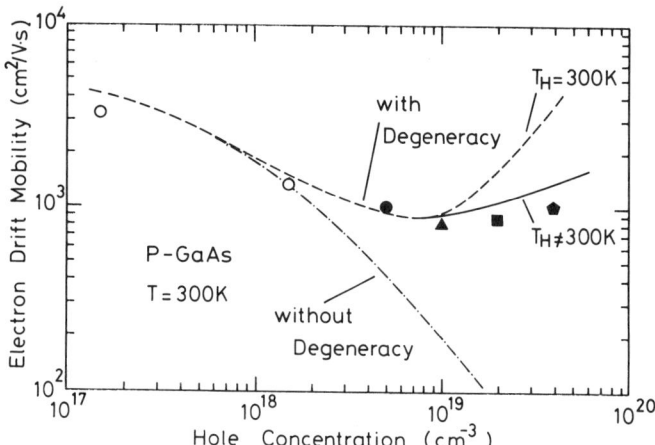

FIG. 20. Low-field drift mobilities for minority electrons in p-GaAs are plotted as a function of hole concentration. Also included are the Monte Carlo calculations. Each line has the same significance as in Fig. 19 (from Furuta and Tomizawa, 1990).

degeneracy and dependence of the hole distribution function on electric field, as well as the electron–hole interaction, show good agreement with the experimental data from low to high doping concentration.

Recently, Lowney and Bennett (1991) calculated the minority-electron mobility in p-GaAs for acceptor densities between 5×10^{16} and 1×10^{20} cm^{-3}. They considered all the important scattering mechanisms such as carrier–carrier and plasmon scattering. The ionized impurity and electron–hole scattering processes were treated with a phase-shift analysis. Their calculated results are also in good agreement with the experimental results presented above. However, in their calculation, the plasmon scattering is the dominant scattering mechanism to explain the experimental data, which is contradictory to my calculation. Therefore, further study must concentrate on improving the physical models of electron–hole and plasmon scattering processes.

c. *Electron Temperature*

The electron–hole interaction was investigated from the standpoint of energy transfer. In Figs. 21a and 21b, the electron temperatures are plotted as a function of electric field for hole concentrations of $p = N_A = 1.5 \times 10^{17}$ and 1.5×10^{18} cm^{-3}, respectively (Furuta *et al.*, 1990; Taniyama *et al.*, 1990). Three kinds of lines have the same significance as in Fig. 18.

The differences of energy transfer between electron–hole and electron–electron interactions are clearly calculated. For electron-hole scattering, the reduction of electron temperature with increasing hole concentration agrees well with the experiments. The large energy transfer rate from the hot electrons to cold holes leads to a much lower electron temperature with an increase of hole concentration. For electron–electron scattering, electrons rapidly heat up with an increasing electric field, because the electron–electron scattering conserves the total electron energy. Therefore, the electron temperature is not appreciably reduced through electron–electron scattering.

If minority electrons are compared with majority electrons at high electric fields, the fraction of carriers that are scattered to the upper valleys is reduced and the occupation is limited to the Γ valley alone in the case of minority electrons. Thus, because of the lower electron temperature, enhanced drift velocity has been reported (Sadra *et al.*, 1988). However, other theoretical (Saito *et al.*, 1989; Taniyama *et al.*, 1990) and experimental (Furuta and Tomizawa, 1990) studies show that the minority-electron drift velocity is not enhanced as much as the

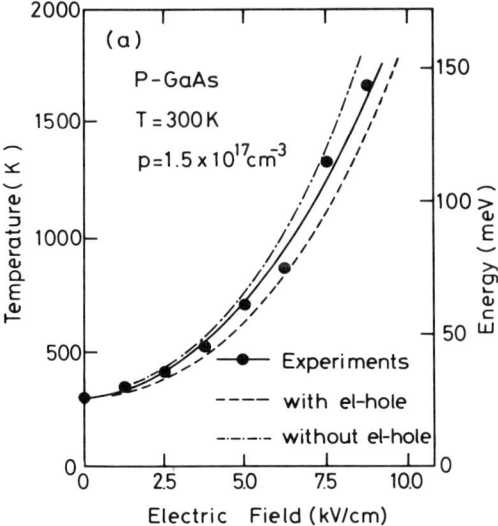
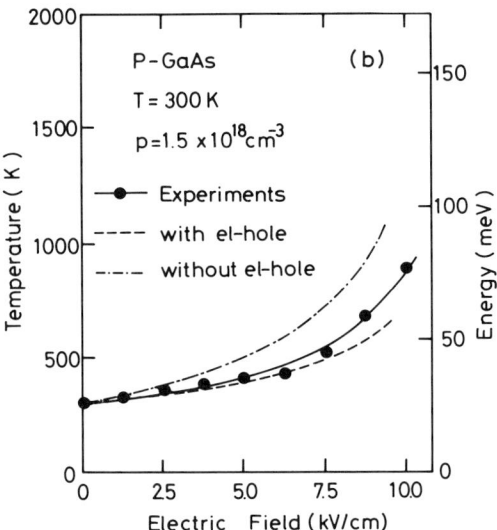

FIG. 21. Monte Carlo calculations for the electron temperature dependence on electric field in p-GaAs with hole concentrations of 1.5×10^{17} (a) and 1.5×10^{18} cm^{-3} (b). Dotted and chain-dotted lines have the same significance as in Fig. 18 (from Furuta et al., 1990).

reported velocities. This may explain why the drift velocity is reduced to a very low level because of strong electron-hole interaction. This reduction increases with increasing carrier concentrations. The velocity is saturated, and develops a lower value even in the Γ valley. The difference between these results and previous work can be attributed to the fact that the cutoff wave number is calculated by including the single-particle excitation of a hole. This inclusion leads to a longer screening length and stronger electron–hole scattering compared with other reports.

VI. Summary

The work reviewed in this chapter describes the transport properties of minority electrons in p-GaAs under high electric fields. Drift velocity, electron temperature, and the energy relaxation processes were investigated by time-of-flight and photoluminescence measurements. The experimental results were different from those of majority electrons and reveal a strong dependence on hole concentration. These data are attributed to the momentum and energy transfer between minority electrons and majority holes by the electron–hole interaction. The discussions of the energy loss rate and momentum and energy relaxation times also play an important role in the electron–hole interaction.

In order to investigate scattering mechanisms in detail, experimental data were compared with calculations by Monte Carlo simulation. The calculated results, which take the electron–hole interaction into account, agree well with experimental results up to a hole concentration of $1 \times 10^{18} \text{cm}^{-3}$. Above this concentration, however, the degeneracy effect and the dependence of hole distribution function on electric field also strongly affect the minority-electron transport. This means that the electron–hole interaction affects the minority-carrier transport. It must thus be included, along with the degeneracy effect and the majority-hole energy distribution dependence on electric field, to obtain accurate Monte Carlo calculations.

This work reveals that the combination of an experimental approach using optical techniques, such as time-of-flight and photoluminescence, with a theoretical approach such as a Monte Carlo simulation is indispensable for a detailed investigation of hot-carrier transport.

Recently, these physical models were applied to the simulation of HBTs. Tomizawa *et al.* (1991) calculated the transit time of electrons in the base-collector depletion layer and investigated the velocity overshoot effects. They pointed out that the initial drift velocity and energy distribution for electrons injected from base to collector regions, which are mainly determined by minority electron transport in the base region, strongly affect the velocity overshoot effect. This means that minority electron transport under high electric field is very important to determine the speed performance of HBTs. Their results also support the characteristics of HBTs reported by Ishibashi and Yamauchi (1988).

Finally, it is worth noting that the work reviewed here is very recent at the time of writing. This makes it difficult to obtain an accurate physical model of minority-carrier transport, and a number of problems still have to be resolved. The screening length is one of the most important parameters for describing carrier transport. In my calculation, the screening length is taken into consideration using a formula similar to that for majority carriers. For minority carriers, however the effect of screening is very different than for majority carriers. The minority carriers and majority carriers can screen both other scattering mechanisms and each other. Therefore, further study must concentrate on improving the physical model of screening length. In addition, the scattering potential of the Coulomb interaction and plasmon scattering at high doping concentration must also be included. Further experimental studies, in conjunction with theoretical studies, may contribute to answering these problems.

Acknowledgments

The author would like to thank to Drs. A. Yoshii, T. Ishibashi, and K. Hirata for their encouragement during the course of this work. The author would also like to thank N. Shigekawa for his valuable input regarding the time-of-flight experiments, and H. Taniyama and Dr. M. Tomizawa for performing the Monte Carlo simulation.

References

Ahrenkiel, R. K., Dunlavy, D. J., Hamaker, H. C., Green, R. T., Lewis, C. R., Hayes, R. E., and Fardi, H. (1986). *Appl. Phys. Lett.* **49,** 725.
Ahrenkiel, R. K., Dunlavy, D. J., Greenberg, D., Sclupmann, J., Hamaker, H. C., and MacMillan, H. F. (1987). *Appl. Phys. Lett.* **51,** 776.
Bradley, C. W. W., Taylor, R. A., and Ryan, J. F. (1989). *Solid State Elec.* **32,** 1173.
Canali, C., Nava, F., and Reggiani, L. (1985). In *Hot-Electron Transport in Semiconductors* (Reggiani, L., ed.). Springer, Berlin.
Chattopadhyay, D., and Quiesser, H. J. (1981). *Rev. Mod. Phys.* **53,** 1745.

Combescot, M., and Combescot, R. (1987). *Phys. Rev.* **B35,** 7989.
Conwell, E. M. (1967). *High Field Transport in Semiconductors.* Academic Press, New York.
Debye, P. P., and Conwell, E. M. (1954). *Phys. Rev.* **93,** 693.
Degani, J., Leheny, R. F., Nahory, R. E., and Heritage, J. P. (1981). *Appl. Phys. Lett.* **39,** 569.
Ehrenreich, H. (1957). *J. Phys. Chem. Solids* **2,** 131.
Ettenberg, M. E., Kressel, H., and Gilbert, S. L. (1973). *J. Appl. Phys.* **44,** 827.
Evans, A. G. R., and Robson, P. N. (1974). *Solid-State Electron.* **17,** 805.
Fawcett, W., Boadman, A. D., and Swain, S. (1970). *J. Phys. Chem. Solids* **31,** 1963.
Furuta, T., and Tomizawa, M. (1990). *Appl. Phys. Lett.* **56,** 824.
Furuta, T., and Yoshii, A. (1991). *Appl. Phys. Lett.* **59,** 3607.
Furuta, T., Taniyama, H., and Tomizawa, M. (1990). *J. Appl. Phys.* **67,** 293.
Goodnick, S., and Lugli, P. (1988). *Phys. Rev.* **B38,** 10135.
Haynes, J. R., and Shockley, W. (1949). *Phys. Rev.* **75,** 691.
Haynes, J. R., and Shockley, W. (1951). *Phys. Rev.* **81,** 835.
Höpfel, R. A., Shah, J., Gossard, A. C., and Wiegmann, W. (1985). *Physica.* **134B,** 509.
Höpfel, R. A., Shah, J., and Gossard, A. C. (1986a). *Appl. Phys. Lett.* **48,** 148.
Höpfel, R. A., Shah, J., and Gossard, A. C. (1986b). *Phys. Rev. Lett.* **56,** 765.
Höpfel, R. A., Shah, J., Wolff, P. A., and Gossard, A. C. (1988). *Phys. Rev.* **B37,** 6941.
Ishibashi, T., and Yamauchi, Y. (1988). *IEEE Trans. Electron Devices* **ED-35,** 401.
Ito, H., and Ishibashi, T. (1986). *Inst. Phys. Conf. Ser.* **79,** 607.
Ito, H., and Ishibashi, T. (1989). *J. Appl. Phys.* **65,** 5197.
Jacoboni, C., and Reggiani, L. (1983). *Rev. Mod. Phys.* **55,** 645.
Joshi, R. P., and Grondin, R. O. (1989). *J. Appl. Phys.* **66,** 4288.
Kim, M. E., Das, A., and Sentuia, S. D. (1987). *Phys. Rev.* **B18,** 6890.
Klausmeirer-Brown, M. E., Melloch, M. R., and Lundstrom, M. S. (1990). *Appl. Phys. Lett.* **56,** 160.
Knox, W. H., Hirlimann, C., Miller, D. A. B., Shah, J., Chemla, D. S., and Shank, C. V. (1986). *Phys. Rev. Lett.* **56,** 1191.
Knox, W. H., Chemla, D. S., Livescu, G., Cunningham, J. E., and Henry, J. E. (1988). *Phys. Rev. Lett.* **61,** 1290.
Lagunova, T. S., Marushchak, V. A., Stepanova, M. N., and Titkov, A. N. (1985). *Sov. Phys. Semicond.* **19,** 71.
Leo, K., and Collet, J. H. (1991). *Phys. Rev.* **B44,** 5535.
Levi, A. F., and Yafet, Y. (1987). *Appl. Phys. Lett.* **51,** 42.
Levine, B. F., Tsang, W. T., Bethea, C. G., and Capasso, F. (1982). *Appl. Phys. Lett.* **41,** 470.
Levine, B. F., Bethea, C. G., Tsang, W. T., Capasso, F., Thornber, K. K., Fulton, R. C., and Kleinmann, D. A. (1983). *Appl. Phys. Lett.* **42,** 769.
Littlejohn, M. A., Hauser, J. R., and Glisson, T. H. (1977). *J. Appl. Phys.* **48,** 4587.
Lovejoy, M. L., Keyes, B. M., Klausmeier-Brown, M. E., Melloch, M. R., Ahrenkiel, R. A., and Lundstrom, M. S. (1990). In *Extended Abstracts for the 22nd International Conference of Solid State Devices and Materials, Sendai, Japan,* pp. 613–616.
Lowney, J. R., and Bennett, H. S. (1991). *J. Appl. Phys.* **69,** 7102.
Lugli, P., and Ferry, D. K. (1985). *IEEE Trans. Electron Devices* **ED-32,** 2431.
Madelung, O. (1978). *Introduction to Solid-State Theory.* Springer, Berlin.
McLean, T. P., and Paige, E. G. S. (1960). *J. Phys. Chem. Solids* **16,** 220.
Moglestue, C. (1984). *IEE Proc.* **131,** 188.
Morohashi, M., Sawaki, N., and Akasaki, I. (1985). *Jpn. J. Appl. Phys.* **24,** 661.

Nag, B. R. (1980). *Electron Transport in Compound Semiconductors*. Springer, Berlin.
Nathan, M. I., Dumke, W. P., Wrenner, K., Tiwari, S., Wright, S. L., and Jenkins, A. (1988). *Appl. Phys. Lett.* **52,** 654.
Nelson, R. J. (1979). *Inst. Phys. Conf. Ser.* **45,** 256.
Neukermans, A., and Kino, G. S. (1973). *Phys. Rev.* **B7,** 2703.
Osman, M. A., and Ferry, D. K. (1987a). *J. Appl. Phys.* **61,** 5330.
Osman, M. A., and Ferry, D. K. (1987b). *Phys. Rev.* **B46,** 6018.
Osman, M. A., and Grubin, H. L. (1987). *Appl. Phys. Lett.* **51,** 1812.
Paige, E. G. S. (1960). *J. Phys. Chem. Solids* **16,** 207.
Pines, D. (1956). *Rev. Mod. Phys.* **28,** 184.
Prince, M. B. (1953). *Phys. Rev.* **92,** 681.
Ridley, B. K. (1982). *Quantum Processes in Semiconductors*. Oxford University Press, New York.
Ruch, J. G., and Kino, G. S. (1967). *Appl. Phys. Lett.* **10,** 40.
Sadra, K., Maziar, C. M., Streetman, B. G., and Tang, D. S. (1988). *Appl. Phys. Lett.* **53,** 2205.
Saito, K., Yamada, T., Akatsuka, T., Fukamachi, T., Tokumitsu, E., Konagai, M., and Takahashi, K. (1989). *Jpn. J. Appl. Phys.* **28,** L.2081.
Seeger, K. (1985). *Semiconductor Physics*. Springer, Berlin.
Shah, J. (1978). *Solid State Elec.* **21,** 43, and references therein.
Shah, J. (1988). *IEEE J. Quantum Electron.* **QE-24,** No. 2, 276.
Shah, J. (1989). *Solid State Elec.* **32,** 1051, and references therein.
Shah, J., and Leite, R. C. C. (1964). *Phys. Rev. Lett.* **22,** 1304.
Shah, J., Nahory, R. E., Leheny, R. F., Degani, J., and DiGiovanni, A. E. (1982). *Appl. Phys. Lett.* **40,** 505.
Shah, J., Pinczuk, A., Störmer, H. L., Gossard, A. C., and Wiegmann, W. (1983). *Appl. Phys. Lett.* **42,** 55.
Shah, J., Pinczuk, A., Störmer, H. L., Gossard, A. C., and Wiegmann, W. (1984). *Appl. Phys. Lett.* **44,** 322.
Shah, J., Pinczuk, A., Gossard, A. C., and Wiegmann, W. (1985). *Phys. Rev. Lett.* **54,** 2045.
Shigekawa, N., Furuta, T., Maezawa, K., and Mizutani, T. (1990). *Appl. Phys. Lett.* **56,** 1146.
Smith, P. M., Inoue, M., and Frey, J. (1980). *Appl. Phys. Lett.* **37,** 797.
Tang, D. D., Fang, F. F., Scheuermann, M., and Chen, T. C. (1986). *Appl. Phys. Lett.* **49,** 1540.
Taniyama, H., Furuta, T., Tomizawa, M., and Yoshii, A. (1990). *J. Appl. Phys.* **68,** 621.
Tiwari, S., and Wright, S. L. (1990). *Appl. Phys. Lett.* **56,** 563.
Tomizawa, M., Taniyama, H., Furuta, T., and Yoshii, A. (1991). In *Computer Aided Innovation of New Materials* (M. Doyama, T. Suzuki, J. Kihara, and R. Yamamoto, eds.). North-Holland, Amsterdam.
Vilims, J., and Spicer, W. E. (1965). *J. Appl. Phys.* **36,** 2815.
Walukiewicz, W., Lagowski, J., Jastrzebski, L., and Gatos, H. C. (1979). *J. Appl. Phys.* **50,** 5040.
Xiaoguang, Wu., Peeters, F. M., and Devreese, J. T. (1985). *Phys. Rev.* **B32,** 6982.
Zhou, X. Q., Cho, G. C., Leummer, K., Kütt, W., Wolter, K., and Kurz, H. (1989). *Solid State Elec.* **32,** 1591.

CHAPTER 4

Minority-Carrier Transport in III–V Semiconductors

Mark S. Lundstrom

SCHOOL OF ELECTRICAL ENGINEERING
PURDUE UNIVERSITY
WEST LAFAYETTE, INDIANA

I. INTRODUCTION	194
II. MINORITY-CARRIER TRANSPORT IN COMPOSITIONALLY NONUNIFORM SEMICONDUCTORS	195
1. *Energy Bands in Compositionally Nonuniform Semiconductors*	195
2. *Solution of the Boltzmann Equation*	197
3. *Drift-Diffusion Current Equations*	198
III. HEAVY DOPING EFFECTS AND MINORITY-CARRIER TRANSPORT	201
4. *Band Structure of Heavily Doped Semiconductors*	201
5. *The $n_0 p_0$ Product*	202
6. *Minority-Carrier Transport*	204
7. *Minority-Carrier Transport in Quasi-neutral Regions*	205
IV. COUPLED PHOTON/MINORITY-CARRIER TRANSPORT	205
8. *Photon Recycling*	205
9. *Mathematical Formulation*	206
10. *Influence on Minority-Carrier Recombination*	209
11. *Influence on Minority-Carrier Transport*	210
12. *Effects on Devices*	212
V. EFFECTS OF HEAVY DOPING ON DEVICE-RELATED MATERIALS PARAMETERS	213
13. *Effective Energy Gap Shrinkage and Injected Currents*	213
14. *Measurements of Effective Energy Gap Shrinkage*	214
15. *Minority-Carrier Mobility*	217
16. *Measurements of Minority-Carriers Mobility*	219
17. *Survey of Measured Minority-Carrier Mobilities for Carrier in* GaAs	226
VI. MINORITY-CARRIER TRANSPORT IN III–V DEVICES	228
18. *Dc Characteristics of* AlGaAs/GaAs *HBTs*	229
19. *All-*GaAs *Pseudo-*HBT's	231
20. *Carrier Drift in Doping Gradients*	234
21. GaAs *Solar Cells*	238
22. *Quantum Efficiency of* GaAs *Photodiodes*	245
23. *Minority Electron Transport in* InP/InGaAs *HBTs*	247
VII. SUMMARY	254
ACKNOWLEDGMENT	254
REFERENCES	255

I. Introduction

The performance of electronic devices such as bipolar transistors and solar cells is often controlled by the injection of minority carriers into quasi-neutral regions and by the transport of minority carriers across such regions. Minority carrier injection is described by Shockley's "law of the junction" (Shockley, 1959; Marshak and Shrivastava, 1979) and minority-carrier transport by drift-diffusion equations (van Roosbroeck, 1950). For modern, compound semiconductor devices, however, minority-carrier transport is complicated by several factors. For example, in devices with a heterojunction launching ramp, minority-carrier transport may be very far from equilibrium. In this chapter, the focus is on near-equilibrium transport with an emphasis on the influence of heavy impurity doping and radiative recombination and reabsorption. So-called off-equilibrium minority carrier transport is, however, briefly examined. This chapter is intended to relate minority-carrier physics to the performance of bipolar devices.

Modern devices often contain layers in which the impurity doping is extremely heavy. An important example is the heterojunction bipolar transistor in which the base is doped, very heavily to lower its resistance (Kroemer, 1982). As discussed by Abram in this volume, many-body effects associated with the heavy impurity doping shrink the energy gap, and disorder associated with the dopants introduces band tails. The high majority-carrier densities also increase the scattering rate for minority carriers, as discussed in the chapter by Yevick and Bardyszewski. Finally, heavy impurity doping lowers the minority-carrier lifetime, as reviewed by Ahrenkiel. These so-called *heavy-doping effects* can have a profound effect on the performance of devices.

For moderately to heavily doped III–V semiconductors, radiative recombination often dominates. If the active layer of the device has a thickness on the order of $1/\alpha$, where α is the absorption coefficient of the emitted photons, then photons will often be reabsorbed and new minority carriers generated. Under such conditions, minority-carrier transport must be described as the coupled flow of minority carriers and photons. These so-called *photon-recycling effects* can be important in devices and in diagnostic measurements.

To put the chapter into perspective, we begin with a brief synopsis. Because the impurity doping density varies with position within a device and because heavy doping perturbs the band structure, all devices should be treated as heterostructures. In Section II, we develop current equations for compositionally nonuniform semiconductors. The focus is on low-field transport under near-equilibrium conditions; hot carrier effects have been discussed by Furuta. In Section III, the effects of heavy impurity doping are reviewed, and the current equations are reformulated in terms of the effective (or apparent) energy gap shrinkage, a parameter commonly used to characterize minority-carrier injection in heavily doped semiconductors. We show that for quasineutral, heavily doped regions, the effective energy-gap shrinkage is the only parameter needed to describe the effects of band structure perturbations on near-equilibrium, minority-carrier transport. In semiconductors such as GaAs, radiative recombination and reabsorption leads to a coupled flow of electrons and holes. The mathematical treatment of this coupled flow and its effect on devices is the subject of Section IV. In Section V we examine how heavy doping affects two important material parameters, the $n_0 p_0$ product and the minority carrier mobility. Theoretical treatments are first reviewed, and experimental techniques to measure the parameters are then described. Finally, in Section VI, the influence of minority-carrier physics on the performance of bipolar devices is examined.

II. Minority-Carrier Transport in Compositionally Nonuniform Semiconductors

With modern epitaxial growth techniques it is possible to vary the material composition within a device almost at will. During the past decade or so, this new degree of freedom has made it possible to explore a variety of new *heterostructure* devices. As discussed by Abram, however, heavy impurity doping perturbs the band structure, so any device in which the impurity doping changes with position is also a heterostructure device. In this section, we begin by briefly examining transport in heterostructures; the results are applied to heavily doped semiconductors in Section III.

1. Energy Bands in Compositionally Nonuniform Semiconductors

Figure 1 shows an energy band diagram we'll use to describe a general heterostructure (Marshak and van Vliet, 1978a). As shown in Fig. 1, the valence band maximum, E_{V0}, and the conduction band minimum, E_{C0}, vary with position for two reasons. One is that a macroscopic potential, due to impurity doping or applied biases, may shift the levels, and the

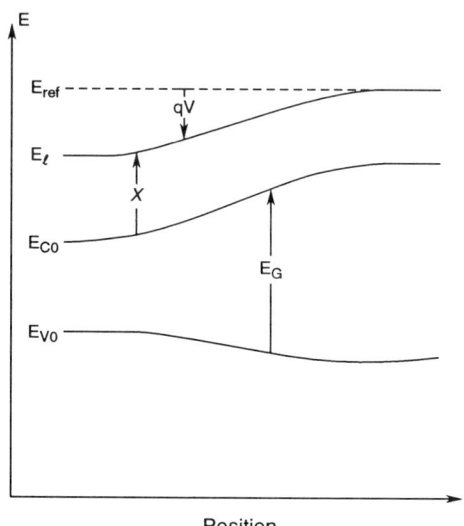

FIG. 1. Energy band diagram for a general, compositionally nonuniform semiconductor. $E_\ell = E_{\text{ref}} - qV(z)$ is a local reference level. (Reprinted with permission from *Solid State Electron.* **21,** A. H. Marshak and K. M. van Vliet, "Electrical Current in Solids with Position-Dependent Band Structure," Copyright © 1978, Pergamon Press plc.)

second is that the material composition (microscopic potential) may vary with position. We denote the macroscopic electrostatic potential by $V(z)$ and measure the material composition by $\chi(z)$, which is the energy difference between the conduction band minimum and a reference level. (It has been common to select the field-free vacuum level as the reference, in which case χ is the electron affinity, but an internal reference is preferable.)

With these definitions, we write the conduction and valence band extrema as

$$E_{C0}(z) = E_{\text{ref}} - qV(z) - \chi(z) \tag{1a}$$

and

$$E_{V0}(z) = E_{\text{ref}} - qV(z) - \chi(z) - E_G(z). \tag{1b}$$

The force on electrons is

$$F_e = -\frac{dE_{C0}}{dz} = -q\left\{\mathscr{E}_z + \frac{d\chi/q}{dz}\right\}, \tag{2}$$

4. MINORITY-CARRIER TRANSPORT IN III–V SEMICONDUCTORS

which we interpret as the sum of forces due to the electric field and to the so-called *quasi-electric field*, which arises from compositional variations (Kroemer, 1957). Similarly, for holes,

$$F_b = \frac{dE_{V0}}{dz} = +q\left\{\mathscr{E}_z - \frac{d}{dz}(\chi/q + E_G/q)\right\}. \tag{3}$$

Although it is not shown in Fig. 1, we also assume that the effective mass varies slowly with position. The assumption is that at $z = z_1$, the band structure is that of homogeneous semiconductor with a composition identical to that at z_1. The concept of a semiconductor with a position-dependent effective mass makes sense when the composition does not vary too rapidly (Marshak and van Vliet, 1978a; van Vliet and Marshak, 1982; Gora and Williams, 1967); it should be valid when the material composition varies slowly over about 100 lattice spacings (Gora and Williams, 1967). We further assume that the quantum density of states is proportional to the square root of kinetic energy. (Use of this approximation ignores transport in delocalized band-tail states.) For such cases, we can define an effective mass and density of states and relate the electron concentration to the Fermi level by

$$n_0 = N_C(z)\mathscr{F}_{1/2}[(E_F - E_{C0})/k_B T] \tag{4}$$

and determine the intrinsic carrier concentration from

$$n_i(z) = \sqrt{N_C(z)N_V(z)}\, e^{-E_G(z)/2k_B T}. \tag{5}$$

2. SOLUTION OF THE BOLTZMANN EQUATION

To describe electron transport in a heterostructure, we begin with the steady-state Boltzmann equation,

$$v_z \frac{\partial f}{\partial z} + \frac{dp_z}{dt}\frac{\partial f}{\partial p_z} = \frac{\partial f}{\partial t}\bigg\}_{\text{coll}}. \tag{6}$$

For a heterostructure, the equation of motion is (Marshak and van Vliet, 1978a)

$$\frac{dp_z}{dt} = -\frac{\partial}{\partial z}\{E_{C0} + E(p)\}, \tag{7}$$

where E_{C0} is the potential energy, and the kinetic energy is $E(p) = p^2/2m^*$ for spherical bands. We decompose f into symmetric and antisymmetric components f_S and f_A, where

$$f_S = \frac{1}{1 + e^{(E_{C0}+E(p)-F_n)/k_B T}}, \qquad (8)$$

and F_n is the electron quasi-Fermi level. Making the relaxation time approximation for the collision term, the Boltzmann equation becomes

$$v_z \frac{\partial f_S}{\partial z} - \frac{\partial(E_{C0} + E(p))}{\partial z} \frac{\partial f_S}{\partial p_z} = -f_A/\tau. \qquad (9)$$

For an isothermal, nondegenerate semiconductor, we find

$$f_A = \frac{\tau}{k_B T_L} f_S v_z \frac{\partial F_n}{\partial z}, \qquad (10)$$

which can be solved for the electron current,

$$J_n = \frac{1}{\Omega} \sum_p (-q) v_z f_A, \qquad (11)$$

(where Ω is a normalization volume) to find

$$J_{nz} = n\mu_n \frac{\partial F_n}{\partial z}. \qquad (12)$$

The mobility is defined as

$$\mu_n = \frac{q \langle\!\langle \tau \rangle\!\rangle}{m^*}, \qquad (13)$$

with $\langle\!\langle \tau \rangle\!\rangle$ being the "average" relaxation time (Lundstrom, 1990). Equation (12) shows that the conventional thermodynamic expression for the current flow (Smith et al., 1967) remains valid for heterostructures.

3. DRIFT-DIFFUSION CURRENT EQUATIONS

For device analysis, it is useful to reexpress the current equation as a drift-diffusion equation. Although it is not a necessary condition, the

4. MINORITY-CARRIER TRANSPORT IN III–V SEMICONDUCTORS

derivation is most easily done for a nondegenerate semiconductor. Differentiating

$$F_n = E_{C0} + k_B T_L \log(n/N_C) \tag{14}$$

and inserting the results in (12), we find

$$J_{nz} = n\mu_n \frac{dE_{C0}}{dz} + qD_n \frac{dn}{dz} - n\mu_n k_B T_L \frac{d(\log N_C)}{dz}. \tag{15}$$

The first term is a drift current, which consists of drift in the electric field (due to gradients in the macroscopic potential) as well as drift in the quasi-electric field (due to gradients in the microscopic potential). In addition to the diffusion current, we also find a *density-of-states effect* associated with gradients in the effective mass. If the effective mass (density-of-states) varies with position, then electrons will tend to diffuse in the direction of increasing effective mass.

For computational work, it is convenient to make Eq. (15) look as much as possible like a conventional drift-diffusion equation. By inserting Eq. (1a) in Eq. (15) and rearranging the result, we find

$$J_{nz} = -nq\mu_n \frac{d}{dz}(V + V_n) + k_B T \mu_n \frac{dn}{dz}, \tag{16}$$

where

$$qV_n = \chi(z) + k_B T \log N_C(z) + \text{constant}. \tag{17a}$$

It is convenient to select the constant so that V_n reduces to zero in a reference, or unperturbed, location (Lundstrom *et al.*, 1981). The result is that Eq. (17a) becomes

$$qV_n = \chi(z) - \chi_{\text{ref}} + k_B T \log(N_C(z)/N_{C\text{ref}}). \tag{17b}$$

A similar development for holes leads to

$$J_{pz} = -pq\mu_p \frac{d}{dz}(V - V_p) - k_B T \mu_p \frac{dp}{dz}, \tag{18}$$

where V_p is given by

$$qV_p = -(\chi(z) - \chi_{\text{ref}}) - (E_G(z) - E_{G\text{ref}}) + k_B T \log(N_V/N_{V\text{ref}}). \tag{19}$$

Equations (16) and (18) show that the drift-diffusion equations for heterostructures are just like the conventional equations, but with additional terms that account for the quasi-electric fields and position-dependent effective mass.

It is interesting to observe that

$$n_i^2(z) = n_i^2(\text{ref}) \times e^{q(V_n(z)+V_p(z))/k_B T}, \tag{20}$$

which shows that the quantity $V_n + V_p$ acts like an effective energy gap shrinkage given by

$$\Delta_G(z) = qV_n + qV_p = \Delta E_G + k_B T \log\left(\frac{N_V(z)N_C(z)}{N_{V\text{ref}}N_{C\text{ref}}}\right). \tag{21}$$

For heavily doped semiconductors, it will be more convenient to recast the drift-diffusion equations in terms of the effective energy gap shrinkage. A little algebra shows that (Lundstrom et al., 1981)

$$J_{nz} = -nq\mu_n \frac{d}{dz}\{V + \gamma \Delta_G/q\} + k_B T \mu_n \frac{dn}{dz}, \tag{22a}$$

and

$$J_{pz} = -pq\mu_p \frac{d}{dz}\{V - (1-\gamma)\Delta_G/q\} - k_B T \mu_p \frac{dp}{dz}, \tag{22b}$$

where

$$\gamma \equiv \frac{V_n}{V_n + V_p} \tag{23}$$

is the *effective asymmetry factor*, a measure of the change in band edge and band structure associated with the conduction band.

The form of Eqs. (22) suggests that they are restricted to nondegenerate semiconductors, but they are easily generalized. When carriers are degenerate, $k_B T \mu \neq qD$, so the final term in the current equations should not be interpreted as a diffusion current. Instead, the influence of carrier degeneracy on the diffusion coefficient, D, can be included in the parameters, γ and Δ_G, if they are generalized for Fermi–Dirac statistics (Lundstrom et al., 1981). We now turn to the perturbations induced by heavy impurity doping and specialize the transport equations for such applications.

III. Heavy Doping Effects and Minority Carrier Transport

In a heavily doped semiconductor, carrier–carrier interactions reduce the energy gap and carrier–dopant interactions distort the bands by introducing band tails (Abram et al., 1978). The resulting position-dependent, perturbed band structure influences carrier transport and has a strong effect on the electrical performance of devices. Modeling transport in such structures is much like modeling transport in any heterostructure, but the details of the perturbed band structure are still not well understood, so a specialized terminology has developed. The specialized terminology is reviewed in this section and is related to the conventional treatment of transport in heterostructures.

4. Band Structure of Heavily Doped Semiconductors

Figure 2 illustrates the band structure perturbations induced by heavy impurity doping (del Alamo and Swanson, 1987). When the semiconductor is lightly doped, the conduction and valence bands are parabolic with

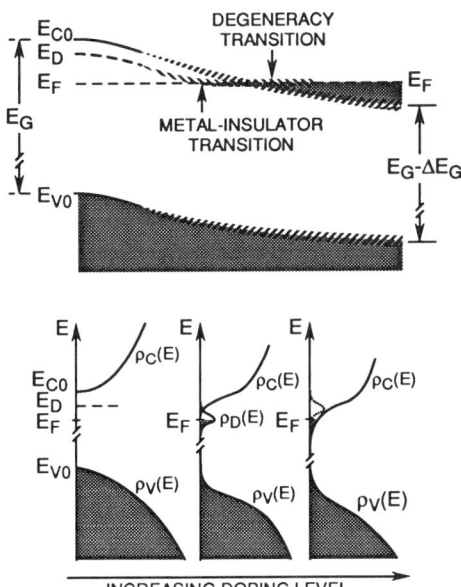

FIG. 2. Illustration of how heavy impurity doping affects the band structure of an n-type semiconductor. (Reprinted with permission from *Solid State Electron.* **30**, J. A. del Alamo and R. M. Swanson, "Modeling of Minority-Carrier Transport in Heavily Doped Silicon Emitters," Copyright © 1987, Pergamon Press plc.)

sharply defined band edges. As the doping density increases, the localized impurity states broaden into an impurity band, and the fluctuating potential associated with the dopants located on randomly situated lattice sites produces tails in the conduction and valence bands. For very heavy impurity doping, the impurity band merges with the conduction band. Under such conditions, electron–hole correlation and exchange effects dominate, and these many-body effects produce a more or less rigid shrinkage of the band gap. The electronic properties of heavily doped semiconductors are profoundly influenced by these perturbations in the band structure.

5. The $n_0 p_0$ Product

In heavily doped semiconductors, the perturbed band structure alters $n_0 p_0$, which must be evaluated from

$$n_0 = \int_{E'_{C0}}^{E_{top}} \rho'_C(E - E'_{C0}) f(E - E_F) \, dE, \tag{24a}$$

and

$$p_0 = \int_{E_{bot}}^{E'_{V0}} \rho'_V(E'_{V0} - E)[1 - f(E - E_F)] \, dE, \tag{24b}$$

where ρ'_C is the perturbed conduction band density-of-states, f is the Fermi function, and E'_{C0} is the conduction band mobility edge (a corresponding set of definitions for the valence band apply to the quantities with a subscript V). For lightly doped semiconductors, the densities-of-states are parabolic, and the product of Eqs. (24a) and (24b) reduces to

$$n_0 p_0 = n_{i0}^2 = N_C N_V e^{-E_G/k_B T}, \tag{25}$$

where N_C and N_V are the effective densities-of-states for the conduction and valence bands. For heavily doped semiconductors, however, the result is much different. The doping-induced perturbation of the energy bands can be viewed at the simplest level as an effective narrowing of the energy gap, from which we conclude that $n_0 p_0$ will increase with doping density. Both detailed many-body calculations and measurements confirm that $n_0 p_0$ does increase in heavily doped p-GaAs (Bennett and Lowney, 1987; Klausmeier-Brown et al., 1988; Harmon et al., 1991).

4. MINORITY-CARRIER TRANSPORT IN III–V SEMICONDUCTORS

The $n_0 p_0$ product is an important factor in device design and analysis, and for such purposes it is convenient to express the product in the simple form (Lundstrom *et al.*, 1981)

$$n_0 p_0 \equiv n_{ie}^2 \equiv n_{i0}^2 e^{\Delta_G^0/k_B T}. \tag{26}$$

The effective intrinsic carrier concentration n_{ie} is greater than n_{i0}, which refers to a lightly doped semiconductor. It is essential to note that Eq. (26) is a defining equation for Δ_G^0, the equilibrium *effective* (or *apparent* or *electrical*) energy gap shrinkage, which is not a physical energy gap shrinkage but is, rather, related to the band structure by equating Eq. (26) to the product of Eqs. (24a) and (24b). If we adopt a simple rigid band model for the band structure of a heavily doped semiconductor, then the effective gap shrinkage for a nondegenerate semiconductor reduces to Eq. (21). In this model, the energy gap shrinks, but the band shape (or effective mass) does not change. When the semiconductor is degenerate, the effective gap shrinkage becomes

$$\Delta_G = \Delta E_G + k_B T \log\left(\frac{N_V(z) N_C(z)}{N_{Vref} N_{Cref}}\right) + k_B T \log\left(\frac{\mathscr{F}_{1/2}(\eta_V)}{e^{\eta_V}} \frac{\mathscr{F}_{1/2}(\eta_C)}{e^{\eta_V}}\right), \tag{27}$$

where $\eta_V = (E_{V0} - E_F)/k_B T$ and $\eta_C = (E_F - E_{C0})/k_B T$, and $\mathscr{F}_{1/2}$ is the Fermi–Dirac integral of order one-half. For a degenerate p-type semiconductor, $\mathscr{F}_{1/2}(\eta_V)/e^{\eta_V} < 1$ and $\mathscr{F}_{1/2}(\eta_C)/e^{\eta_C} \simeq 1$, so the effect of majority carrier degeneracy is to widen the effective energy gap, which is analogous to the well-known Burstein–Moss shift in the optical gap. Equation (27) is valid only when the energy bands are parabolic, but it does illustrate how band gap shrinkage and Fermi–Dirac statistics influence the $n_0 p_0$ product.

It must be emphasized that a variety of effective or apparent energy gap shrinkages have been defined and that these nonphysical, defined quantities should not be confused (Lundstrom *et al.*, 1981; del Alamo, and Swanson, 1987; Marshak, 1987). This distinction is especially important when comparing the results of electrical measurements of devices to optical absorption or photoluminescence measurements (Pantelides, *et al.* 1985). The effective gap shrinkage (or equivalently, the effective intrinsic carrier concentration) is the important quantity for modeling bipolar devices because it determines the minority carrier concentration and, as we demonstrate below, the minority carrier current in a heavily doped semiconductor.

6. MINORITY-CARRIER TRANSPORT

If we restrict our attention to low-field transport in isothermal semiconductors, then, as shown in Section II.6, the electron and hole current equations are

$$\vec{J}_n = n\mu_n \nabla F_n \tag{28a}$$

and

$$\vec{J}_p = p\mu_p \nabla F_p, \tag{28b}$$

where F_n and F_p are the quasi-Fermi levels for electrons and holes. The nonequilibrium carrier densities are evaluated from Eqs. (24a) and (24b) with the Fermi levels replaced by the appropriate quasi-Fermi levels. From the resulting expressions, we can solve for ∇F_n and ∇F_p, and insert the results in Eq. (28) to express the current equations as the sum of conventional drift and diffusion components plus a term that involves the gradient of the perturbed band structure (Marshak and van Vliet, 1978a, b). For nondegenerate semiconductors with parabolic bands, Marshak's equations reduce to Eq. (15), which is commonly used to model transport in semiconductor heterostructures (Sutherland and Hauser, 1977). For use in modeling heavily doped devices, however, we find it most convenient to express the current equations in the form of Eq. (22). The perturbed band structure introduces two additional parameters into the current equations. The first, Δ_G, is precisely the effective gap shrinkage defined in equilibrium by Eq. (26), and the second, the effective asymmetry factor, γ, is a measure of how much of the perturbation is associated with the conduction band. Equation (26) defines the effective gap shrinkage in equilibrium. Under low injection conditions, it is common to assume that the effective gap shrinkage is equal to its value in equilibrium, but high densities of injected carriers may perturb the band structure.

Equations (22a) and (22b) are equivalent to those commonly used for analyzing semiconductor heterostructures (Sutherland and Hauser, 1977; Schuelke and Lundstrom, 1983) but are expressed in the form most useful when the compositional variation is due to heavy doping effects (Lundstrom et al., 1981). Detailed expressions for Δ_G and γ are available (Lundstrom et al., 1981), but it may be helpful to note that if the semiconductor is nondegenerate and the band structure perturbation is a simple, rigid shift of the bands without any change in band shape, then Δ_G is the actual energy gap shrinkage and γ is the fraction of the shrinkage associated with a perturbation in the conduction band edge.

7. Minority Carrier Transport in Quasi-Neutral Regions

Many minority carrier devices are controlled by diffusion of carriers across a quasi-neutral region in low-level injection. Consider a quasi-neutral p-type region in low level injection. The equilibrium field is found from Eq. (22b) by setting $\vec{J}_p = 0$, which is then inserted in Eq. (22a) to find

$$\vec{J}_n = -nq\mu_n \left[-(k_B T/q)\frac{\nabla p_0}{p_0} + \nabla(\Delta_G^0/q) \right] + k_B T \mu_n \nabla n, \qquad (29)$$

where a subscript or superscript 0 denotes the equilibrium value of a quantity. Equation (29) should accurately describe low-injection electron transport in p^+-GaAs because under such conditions, the electrostatic potential, the effective gap shrinkage, and the majority carrier hole concentration are all very near their equilibrium values. The important point is that to describe low-injection electron transport in quasi-neutral p^+-GaAs, the only information required about the perturbed band structure is the quantity, Δ_G—the effective gap shrinkage, which explains its widespread use in device modeling. Conversely, electrical measurements of minority carrier currents only provide information on Δ_G; they are insensitive to γ. When the simplifying assumptions of minority carrier diffusion in quasi-neutral regins under low-level injection are not met, then a detailed understanding of the perturbed band structure is necessary. For example, just to compute the built-in potential of a p–n junction requires knowledge of both Δ_G^0 and γ^0 (Lundstrom et al., 1981).

IV. Coupled Photon/Minority-Carrier Transport

8. Photon Recycling

Radiative recombination often dominates in III–V semiconductors, but one shouldn't assume that the minority-carrier lifetime is the radiative lifetime. Because III–V semiconductors have a high absorption coefficient near the band edge, photons emitted when electrons and holes recombine are very often absorbed within the semiconductor. The result is that observed minority-carrier lifetimes are often several times the radiative lifetime. In III–V semiconductors, minority-carrier transport is accompanied by the transport of photons, and radiative recombination and reabsorption (so-called photon recycling) couple the two flows.

Recombination/reabsorption effects can have a profound influence on minority-carrier lifetimes and can affect the transport of minority carriers as well (Dumke, 1957). Our purposes in this section are to explain how these effects are mathematically modeled and to examine some of the consequences for III–V semiconductors.

9. MATHEMATICAL FORMULATION

Figure 3a illustrates the physical problem we wish to describe: a distribution of excess minority carriers drifting and diffusing through a semiconductor under the influence of radiative recombination and reabsorption. At position z, the radiative recombination rate is simply

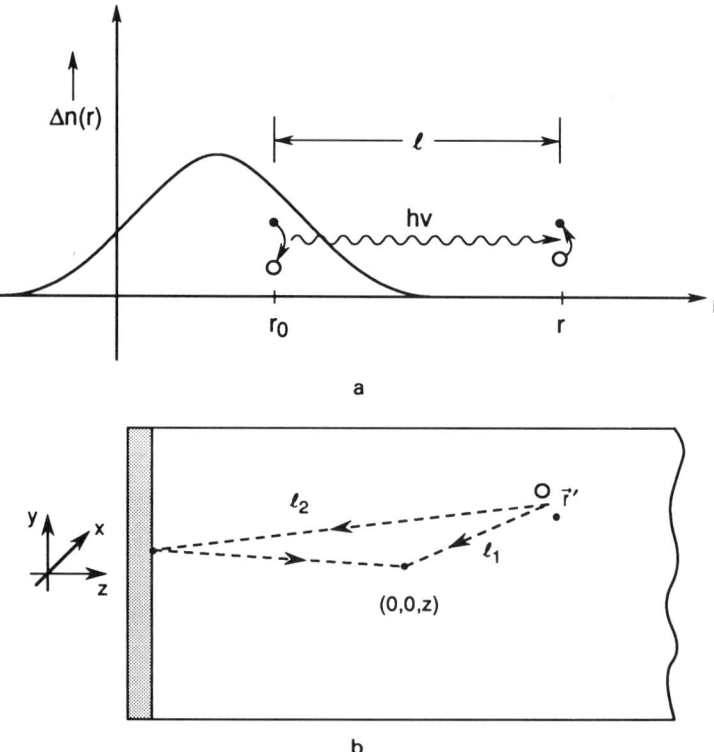

FIG. 3. (a) Illustration of how radiative recombination and reabsorption are related in an infinite semiconductor sample. (b) Illustration of two possible paths for generation of minority carriers at position z due to radiative recombination at $\vec{r}'$. A semi-infinite sample with the surface at $z = 0$ is assumed.

4. MINORITY-CARRIER TRANSPORT IN III–V SEMICONDUCTORS

$\Delta n/\tau_r$, where τ_r is the radiative lifetime. Recombination at position $\vec{r}_0$ gives rise to near band-edge photons, which propagate through the semiconductor. Because the absorption edge is steep in III–V semiconductors, however, the absorption depth $1/\alpha$ is typically only a few micrometers. Photons produced at $\vec{r}'$ and absorbed at $\vec{r}_0$ generate carriers at a rate of $g_{pr}(\vec{r}_0, \vec{r}', \omega) = \alpha(\omega)\exp(-\alpha\ell)$, where ℓ is the optical path length between $\vec{r}_0$ and $\vec{r}'$. To describe the overall effect, we need to include photons generated throughout the structure by integrating over position and over the spectral distribution of the emitted photons.

These ideas can be specified to formulate a minority carrier diffusion equation that describes the effects of radiative recombination and reabsorption. By adding radiative recombination and generation terms to the minority carrier diffusion equation, we find

$$\frac{\partial \Delta n}{\partial t} = D_n \frac{\partial^2 \Delta n}{\partial z^2} - \frac{\Delta n}{\tau_{nr}} - \frac{\Delta n}{\tau_r} + G_{pr}(z, t), \qquad (30)$$

where

$$G_{pr}(z, t) = \int_0^\infty d\omega \int d\vec{r}' g_{pr}(z, \vec{r}', \omega). \qquad (31)$$

We can express the generation rate, g_{pr}, as

$$g_{pr}(z, \vec{r}', \omega) = \frac{\Delta n(\vec{r}', t)}{\tau_r} \sum_i \frac{R_i \alpha(\omega) e^{-\alpha(\omega)\ell_i}}{4\pi \ell_i^2} \hat{S}(\omega), \qquad (32)$$

where $\hat{S}(\omega)$ is the normalized spectral distribution of the emitted radiation. The sum in Eq. (32) is over the various optical paths. For example, Fig. 3b shows a semi-infinite sample for which there is a direct path as well as one involving a reflection from the front surface. If the path involves a reflection, then R_i is the reflection coefficient. For the direct path, $R_i = 1$.

To find $\hat{S}$, we begin with the spontaneous emission rate as given by the von Roosbroeck–Shockley relation (van Roosbroeck and Shockley, 1954; von Roos, 1983),

$$S(\vec{r}, t, \omega) = \left(\frac{n\omega}{\pi c}\right)^2 \frac{\alpha(\omega)}{[e^{(\hbar\omega - \Delta E_F)/k_B T} - 1]}. \qquad (33)$$

The emitted radiation is centered near the band edge, so for low-level injection, the separation between the quasi-Fermi levels is much smaller than $\hbar\omega$, and Eq. (33) can be simplified to

$$S(\omega) = \frac{\Delta n(\vec{r})}{n_0} \times \left(\frac{n(\omega)\omega}{\pi c}\right)^2 \alpha(\omega) e^{-\hbar\omega/k_B T}, \qquad (34)$$

from which we find the normalized spectral distribution as

$$\hat{S}(\omega) = \frac{\omega^2 \alpha(\omega) e^{-\hbar\omega/k_B T}}{\int_0^\infty \omega^2 \alpha(\omega) e^{-\hbar\omega/k_B T}}. \qquad (35)$$

In Eq. (34), Δn is the excess minority carrier concentration, n_0 is the equilibrium minority carrier concentration, and $n(\omega)$ is the index of refraction. In practice, it is important that the radiative recombination and reabsorption terms be consistent. Since the reabsorption term G_{pr} is evaluated from the measured absorption coefficient (after subtracting out the contribution from the free carriers), we evaluate the radiative lifetime from

$$1/\tau_r = \frac{1}{n_0} \int_0^\infty (n\omega/\pi c)^2 e^{-\hbar\omega/k_B T} \alpha(\omega) \, d\omega, \qquad (36)$$

which is derived from the von Roosbroeck–Shockley relation (van Roosbroeck and Shockley, 1954; von Roos, 1983). (In Eq. (36), n is the index of refraction, c is the speed of light in vacuum, and $n_0 = n_{ie}^2/N_A$ is the equilibrium minority-carrier concentration.)

Both analytical and numerical solutions to Eq. (30) have been reported (Asbeck, 1977; Kuriyama et al., 1977). The analytical solutions require a number of simplifying assumptions, but they have proven useful for interpreting experiments (Nelson and Sobers, 1978). Standard numerical techniques also work and can treat a broader range of problems. A numerical solution can be obtained by using finite difference techniques to solve Eq. (30). Initial conditions are specified, and the minority-carrier distribution is updated after a small time step. At the end of the time step, the optical generation term, G_{pr}, is evaluated from the updated minority-carrier distribution. This generation term is then used in the minority-carrier diffusion equation to update the minority-carrier profile after the next time step. The process continues until steady-state conditions are achieved. In the following sections, we'll examine the general consequences of radiative recombination and reabsorption using simple, analytical solutions. Numerical techniques will be employed in Section VI in order to examine the influence on devices.

4. MINORITY-CARRIER TRANSPORT IN III–V SEMICONDUCTORS 209

10. INFLUENCE ON MINORITY-CARRIER RECOMBINATION

When nonradiative and radiative recombination occurs, one might be tempted to write the minority-carrier lifetime as

$$1/\tau = 1/\tau_{nr} + 1/\tau_r, \tag{37}$$

but photon recycling can significantly increase the observed lifetimes. The effects are easiest to examine for an infinite semiconductor with a uniform excess minority carrier concentration because there is only a single optical path between a photon emitted at $\vec{r}'$ and position $(0, 0, z)$. For such conditions, the generation term to reabsorption becomes

$$G_{pr} = \frac{\Delta n}{\tau_r} \int_0^\infty \hat{S}(\omega)\, d\omega \int d^3r \frac{\alpha e^{-\alpha r}}{4\pi r^2} = \frac{\Delta n}{\tau_r}. \tag{38}$$

By inserting the result in Eq. (30), we find that carrier generation due to reabsorption precisely balances the loss due to radiative recombination, and the net recombination rate is

$$1/\tau = /\tau_{nr}, \tag{39}$$

so for an infinite sample with a uniform excess carrier concentration, radiative recombination has no effect on the observed lifetime of minority carriers.

For a finite-sized sample, some phonons escape without generating electron–hole pairs, so the observed lifetimes are reduced. For this case, we must integrate over the volume of the sample while accounting for the various optical paths and the reflection coefficients from the surfaces. Consider, for example, a thin film of thickness d with a uniform concentration of excess minority carriers. Equation (31) becomes

$$G_{pr} = \frac{\Delta n}{\tau_r} \int_0^\infty \hat{S}(\omega)\, d\omega \int_0^{2\pi} \sin\theta\, d\theta \int_0^\infty r\, dr \int_0^d dz \sum_i \frac{R_i \alpha e^{-\alpha \ell_i}}{4\pi \ell_i^2}, \tag{40}$$

which can be written as

$$G_{pr} = \frac{\Delta n}{\tau_r} F(d), \tag{41}$$

where $0 \le F(d) \le 1$ is the probability that an emitted photon is reabsorbed within the thin film. $F(d)$ obviously increases as d increases, but it also depends on the reflection coefficients at the top surface and at the film/substrate interface because they affect the degree of photon confinement. Expressions for $F(d)$ have been presented by Asbeck (1977).

If we insert Eq. (41) into the minority-carrier diffusion equation, (30), we find that the minority-carrier lifetime is

$$1/\tau = 1/\tau_{nr} + (1 - F)/\tau_r, \qquad (42)$$

which we write as

$$1/\tau = 1/\tau_{nr} + 1/\phi\tau_r, \qquad (43)$$

where $\phi = 1/(1 - F)$ is the recycling factor (Asbeck, 1977). When all of the emitted photons are reabsorbed, the recycling factor approaches infinity, and the lifetime is the nonradiative lifetime. For thin films with active layers a few micrometers thick, the recycling factor can be on the order of 10 (Asbeck, 1977). Photon recycling clearly has a profound influence on the minority-carrier lifetimes; the measured lifetimes bear no simple relation to the radiative lifetime. As discussed by Ahrenkiel in this volume, experimental measurements of recombination lifetimes need to be carefully interpreted to account for these recycling effects.

11. Influence on Minority-Carrier Transport

The effects of photon recycling on minority-carrier recombination are easy to appreciate, but recombination and reabsorption can also affect minority carrier transport (Dumke, 1957; von Roos, 1983). To examine the effects on transport, we assume an infinite semiconductor and expand $\Delta n(z)$ as

$$\Delta n(z) = \Delta n(0) + \frac{\partial \Delta n}{\partial z} z + \frac{1}{2} \frac{\partial^2 \Delta n}{\partial z^2} z^2 + \cdots, \qquad (44)$$

insert the result in Eq. (31) and integrate to write the generation rate as

$$G_{pr} = G_{pr}^0 + G_{pr}' + G_{pr}'' + \cdots. \qquad (45)$$

The first term integrates to $\Delta n(0)/\tau_r$, which just cancels the radiative recombination term in the minority carrier diffusion equation. If we assume that $\partial \Delta n/\partial z$ is roughly constant, then the second term integrates to zero. Finally, if we assume that $\partial^2 \Delta n/\partial z^2$ is nearly constant, then the third term in the expansion gives

$$G'''_{pr} = \frac{1}{2\tau_r} \frac{\partial^2 \Delta n}{\partial z^2} \int_0^\infty \alpha(\omega)\hat{S}(\omega)\,d\omega \int_0^{2\pi} d\phi \int_0^\pi \int_0^\infty r^2 \sin\theta\, d\theta\, dr \times \frac{z^2 e^{-\alpha r}}{4\pi r^2}, \tag{46}$$

which can be integrated to find

$$G''_{pr} = \frac{1}{\tau_r} \int_0^\infty \frac{\hat{S}(\omega)}{3\alpha^2}\,d\omega \times \frac{\partial^2 \Delta n}{\partial z^2} = \frac{\langle 1/\alpha^2 \rangle}{3\tau_r} \times \frac{\partial^2 \Delta n}{\partial z^2}. \tag{47}$$

By inserting the result for G_{pr} into Eq. (30), we find an equation that describes the transport of minority carriers near $z = 0$. The results can be expressed as

$$\frac{\partial \Delta n}{\partial t} = (D_n + D_R)\frac{\partial^2 \Delta n}{\partial z^2} - \frac{\Delta n}{\tau_{nr}}, \tag{48}$$

where

$$D_R \equiv \frac{1}{\tau_r}\int_0^\infty \frac{\hat{S}(\omega)}{3\alpha^2}\,d\omega = \frac{\langle 1/\alpha^2 \rangle}{3\tau_r}. \tag{49}$$

Equation (48) is the desired result; it shows that photon recycling affects minority carriers in an infinite semiconductor in two different ways. The first effect is that photon recycling eliminates the influence of radiative recombination on the minority-carrier lifetime, and the second is an enhancement of the minority-carrier diffusion coefficient. Since $1/\tau_r = BN_A$, we expect D_R to increase with doping. Minority-carrier diffusion coefficients generally decrease with doping, so for heavily doped semiconductors, D_R may become important. The enhancement of D is of special concern for n-type semiconductors, because the minority hole diffusion coefficient tends to be small.

To estimate the importance of D_R, we should evaluate it from Eq. (49), but $\langle 1/\alpha^2 \rangle$ diverges as α approaches zero. In practice, however, the lower limit of the integral in Eq. (49) should be set by $\alpha^{-1} \geq \mathscr{L}$, where $\mathscr{L}$ is some characteristic distance over which the minority carrier concentration varies (e.g., the diffusion length for a long sample or the sample dimension for a short one). Data for the doping-dependent absorption coefficient are available (Casey *et al.*, 1975), but for a simple estimate

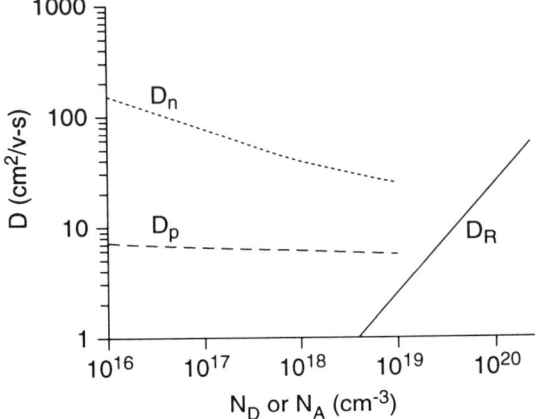

FIG. 4. Comparison of D_R and the diffusion coefficients for minority carrier electrons (D_n) and holes (D_p) in GaAs. D_R is estimated from $D_R = 1/(3\tau_r \alpha_0^2)$, which assumes that $\alpha(\omega) = \alpha_0 \theta(E - E_G)$. A value of $\alpha_0 = 2 \times 10^4 \text{ cm}^{-1}$ was assumed. The minority carrier diffusion coefficients are a summary of the available data presented in Section V.

of the importance of D_R, we may assume that $\alpha(\omega) = \alpha_0 \theta(E - E_G)$, where θ is the unit step function. With this assumption, $D_R = 1/(3\tau_r \alpha_0^2)$. Figure 4 shows the resulting D_R vs. doping density and compares it with the available data for the minority-carrier diffusion coefficients in GaAs. This calculation demonstrates that for heavily doped GaAs, enhanced diffusion due to photon recycling can be significant. For n-type GaAs doped above $\sim 10^{18} \text{ cm}^{-3}$ and for p-type GaAs doped above $\sim 10^{19} \text{ cm}^{-3}$, D_R may be comparable to the minority-carrier diffusion coefficient.

12. Effects on Devices

For infinite semiconductors, the effects of photon recycling can be treated analytically, and the results show that both the minority-carrier lifetime and the diffusion coefficient are enhanced. Similar effects are expected for devices, but they are more difficult to treat quantitatively. The device problem is complicated because of the finite size of the sample, which means that some photons escape. Of more concern, however, is the fact that the excess carrier profiles are nonuniform, so recycling effects cannot be described by a simple minority-carrier diffusion equation as in Eq. (48). Photon recycling in devices is best treated numerically by solving Eqs. (30)–(32). For devices, the effects of

4. MINORITY-CARRIER TRANSPORT IN III–V SEMICONDUCTORS

photon recycling can be much more complex than a simple increase in D and τ. Numerical computations that illustrate the effects in devices are presented in Section VI.

V. Effects of Heavy Doping on Device-Related Materials Parameters

The heavy impurity doping commonly employed in devices can alter important semiconductor materials properties such as the minority-carrier lifetime, the diffusion coefficient, and the equilibrium np product. The effects of heavy doping on the minority-carrier lifetime are reviewed by Ahrenkiel in this volume. The focus in this section is on how the mobility and np product are affected. Techniques for measuring these parameters are described, and results for GaAs are presented and discussed.

13. Effective Energy Gap Shrinkage and Injected Currents

The injected electron current in an np diode is

$$J_n = qn_{p0}\frac{D_n}{L_n}(e^{qV_A/k_BT} - 1), \tag{50}$$

where n_{p0} is the equilibrium minority electron concentration, which can be written as

$$n_{p0} = \frac{n_{ie}^2}{N_A} = \frac{n_{i0}^2}{N_A}e^{\Delta G_G^0/k_BT}. \tag{51}$$

The equilibrium np product, n_{ie}^2, is a key parameter for bipolar devices; we are interested in how it varies with doping. Alternately, we can focus on ΔG_G^0, the effective, or apparent, energy gap shrinkage, which is a nonphysical quantity defined to produce the correct $n_0 p_0$ product.

Recall that heavy doping perturbs a semiconductor's band structure. If the perturbed band structure is known, then n_{ie}^2 (and, equivalently, ΔG_G^0) can be evaluated from Eq. (24). The perturbed band structure is difficult to measure directly. Optical absorption and luminescence studies can provide information about the location of the band edges and Fermi level, but not detailed information about the perturbed band structure, which is needed to evaluate n_{ie}^2. Electrical measurements can provide information about n_{ie}^2 (or Δ_G^0), but not on the details of the band structure. Nevertheless, electrical measurements are quite sensitive to band structure changes, and they provide tests for detailed theoretical models as well as information that is directly useful for device analysis and design. Such measurements are considered next.

14. Measurements of Effective Energy Gap Shrinkage

Equation (50) suggests that n_{ie}^2 can be inferred from the measured I–V characteristic of a *pn* diode. Such measurements are feasible, but they are complicated by the need to subtract out extraneous current components (Klausmeier-Brown et al., 1988). The bipolar transistor, however, is an ideal device for characterizing electron injection currents because only the current of interest exits through the collector. For heavily doped silicon, such measurements have proven to be reliable and consistent — if properly interpreted (Slotboom and de Graaff, 1976; del Alamo and Swanson, 1987). The transistor-based approach consists of measuring the steady-state collector current with the emitter-base junction forward-biased and the base-collector junction short-circuited. Under these conditions the collector current in an *n*–*p*–*n* transistor is given by an expression similar to Eq. (50), which may be written as

$$\ln I_C = \ln\left[\frac{qA_E(n_{ie}^2 D_n)}{N_A W_B}\right] + \frac{qV_{BE}}{k_B T}. \tag{52}$$

Given a plot of $\ln I_C$ versus V_{BE}, $(n_{ie}^2 D_n)$ is determined from the intercept and the device temperature from the slope. It is essential that both the doping density and width of the base layer be accurately known, so test structures to determine the base layer resistivity by Hall effect measurements are placed adjacent to the transistors under test, and the base layer thickness is measured.

When designing the transistors for such experiments, care must be taken to be sure that Eq. (52) is a valid description of the collector current characteristics. To minimize excessive base recombination, which would invalidate Eq. (52), the base width must be much less than a minority-carrier diffusion length. The minority-carrier lifetime can be measured by photoluminescence decay on a diode test structure grown, with identical *p*-layer doping, immediately after growth of the transistor film. For valid measurements, we require that

$$L_n^2 = D_n \tau_n \gg W_B^2. \tag{53}$$

4. MINORITY-CARRIER TRANSPORT IN III–V SEMICONDUCTORS

Since we expect L_n to vary from about 1 to 10 μm for the doping densities to be considered, and W_B is typically 0.1 to 0.2 μm, Eq. (53) is readily satisfied.

For transistor-based characterization of heavy doping effects, it is preferable to work with homojunction transistors rather than with heterojunction transistors. This choice is dictated by the fact that improper compositional grading or unintended offsets between the doping and compositional junctions in a heterojunction bipolar transistor can give rise to an energy barrier that would invalidate Eq. (52), but Eq. (52) should always describe homojunction transistors accurately. The fact that the gain of the homojunction transistor may be low is of little consequence because only the collector current versus base-emitter voltage is of interest. Because no AlGaAs growth is involved, the growth temperatures are also lower, so uniform dopant profiles with very steep shoulders result (Klausmeier-Brown, 1990a).

Measurements of minority carrier injection in p- and n-type GaAs have recently been reported (Klausmeier-Brown *et al.*, 1990b; Harmon *et al.*, 1991; Patkar *et al.*, 1991). By measuring the steady-state I_C vs. V_{BE} characteristic for transistors with various base dopings, $n_{ie}^2 D$ vs. doping was characterized for $10^{18}\,\text{cm}^{-3} \le N_A \le 2 \times 10^{20}\,\text{cm}^{-3}$ for p-GaAs (Klausmeier-Brown *et al.*, 1990b; Harmon *et al.*, 1991), and from $1 \times 10^{17}\,\text{cm}^{-3}$ to $4 \times 10^{18}\,\text{cm}^{-3}$ for n-GaAs (Patkar *et al.*, 1991). Figures 5a and 5b summarize the results. To deduce the np product from these measurements, we need to know the minority-carrier diffusion coefficient. Comprehensive data are still lacking, but if we use the available data summarized in Section V.16, we obtain the results displayed in Fig. 6. For p^+-GaAs doped at $N_A \simeq 10^{19}\,\text{cm}^{-3}$, n_{ie}^2 is more than 10 times its value in lightly doped GaAs. Because the gain of a bipolar transistor is proportional to n_{ie}^2, it is clear that these results have important implications for devices. As shown in Fig. 6a, n_{ie}^2 begins to decrease for doping densities above $\simeq 5 \times 10^{19}\,\text{cm}^{-3}$. This effect is analogous to the Burstein–Moss shift of the optical gap and can be viewed as an effective widening of the energy gap. This effective gap widening sets an upper limit to the base doping for AlGaAs/GaAs HBTs because it decreases the emitter injection efficiency.

Figure 6 also shows that effective energy gap shrinkage effects are strikingly different for n-GaAs. In n-GaAs, energy gap shrinkage is offset by effective gap widening due to degenerate carrier statistics, so the np product decreases with doping (refer back to Eq. (26)). These results show that for an n–p^+ GaAs pn junction, the effective energy gap of the p^+ layer is smaller than that of the n-type layer.

These results indicate that for p^+-GaAs, the $n_0 p_0$ product is substan-

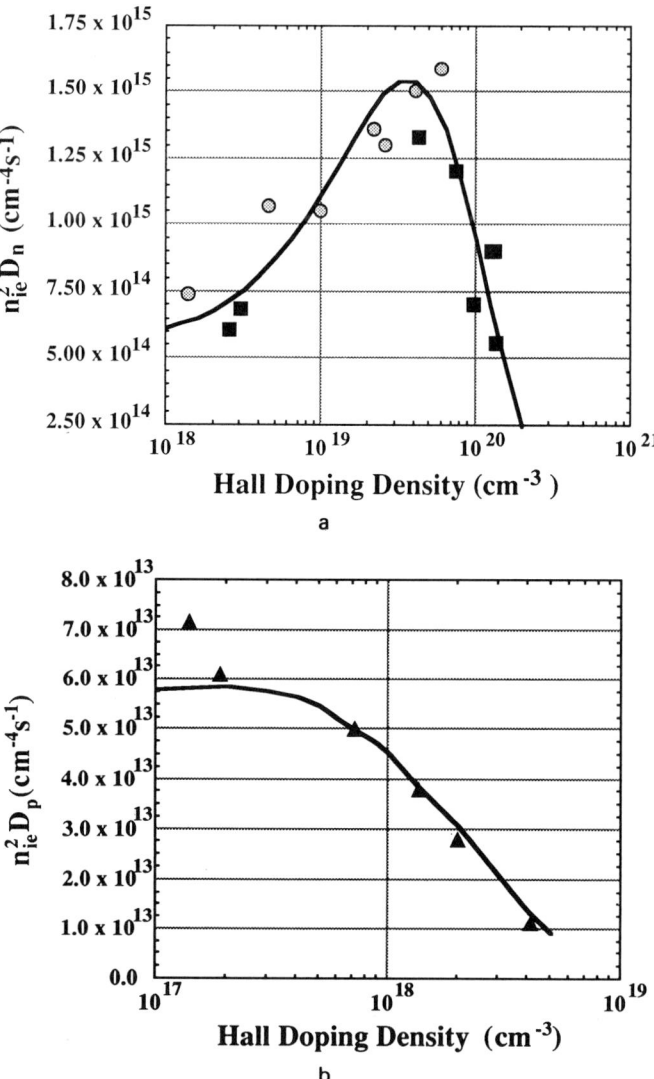

FIG. 5. (a) Measured $n_{ie}^2 D_n$ product for MBE-grown, p-type GaAs (from Harmon et al., 1991). (b) Measured $n_{ie}^2 D_p$ product for MBE-grown, n-type GaAs (from Patkar et al., 1991). (The doping density for the most heavily doped sample was lowered to correct a measurement error.)

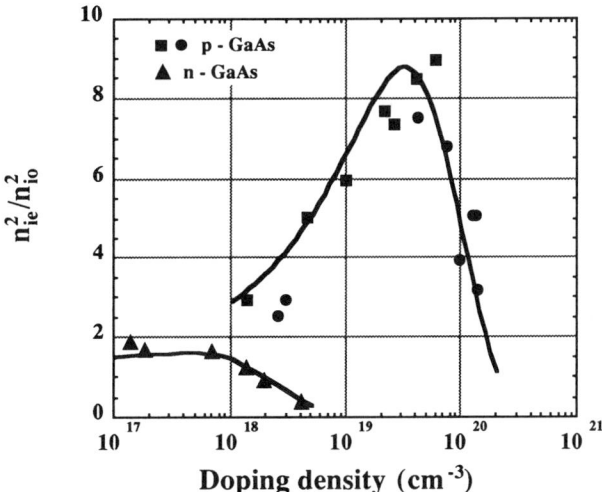

FIG. 6. Estimated value of n_{ie}^2 vs. doping density for p- and n-type GaAs. The p-type results are from the data, displayed in Fig. 5a using reported minority electron mobilities (Lovejoy, 1992), and the n-type results were extracted from the data of Fig. 5b using reported minority hole mobilities (Lovejoy et al., 1992b). $D_n(N_A)$ was assumed to saturate for $N_A > 10^{19}$ cm^{-3}.

tially greater than that predicted by the perturbed band structure model used to treat the optical effects of heavy doping (Casey and Stern, 1976). The results are, however, in general agreement with the recent many-body treatment of Bennett and Lowney (Bennett, 1986; Bennett and Lowney, 1987), although the energy gap shrinkage we observe for p^+-GaAs seems to be somewhat stronger. The distinct difference observed for n_{ie} in p^+- and n^+-GaAs is also consistent with theoretical predictions (Bennett and Lowney, 1987). The data shown in Fig. 6 can be approximated by simple, closed-form expressions for the effective energy-gap narrowing in p- and n-type GaAs (Jain and Roulston, 1991; Jain et al., 1992). Comprehensive measurements are not available for other III–V semiconductors, but similar effects are expected. Recent estimates of n_{ie} in n-type InP based on optical luminescence studies are quite similar to the results for n-type GaAs (Tyagi et al., 1991).

15. Minority-Carrier Mobility

A proper theoretical treatment of minority-carrier transport must account for the short-range electron–hole scattering events, the long-range excitations of the hole plasma, and the effects of majority carrier degeneracy (Lowney and Bennett, 1991). A rigorous treatment is not yet available, but some simplified approaches have been reported. Walukiewicz et al. (1979) evaluated the minority electron mobility by solving the Boltzmann equation using a variational approach. In addition to screened polar optical phonon and ionized impurity scattering, they added electron–heavy hole scattering. Because heavy holes are so much more massive than electrons, Walukiewicz treated the scattering process as an elastic one and used the same formalism as for ionized impurity scattering. For p^+-GaAs, Walukiewicz found that the minority electron mobilities were roughly one-half of the corresponding majority electron mobilities, and that the minority electron mobility increased at low temperatures, in contrast to the roughly constant mobility for majority electrons in n^+-GaAs.

What's missing in Walukiewicz's treatment is the fact that electrons can excite long-range oscillations in the hole plasma. Lowney and Bennett added a simple treatment of plasmon scattering and evaluated the mobility for minority electrons and holes in GaAs. They also treated electron-hole scattering, in much the same way Walukiewicz did, but included the effects of degeneracy. For heavily doped material, empty states are not available for the scattered majority carriers to occupy, so the scattering rate decreases. Lowney and Bennett treated degeneracy by assuming that electrons could not scatter from holes located below the Fermi energy. For electrons in p^+-GaAs doped at $\sim 10^{19}\,\mathrm{cm}^{-3}$, they found, in agreement with Walukiewicz, that the minority electron mobility was less than one-half of the corresponding majority mobility, but for more heavily doped material, a very sizable increase in the minority electron mobility was predicted.

Lowney and Bennett also evaluated the minority electron mobility for n-type GaAs. For minority holes, plasmon scattering is much less important because the small electron effective mass leads to a higher plasma frequency. As for minority electrons, however, the minority hole mobilities are predicted to be less than the corresponding majority hole mobilities, and an increase in mobility for heavily doped materials due to degeneracy effects is predicted. At a doping density of $\sim 10^{18}\,\mathrm{cm}^{-3}$, the minority hole mobility is predicted to be only one-third of the corresponding majority hole mobility.

The work of Walukiewicz et al. and Lowney and Bennett provides insight into minority-carrier scattering physics, but both treatments are simple approaches to a very complicated problem. For example, Lowney and Bennett treated plasmon scattering, but in a polar semiconductor the plasmon and optical phonon modes couple. The response of the coupled

4. MINORITY-CARRIER TRANSPORT IN III–V SEMICONDUCTORS

majority carrier/phonon system is described by the dielectric function (Levi and Yafet, 1987),

$$\varepsilon(q, \omega) = \varepsilon_\infty \left(\frac{\omega^2 - \omega_{LO}^2}{\omega^2 - \omega_{TO}^2} \right) + \chi(q, \omega), \tag{54}$$

where ω_{LO} and ω_{TO} refer to the longitudinal and transverse optical phonon frequencies and ε_∞ is the high-frequency dielectric constant. The first term in Eq. (54) is the long-wavelength phonon contribution, and the second derives from the mobile holes and includes intraband contributions from heavy and light holes and light–heavy hole interband excitations. From the dielectric function, which has been evaluated in the random phase approximation (Bardyszewski and Yevick, 1989) the electron scattering rate can be obtained from

$$\frac{1}{\tau} = \frac{2q^2 m_n}{\pi \hbar^2 k} \int \frac{dq}{q} d\omega [1 + P(\omega)] \, \text{Im} \left[-\frac{1}{\varepsilon} \right], \tag{55}$$

where k is the wavevector of the electron and $P(\omega)$ is the Bose-Einstein factor. Equation (55) has been evaluated to assess the potential of hot electron transistors, and the results showed that the electron–plasmon scattering rate is very high (Levi, 1987), but it has not yet been applied to the treatment of near-equilibrium, minority carrier transport.

16. MEASUREMENTS OF MINORITY-CARRIER MOBILITY

Several different techniques for measuring the minority carrier mobility are available, including various time-of-flight techniques (Ahrenkiel *et al.*, 1986; Furuta and Tomizawa, 1990), illuminated Hall effect measurements (Ito and Ishibashi, 1987), and analysis of the high-frequency characteristics of bipolar transistors (Nathan *et al.*, 1988). The diffusion, or zero-field, time-of-flight technique (Ahrenkiel *et al.*, 1986, 1987) and the bipolar transistor techniques both measure the minority carrier diffusion coefficient under field-free, low-injection conditions. The two techniques are closely related, but the first is done in the time domain and the second in the frequency domain. Our discussion will focus on carrier transport under field-free conditions; techniques for measuring the high-field mobility of minority carriers have been discussed by Furuta in another chapter of this volume.

Figure 7a illustrates the device structure and measurement system for zero-field, time-of-flight (ZFTOF) studies. The device under test is a *pn*

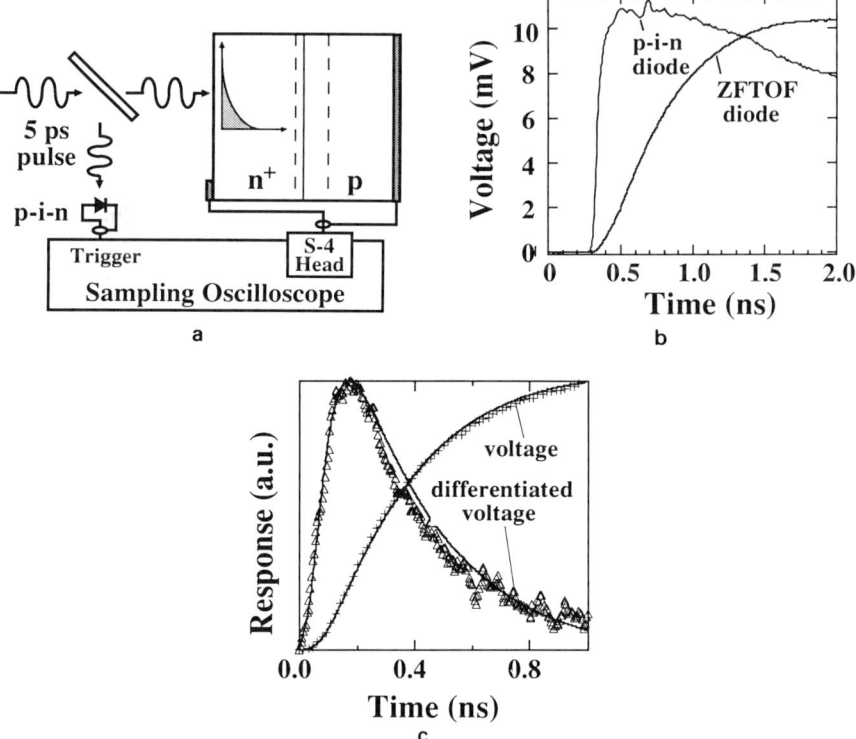

FIG. 7. (a) Schematic illustration of the zero-field, time-of-flight technique for studying minority carrier diffusion. (b) Typical measured transient voltage vs. time characteristic for an n on p GaAs diode. The n-type layer is 1.0 µm thick and doped at 1.8×10^{18} cm^{-3}. Also shown is the response of an identically packed p–i–n diode, which is used to establish $t = 0$. (c) Measured response of a diode with an n-GaAs top layer 1.0 µm thick doped at 1.8×10^{18} cm^{-3}. Also shown is the fitted response assuming $D_p = 5.8$ cm^2/s and $\tau_p = 3.5$ ns. (Figs. 7a, b and c reprinted with permission from *Solid State Electron* **35**, M. L. Lovejoy, M. R. Melloch, R. K. Ahrenkiel, and M. S. Lundstrom, "Measurement Considerations for Zero-Field Time-of-Flight Studies of Minority Carrier Diffusion in III–V Semiconductors," Copyright © 1992, Pergamon Press plc.)

diode mounted in a microwave package with only a small portion of the diode's top surface obscured by a metal contact. The remainder of the illuminated surface is passivated with a wide energy gap window. A high-speed laser photoexcites minority carriers that diffuse across the quasi-neutral p- or n-layer and are collected by the junction. The wavelength of the laser source is selected so that minority carriers are absorbed very near the top surface but so that few, if any, carriers are generated within the window or base layer of the diode. The size of the diode should be large enough so that the influence of photogeneration along the junction perimeter is small. A junction area of ~500 μm by 500 μm is typically adequate, but for some devices, especially n on p diodes, the junction perimeter must not be illuminated. The electrical response of the diode is monitored with a high-speed sampling oscilloscope.

Typical measured data are displayed in Figure 7b, which shows that the measured rise times are commonly about 1 ns. The rise time of the sampling oscilloscope has a negligible effect on such signals, and the effects of the circuit parasitics can also be minimal if the diode is properly packaged. A simple equivalent circuit for the diode consists of a current source representing the photocurrent, the junction capacitance, and the 50 Ω input impedance of the sampling oscilloscope. For short times, the diode acts as an integrator so the meaured voltage is

$$v_m(t) = \int_0^t \frac{i_{ph}(t)\,dt}{C_j}. \tag{56}$$

After the junction capacitance is charged, the voltage simply decays with a time constant of $R_{in}C_j$. The experiment is easiest to analyze when the junction capacitance is constant, which can be insured by limiting the change in voltage to a few millivolts. The laser power is also limited to ensure that low-level injection conditions are maintained in the photoexcited region.

To analyze the results and extract the minority carrier diffusion coefficient, we simulate the photocurrent by solving the minority carrier diffusion equation for an assumed D and τ then compute the $v_m(t)$ characteristic. After comparing the measured and simulated characteristics, the values of D and τ are adjusted until agreement is obtained. In practice we find it easier to compare the simulated photocurrent with the value deduced from the measurement, which is easily obtained by differentiating the measured voltage vs. time characteristic. Figure 7c shows typical measured and simulated photocurrent characteristics.

The ZFTOF technique is conceptually very simple, but there are a number of factors that need to be carefully considered in order to extract accurate parameters (Lovejoy et al., 1992). Devices must be designed and packaged to minimize the effect of photon recycling, grid shadowing, and circuit effects (Lovejoy et al., 1992).

Figure 8 summarizes the transistor method for deducing the minority carrier diffusion coefficient. The technique makes use of a bipolar transistor with an unusually wide base, perhaps 1.0 μm for a GaAs HBT, which is ~10 times the base width for a high-frequency transistor. The

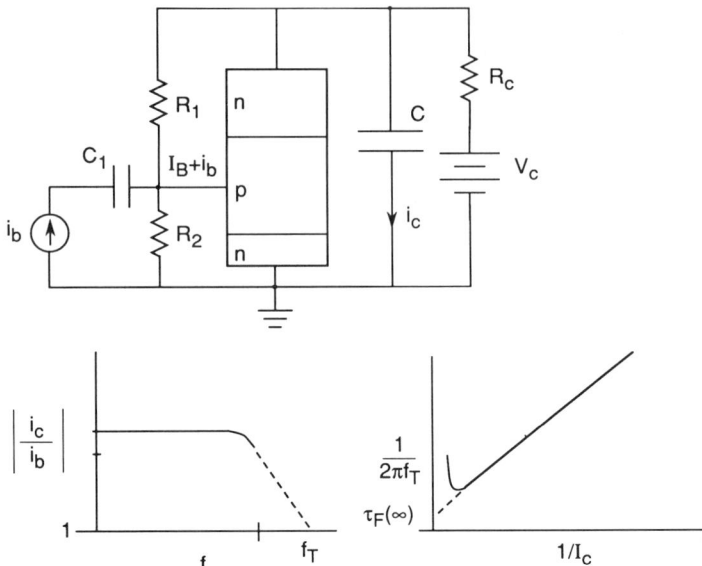

FIG. 8. Illustration of the transistor technique for measuring the minority electron diffusion coefficient in p^+-GaAs. The Npn AlGaAs/GaAs/GaAs bipolar transistor is designed with a wide base, so the extrapolated f_T is controlled by the base transit time.

frequency response of the transistor is characterized by s-parameter measurements, and a plot of $|\beta(f)|$ vs. frequency is constructed. Above f_β, where $|\beta|$ falls by 3 dB, $|\beta|$ drops at a rate of 20 dB per decade. The extrapolated gain–bandwidth product, f_T, is obtained by extrapolating $|\beta|$ to $|\beta| = 1$ at a rate of 20 dB/decade.

For a bipolar transistor, the gain–bandwidth product is given by

$$\frac{1}{2\pi f_T} = \frac{k_B T}{q I_C} C_\pi + \tau_B + \tau_C = \tau_F, \tag{57}$$

where C_π is the capacitance of the emitter-base junction, τ_B the base transit time, and τ_C the collector delay. By constructing a plot of τ_F vs. $1/I_C$ and extrapolating to $1/I_C = 0$, the effects of charging C_π can be eliminated and the intrinsic delay, $\tau_B + \tau_C$, inferred. To avoid high-current effects that would invalidate Eq. (57), it is important that the magnitude of I_C be limited and that the intrinsic delay be deduced by extrapolation. Because the base is wide, the intrinsic delay is dominated

4. MINORITY-CARRIER TRANSPORT IN III–V SEMICONDUCTORS   223

by the base delay, so

$$\tau_F(1/I_C \to 0) = \tau_B = W_B^2/2D_n, \tag{58}$$

and D_n can be determined.

a. Mathematical Analysis of the Zero-Field, Time-of-Flight Technique

The close relation between the zero-field time-of-flight and transistor techniques can be illustrated by mathematically analyzing the two experiments. To simulate the photocurrent vs. time characteristic, we solve the minority-carrier diffusion equation,

$$\frac{\partial \Delta p}{\partial t} = D_p \frac{\partial^2 \Delta p}{\partial z^2} - \frac{\Delta p(z, t)}{\tau_p} + G_{op}(t) \tag{59a}$$

with the boundary conditions,

$$\Delta p(W_B, t) = 0 \tag{59b}$$

at the collecting junction and

$$-D_p \left.\frac{\partial \Delta p}{\partial z}\right)_{z=0} = S_F \Delta p(0, t) \tag{59c}$$

at the illuminated surface. The initial condition is

$$\Delta p(z, 0) = 0, \tag{59d}$$

and the photogeneration rate is given by

$$G_{op}(z, t) = F_0(t)\alpha(\lambda)e^{-\alpha(\lambda)z}, \tag{59e}$$

where $F_0(t)$ is the incident optical flux, and λ is the wavelength of the laser. Equations (59) are easy to solve by standard numerical techniques, but it is useful to consider a simplified, analytical solution (Ahrenkiel *et al.*, 1986). If the photogeneration is taken to be a δ function in time of flux injected at $z = 0$, then we solve Eq. (59a) with $G_{op} = 0$ subject to the boundary conditions,

$$\Delta p(W_B, t) = 0 \tag{60a}$$

and

$$-D_p \left.\frac{\partial \Delta p}{\partial z}\right)_{z=0} = N_0 \delta(t), \tag{60b}$$

where N_0 is the number of photogenerated electron–hole pairs per square centimetre. To solve Eq. (60), we Fourier transform it to find

$$\frac{\partial^2 \Delta P(\omega, z)}{\partial z^2} - \frac{\Delta P(\omega, z)}{L_p^*} = 0, \tag{61a}$$

$$\Delta P(W_B, t) = 0, \tag{61b}$$

$$-D_p \left.\frac{\partial \Delta P}{\partial z}\right)_{z=0} = N_0, \tag{61c}$$

where

$$L_p^* = \frac{L_p}{\sqrt{1 + j\omega\tau_n}}. \tag{62}$$

Equation (61) is readily solved for

$$I_{\text{ph}}(\omega) = -qD_p \left.\frac{\partial \Delta P}{\partial z}\right)_{W_B} = \frac{qN_0}{\cosh(W_B/L_p^*)}. \tag{63}$$

Finally, we obtain the normalized output current in the time domain as

$$\frac{i_{\text{ph}}(t)}{qN_0} = \mathscr{F}^{-1}\left\{\frac{1}{\cosh(W_B/L_p^*)}\right\}. \tag{64}$$

In practice, the optical generation has a finite spatial extent, and surface recombination may occur. To simulate the photocurrent under such conditions, numerical solution techniques are required, but devices are designed for low surface recombination and so that $1/\alpha \ll W$, so Eqs. (63) and (64) provide a good description of the current response in the frequency and time domains.

b. Mathematical Analysis of the Transistor Technique

We now perform a similar analysis of the transistor measurement technique. In a bipolar transistor, we inject a small signal current $\hat{j}_e$ into the base and find the small signal collector current by solving the minority carrier diffusion equation. If we look for solutions of the form

$$\Delta p(z, t) = \Delta \hat{p} e^{j\omega t}, \tag{65}$$

the minority carrier diffusion equation, Eq. (59), becomes

$$\frac{\partial^2 \Delta \hat{p}(\omega, z)}{\partial z^2} - \frac{\Delta \hat{p}(\omega, z)}{L_p^*} = 0, \tag{66a}$$

$$\Delta \hat{p}(W_B, t) = 0, \tag{66b}$$

$$D_p \left.\frac{\partial \Delta \hat{p}}{\partial z}\right)_{z=0} = \hat{j}_e/q, \tag{66c}$$

which is identical in form to Eq. (61). The solution is readily obtained as (Pritchard, 1967)

$$\frac{\hat{j}_c}{\hat{j}_e} = \alpha_T(\omega) = \frac{1}{\cosh(W_B/L_p^*)}. \tag{67}$$

The transistor's response is usually analyzed from the common-emitter current gain,

$$\beta(\omega) = \frac{\alpha(\omega)}{1 - \alpha(\omega)}. \tag{68}$$

If $|W_B/L_p^*| \ll 1$, then we can expand the hyperbolic cosine to obtain

$$|\beta| = \frac{\beta_0}{\sqrt{1 + \omega^2 \tau_p^2}} = \frac{\tau_p/(W_B^2/2D_p)}{\sqrt{1 + \omega^2 \tau_p^2}}. \tag{69}$$

By adjusting D_p and τ_p to fit the measured $|\beta|$ vs. ω characteristic, the minority-carrier diffusion coefficient and lifetime are determined (Nathan et al., 1988). If the base is thin, then the transistor will display gain for high frequencies such that $\omega \tau_p \gg 1$, and Eq. (69) simplifies to

$$|\beta| \simeq \frac{\beta_0}{\omega \tau_p}, \tag{70}$$

which shows that $|\beta|$ drops with frequency at 20 dB/decade. Under such conditions, the gain–bandwidth product is

$$2\pi f_T = \frac{1}{W_B^2/2D_p}, \tag{71}$$

and the minority carrier diffusion coefficient can be obtained from a simple measurement of f_T (Beyzavi et al., 1991). In practice, the base thickness should be selected so that the base delay dominates, but the base should be thin enough so that Eq. (71) applies (Lee et al., 1991).

Comparison of Eqs. (64) and (67) shows that the base transport factor of the bipolar transistor is the Fourier transform of the normalized photocurrent response of the time-of-flight diode. One could, in fact, Fourier-transform the measured time-of-flight current response to obtain the complex transport factor, $\alpha_T(\omega)$, and from that create a plot of β vs. ω for the time-of-flight diode. The difficulty is that the number of electron–hole pairs created by the laser pulse, N_0, is not known precisely, and the simplifying assumptions, such as the assumption that the electron–hole pairs are created in a δ-function at the surface, are not met in practice. Nevertheless, the analysis does show that the two techniques are closely related and that many of the measurement considerations that apply to the time-of-flight technique should apply to the transistor technique as well.

17. Survey of Measured Minority-Carrier Mobilities for Carriers in GaAs

Measurements of the doping- and temperature-dependent mobility for electrons in p^+-GaAs are still rather scarce, and there are even fewer data for the hole mobility in n^+-GaAs. Reported measurements for minority electrons and holes in GaAs are summarized in Figs. 9 and 10. In accord with the theoretical prediction, the measured minority electron mobility is found to be distinctly less than the electron mobility in comparably doped n-GaAs. The increase in minority electron mobility predicted to occur in very heavily doped material has not, however, been observed. More measurements at doping densities above 10^{19} cm^{-3} are still needed to test our understanding of degeneracy effects in p^{++}-GaAs.

Figure 9b shows the measured minority electron mobility versus temperature (Beyzavi et al., 1991). The strong temperature dependence of the minority electron mobility in heavily doped p-type GaAs contrasts with the weak temperature dependence observed for majority carriers in

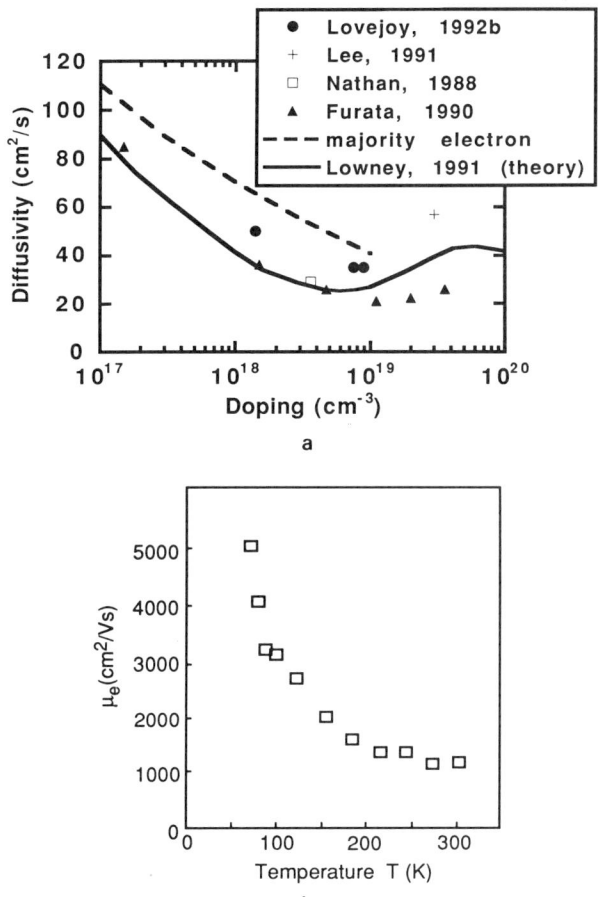

FIG. 9. (a) Measured mobilities for minority carrier electrons in GaAs at room temperature. (b) Temperature-dependent minority electron mobility as measured by the transistor technique for p^+-GaAs doped at 4×10^{18} cm^{-3}. (From Beyzavi et al., 1991.)

heavily doped GaAs. For majority electrons in n^+-GaAs, polar optical phonon and ionized impurity scattering dominate. Ionized impurity scattering decreases with temperature, while phonon scattering increases. The two effects largely offset each other, so the net temperature dependence is a weak one. The distinctly different temperature dependence for minority electrons reflects the difference in the scattering physics for minority carriers. The decrease as temperature increases may be due to increased scattering by the emission of hole plasmons. As the thermal

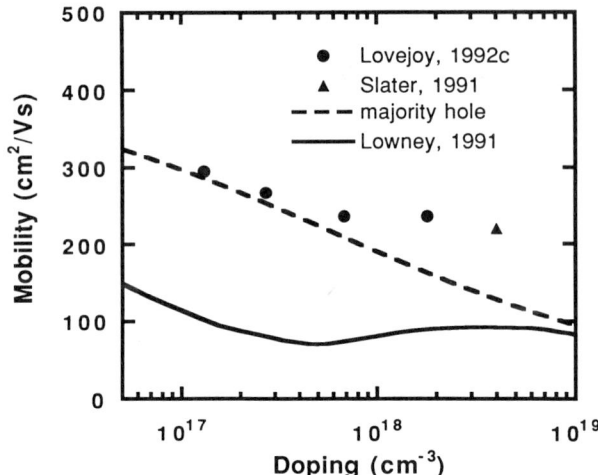

FIG. 10. Measured mobilities for minority carrier holes in GaAs at room temperature. (From Lovejoy et al., 1992b).

energy increases, the fraction of carrier with a thermal energy exceeding the plasmon energy increases, which leads to increased scattering and reduced mobility.

Figure 10 displays the few measured results available for minority carrier holes in n-GaAs. Slater et al. (1991) deduced the minority hole mobility from the high-frequency characteristics of a pnp heterojunction bipolar transistor. Lovejoy et al. (1992b) employed the zero-field time-of-flight technique. In contrast to minority electrons, both Slater and Lovejoy find the minority hole mobility at $N_D \simeq 1.8 \times 10^{18}$ cm^{-3} to be ~30% greater than the corresponding majority hole mobility. Their results appear to contradict available theory, which predicts that the minority hole mobility should be ~60% less than the corresponding majority hole mobility (Lowney and Bennett, 1991), but more measurements are clearly needed.

VI. Minority-Carrier Transport in III–V Devices

The focus in this volume has been on the physics of minority carriers in III–V semiconductors. The objective in this section is to examine how these effects influence the performance of electronic and optoelectronic devices by considering several specific device examples. The examples demonstrate that a detailed understanding of minority carrier physics is essential for predicting device performance. At the same time, however, we demonstrate that these effects can be exploited to realize new kinds of devices.

18. Dc Characteristic of AlGaAs/GaAs HBTs

The base of an AlGaAs/GaAs heterojunction bipolar transistor (HBT) is heavily doped, so the energy-gap narrowing effects have a strong influence on device performance. The main effect is on the common emitter current gain, $\beta_{dc} = I_C/I_B$. In this section, we'll use the measurements presented in Section V.14 to illustrate how heavy base doping affects β_{dc}. For an *Npn* AlGaAs/GaAs HBT, we show that neglecting these effects would result in underestimating the gain by more than one order of magnitude for a typical transistor (Klausmeier-Brown et al., 1989).

The Gummel plot for a typical *Npn* AlGaAs/GaAs HBT is shown in Fig. 11. (A Gummel plot displays the collector and base currents versus the base-emitter voltage.) The epitaxial layers of this device were grown by MOCVD, and the emitter dimensions were 100 μm × 100 μm. The Al mole fraction in the bulk emitter was 0.3 and was graded to 0.0 over a distance of 300 Å on the emitter side of the emitter–base junction. For this transistor, the emitter's energy gap is ~0.4 eV wider than that of the base. The base was Zn-doped to a concentration of 10^{19} cm^{-3}. Secondary ion mass spectroscopy (SIMS) profiling showed the Zn profile in the base to be flat, with a base width of 1,500 Å.

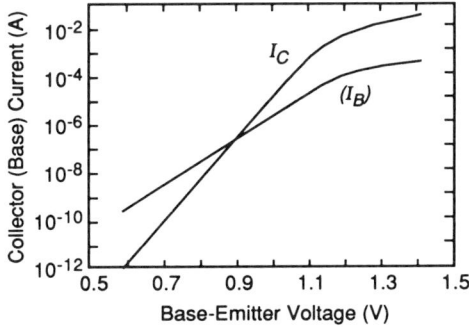

FIG. 11. The measured collector and base current characteristic for a typical *Npn* AlGaAs/GaAs HBT operating at room temperature. The emitter size is 100 μm by 100 μm, the base width 1,500 Å, and base doping 1.0×10^{19} cm^{-3}. (From Klausmeier-Brown et al. © 1989 IEEE.)

As shown in Fig. 11, the collector current characteristic displayed an ideality factor of $n = 1$. The base current versus base-emitter voltage characteristic was fitted by the expression

$$I_B = I_{B0} e^{qV_{BE}/nk_BT}, \qquad (72)$$

with $I_{B0} = 6.57 \times 10^{-16}$ A and $n = 1.78$. To illustrate the importance of effective energy gap narrowing in the quasi-neutral p^+-base, we compute β_{dc} using the measured I_B and a theoretically computed I_C obtained from

$$I_C = \frac{qA_E(n_{ie}^2 D_n)}{N_A W_B} e^{qV_{BE}/k_BT}. \qquad (73)$$

We make use of the measured base current because it tends to be dominated by the recombination of carriers along the perimeter of the device. Perimeter recombination is a strong function of properties of the exposed GaAs surface and is difficult to evaluate theoretically (Dodd et al., 1991a). Our focus is on the collector current, because it is influenced by energy-gap narrowing in the base.

The dc collector current is controlled by $n_{ie}^2 D_n$, which can be obtained from the data plotted in Fig. 5a. At $N_A \simeq 3 \times 10^{19}$ cm^{-3}, we find $n_{ie}^2 D_n \simeq 1.55 \times 10^{15}$ cm^{-4} s^{-1}, which is more than 10 times the value we would have obtained using $n_0 p_0 = n_{i0}^2$. Using the modeled collector current from Eq. (73) and the measured base current fitted as in Eq. (72). we can evaluate the β vs. I_C and β vs. N_A characteristics.

Bandgap narrowing effects can be an important factor in the design of an AlGaAs/GaAs HBT. For example, it is desirable to dope the base as heavily as possible in order to reduce its resistance. As shown in Fig. 5a, however, effective energy gap widening due to majority carrier degeneracy begins to reduce $n_{ie}^2 D_n$ when the doping density exceeds $\sim 5 \times 10^{19}$ cm^{-3}. Figure 12 shows the effect on β_{dc} and compares the calculated gains with and without effective energy-gap narrowing. For moderate base doping densities, effective energy-gap narrowing enhances the gain, but for heavy doping, effective energy-gap widening leads to a very rapid reduction in gain as the emitter injection efficiency degrades. It is possible to improve the emitter injection efficiency by increasing the energy gap of the emitter, but the AlGaAs material quality can degrade if the mole fraction is too high. As a result of effective energy-gap widening, the maximum base doping for an AlGaAs/GaAs HBT is limited in practice to the mid-10^{19} cm^{-3} range.

4. MINORITY-CARRIER TRANSPORT IN III–V SEMICONDUCTORS

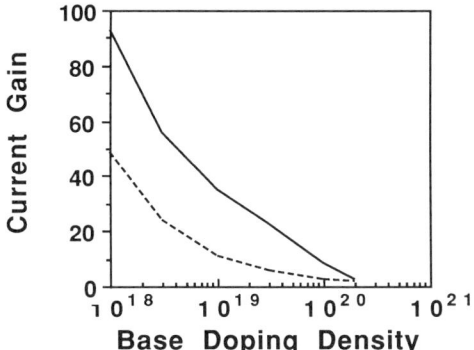

FIG. 12. β vs. N_A for a typical AlGaAs/GaAs HBT. The device is the one shown in Fig. 7.11 at $I_C = 1$ mA. β was evaluated from Eqs. (72) and (73) using the base current data displayed in Fig. 11. The dashed line shows the corresponding result if bandgap narrowing effects are ignored.

Figure 12 shows that at $N_A \simeq 3 \times 10^{19}$ cm^{-3}, effective energy-gap shrinkage increases the gain by a factor of four over what it would have been had no shrinkage been present. Similar results have been reported by Bailbe et al., who observed a current gain greater than 10 in a homojunction transistor with a base doped much more heavily from the emitter (Bailbe et al., 1984). They also attributed their results to effective energy gap shrinkage in the base. In contrast to the case for n–p–n silicon transistors, where energy-gap shrinkage in the heavily doped emitter reduces the gain (Slotboom and de Graaff, 1976), effective energy-gap narrowing enhances the gain of Npn AlGaAs/GaAs HBTs. The difference is that for a silicon transistor, the emitter is heavily doped, so energy gap shrinkage reduces the emitter injection efficiency. For an Npn AlGaAs/GaAs HBT, however, the base is most heavily doped, and energy gap shrinkage in the base improves the emitter injection efficiency.

19. ALL-GaAs PSEUDO-HBTs

A heterojunction bipolar transistor (HBT) is a bipolar transistor whose emitter has a energy gap that is wider than that of the base (Kroemer, 1982). Conventional p–n junction theory gives the collector current as

$$J_C = q \frac{n_{iB}^2}{N_A} \frac{D_n}{W_B} (e^{qV_{BE}/k_B T} - 1), \qquad (74)$$

where n_{iB}^2 is the $n_0 p_0$ product in the base. If the base current is limited by the back-injection of holes into the emitter, then

$$J_B = q \frac{n_{iE}^2}{N_{DE}} \frac{D_p}{W_E} (e^{qV_{BE}/k_B T} - 1). \tag{75}$$

From these expressions, we find the common emitter current gain as

$$\beta_{dc} = \frac{n_{iE}^2 D_n N_{DE} W_E}{n_{iE}^2 D_p N_{AB} W_B}. \tag{76}$$

In a homojunction transistor, the emitter is doped more heavily than the base ($N_{DE} \gg N_{AB}$) in order to maintain high gain, but it is desirable to raise the base doping density in order to lower its resistance. In a heterojunction bipolar transistor, $n_{iB} \gg n_{iE}$, so the doping can actually be inverted ($N_{DE} \ll N_{AB}$), which improves device speed by lowering the base resistance without sacrificing gain. In an $Al_{0.3}Ga_{0.7}As/GaAs$ HBT, the energy gap of the emitter is more than $10 k_B T$ wider than that of the base, so the back-injected hole current is completely suppressed. In this case, Eq. (76) greatly overestimates β_{dc} because the base current is actually due to recombination within the space-charge regions, or within the quasi-neutral base.

Given the fact that the energy gap effectively widens in n^+-GaAs and that it narrows in p^+-GaAs, the possibility exists that a n^+-p^+-n GaAs bipolar transistor will act as an HBT. If we assume a 0.5 μm thick emitter doped at $N_{DE} = 5 \times 10^{17}$ cm^{-3} and a 0.10 μm wide base doped at 5×10^{18} cm^{-3} and use the data displayed in Figs. 5, we estimate $\beta_{dc} \approx 10$. This value of gain is marginal because other sources of base recombination, such as in the perimeter and bulk space-charge regions, would result in a much lower gain. According to the data plotted in Fig. 6, $n_{iB}^2/n_{iE}^2 \approx 5$, so Δ_G is only about $2 k_B T$ at room temperature. At reduced temperatures, however, the effective energy gap difference between the emitter and base becomes large compared to $k_B T$, so the device should begin to behave as an HBT. Figure 13 shows the measured dc characteristics of an $n^+ p^+ n$ all-GaAs transistor (Dodd et al., 1991b). At room temperature, the common emitter current gain is as high as 10 — even though the base is doped more heavily than the emitter. As the temperature is reduced, the device behaves more like an HBT, and the gain rises rapidly, as the data in Fig. 14 demonstrate. For very low temperatures, we find a gain of over 100!

4. MINORITY-CARRIER TRANSPORT IN III–V SEMICONDUCTORS   233

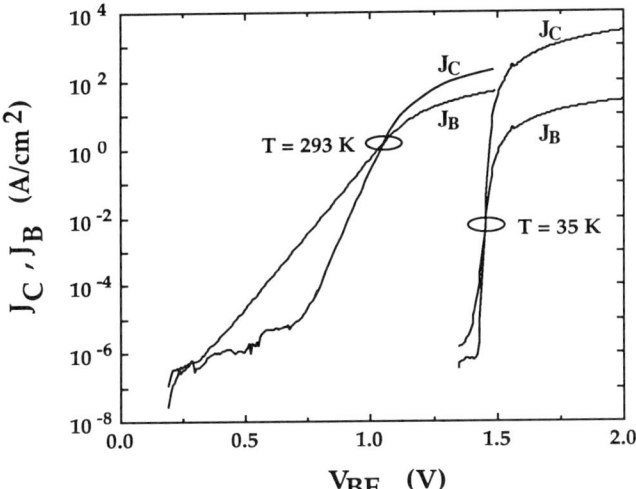

FIG. 13. The measured collector and base current characteristic for a n^+–$p^+$$n$ GaAs pseudo-HBT. The emitter is doped to $N_D = 5 \times 10^{17}$ cm^{-3} and the base to 5×10^{18} cm^{-3}. (From Dodd et al., 1991b, © 1991 IEEE.)

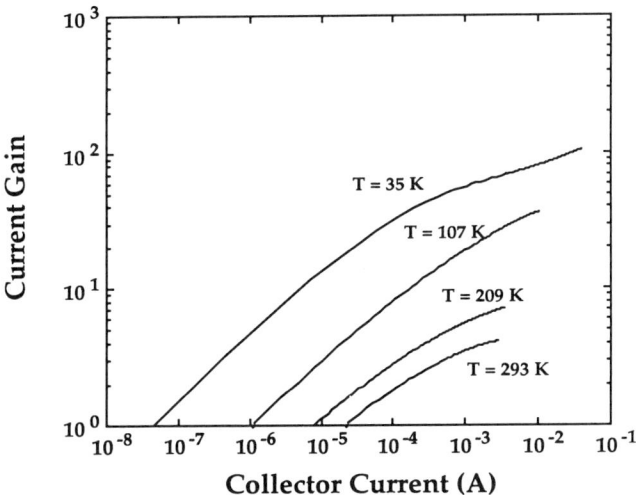

FIG. 14. β_{dc} versus I_C for the GaAs pseudo-HBT with temperature as a parameter. (From Dodd et al., 1991b, © 1991 IEEE.)

This device concept, termed the pseudo-HBT, has also been demonstrated in silicon (Yano et al., 1989). For silicon, however, the effective energy gap shrinkage for n- and p-type material is very similar. The strong shrinkage in p^+-GaAs and the effective widening for n^+-GaAs is especially useful because it produces a large energy gap difference for GaAs transistors. An advantage of the pseudo-HBT is the fact that the doping and compositional junctions are exactly aligned, and there is no need to compositionally grade the emitter–base junction to remove a band spike (Hayes et al., 1983). The concept itself might also prove useful for other III–V semiconductors, such as InAs, which has excellent transport properties, but a compatible wide energy-gap emitter for an InAs base does not exist. In this case, the pseudo-HBT concept would permit the achievement of HBT-like performance in this material system.

20. Carrier Drift in Doping Gradients

A graded doping profile in the base of a bipolar transistor produces an electric field that accelerates minority carriers and reduces the base transit time. When designing and analyzing such structures, it is important to consider the effects of heavy impurity doping. To discuss the problem, we'll adopt a simple, rigid band model for the perturbed band structure. This simple model will illustrate how the doping gradient, Fermi–Dirac statistics, and energy-gap narrowing influence minority-carrier transport, but we'll conclude by treating nondegenerate semiconductors with arbitrary band structure.

Figure 15a shows the energy band diagram for a quasi-neutral p-type region with the doping profile exponentially graded according to

$$p_0(z) = p_0(0)e^{-z/L}. \tag{77}$$

The corresponding energy band diagram sketched in Fig. 15b clearly shows that electrons will be accelerated across this graded region. The force on electrons due to electric and quasi-electric fields is obtained from

$$F_e = -\frac{dE_{C0}}{dz} = -q\mathscr{E}_{\text{eff}}, \tag{78}$$

where $\mathscr{E}_{\text{eff}}$ is the effective field acting on electrons. If the semiconductor is nondegenerate, then

$$p_0(z) = N_V e^{(E_{V0} - E_F)/k_B T} = N_V e^{(-E_G + E_{C0} - E_F)/k_B T}, \tag{79}$$

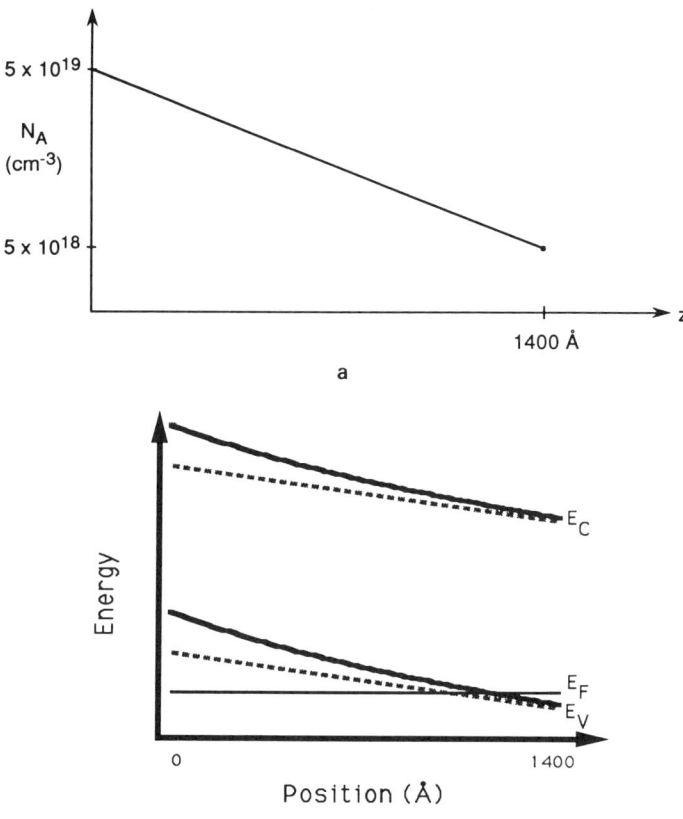

FIG. 15. (a) Exponentially graded, p-type, quasi-neutral region. (b) Energy-band diagram for the p-region shown in (a). The dashed lines assume Boltzmann statistics, and the solid lines use Fermi–Dirac statistics. Bandgap narrowing effects are not considered.

which can be solved for

$$E_{C0} = E_G + E_F + k_B T \log(p_0/N_V). \tag{80}$$

From Eq. (78) we find the effective field to be

$$\mathscr{E}_{\text{eff}} = \frac{d(E_G/q)}{dz} - \frac{k_B T}{q} \frac{1}{p_0} \frac{dp_0}{dz}. \tag{81}$$

(To derive this result, we assumed that the density of states was uniform.)

Equation (81) shows that the effective field acting on minority carriers has two components. The second term is the electric field that arises to maintain quasi-neutrality by keeping $p_0(z)$ nearly equal to $N_A(z)$. For a nondegenerate semiconductor with an exponential grading, the electric field is

$$\mathscr{E} = -\frac{k_B T}{qL}, \qquad (82)$$

where L is the grading parameter in Eq. (77).

The first term in Eq. (81) is due to energy gap narrowing. Because the doping decreases with distance, the energy gap increases. The result is that the heavier doping near $z = 0$ shrinks the energy gap, which produces a quasi-electric field that opposes the electric field. For p^+-GaAs, the electric field produced by the doping gradient can be substantially offset by the quasi-electric field caused by energy-gap shrinkage.

In heterojunction bipolar transistors, graded bases tend to be heavily doped, so the effects of majority carrier degeneracy must also be treated. Instead of Eq. (79), we have

$$p_0 = N_V \mathscr{F}_{1/2}(\eta_V), \qquad (83)$$

which can be solved for the electric field to find

$$\mathscr{E}(z) = -\frac{(k_B T)/q}{L} \times \frac{\mathscr{F}_{1/2}(\eta_V(z))}{\mathscr{F}_{-1/2}(\eta_V(z))}. \qquad (84)$$

For a nondegenerate semiconductor, the ratio of Fermi–Dirac integrals reduces to unity, so Eq. (84) reduces to Eq. (82), the commonly used expression.

For a degenerate semiconductor, the ratio of Fermi–Dirac integrals in Eq. (84) is greater than unity, so majority carrier degeneracy increases the electric field. This result is easy to understand with the energy band sketched in Fig. 15b. This figure shows that when the semiconductor is degenerate, the Fermi level is pushed deeper into the band than would be predicted by the use of Boltzmann statistics. The reason is simply that as the states fill, the Fermi level must penetrate more deeply to fill new states. The result is that the gradient of the energy bands increases, which increases the electric field.

4. MINORITY-CARRIER TRANSPORT IN III–V SEMICONDUCTORS

To summarize, a gradient in the doping profile produces an electric field that accelerates minority carriers. The magnitude of the resulting electric field is affected by two factors, majority carrier degeneracy and energy gap shrinkage. Majority carrier degeneracy acts to increase the electric field, while energy gap narrowing opposes the electric field. We reached these conclusions by using a simple, rigid band model for the perturbed energy bands. As it turns out, it is easy to evaluate the effective field on carriers without assuming a band structure. To do so, we begin with Eq. (22b) and set the majority carrier current to zero to find the electric field as

$$\mathscr{E} = -\frac{d((1-\gamma^0)\Delta_G^0/q)}{dz} - \frac{k_B T}{q}\frac{1}{p_0}\frac{dp_0}{dz}. \tag{85}$$

Equation (85) gives the electric field in terms of the effective energy gap shrinkage and the effective asymmetry factor. While it is easy to measure Δ_G, measurements of γ do not exist. Nevertheless, we can insert Eq. (85) into the minority-carrier current equation, Eq. (22a) to find Eq. (29), from which we observe that the effective field acting on minority carriers is

$$\mathscr{E}_{\text{eff}} = \frac{d(\Delta_G^0/q)}{dz} - \frac{k_B T}{q}\frac{1}{p_0}\frac{dp_0}{dz}, \tag{86}$$

which should be compared with Eq. (81), the result for a nondegenerate semiconductor with parabolic bands. Equation (86) is valid for degenerate or nondegenerate semiconductors with arbitrary band structure. The effective energy gap shrinkage, Δ_G, includes energy-gap narrowing effects as well as the effects of majority-carrier degeneracy. When the semiconductor is degenerate, the second term in Eq. (86) is not the electric field; part of the electric field is due to the degeneracy effects included in Δ_G. We cannot determine the *electric* field without knowing γ, but the *effective* field acting on minority carriers is independent of γ.

It is instructive to evaluate the electric and effective fields for a typical example. Consider an AlGaAs/GaAS HBT with a *p*-type base 1,400 Å wide exponentially graded from $p_0(0) = 5 \times 10^{19}$ cm^{-3} to $p_0(W_b) = 5 \times 10^{18}$ cm^{-3} (Streit *et al.*, 1991). Table I shows that results of some computations of the electric, quasi, and effective fields for minority-carrier electrons. From Eq. (82), we find that the use of Boltzmann statistics predicts a constant electric field of -3.33 kV/cm within the base. Because

TABLE I

ESTIMATED ELECTRIC AND QUASI-ELECTRIC FIELDS FOR A p-TYPE GaAs BASE 1,400 Å WIDE EXPONENTIALLY GRADED FROM $p_0(0) = 5 \times 10^{19}$ cm^{-3} TO $p_0(W_B) = 5 \times 10^{18}$ cm^{-3}

Quantity	Result
Electric field (Boltzmann statistics)	−3.33 kV/cm
Electric field (F–D statistics)	−7.7 kV/cm
Quasi-electric	+3.33 kV/cm
Effective field	−2.6 kV/cm

the base is heavily doped, however, Fermi–Dirac statistics should be used. In this case, the field is stronger, but it becomes position dependent. In Table I, we display the average electric field evaluated from

$$\bar{\mathscr{E}} = -\frac{(E_{C0}(0) - E_{C0}(W_B))}{qW_B}. \qquad (87)$$

The use of Fermi–Dirac statistics shows that the electric field is substantially stronger than the field estimated using Eq. (82), which is based on Boltzmann statistics. To estimate the quasi-electric field that opposes the electric field, we make use of the rigid band model using $\Delta E_G = A N_A^{1/3}$. Again, the effective field is position-dependent, but we evaluate the average quasi-electric field from

$$\bar{\mathscr{E}}_{\text{eff}} = \frac{(\Delta E_G(0) - \Delta E_G(W_B))}{qW_B}. \qquad (88)$$

The result, displayed in Table I, shows that the quasi-electric field is substantial when compared to the electric field. Finally, we evaluate the effective field on minority carrier electrons by adding the electric and quasi-electric fields to find the result listed in Table I. We see that energy gap shrinkage effects reduce the electric field by about 50% in this case. Although the field is reduced, it is still sufficient to produce a sizable reduction in the base transit time (Streit *et al.*, 1991).

21. GaAs SOLAR CELLS

Heavy doping effects are also an important factor in the design of AlGaAs/GaAs solar cells (DeMoulin and Lundstrom, 1989). The structure of a typical high-efficiency GaAs solar cell is displayed in Fig. 16a

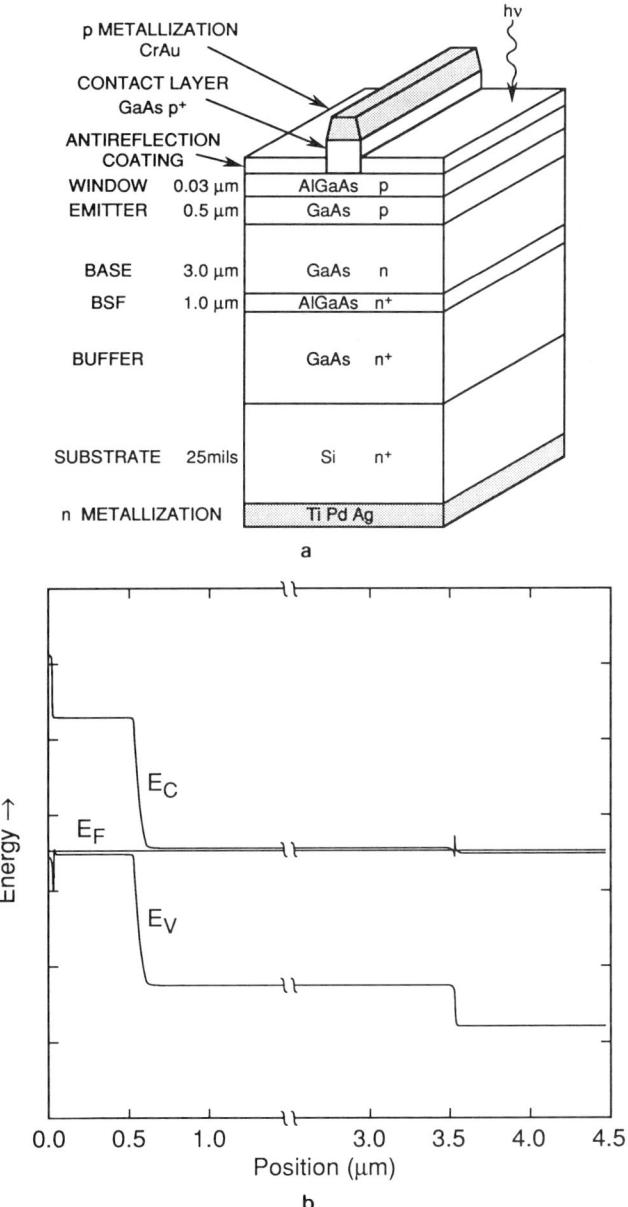

FIG. 16. (a) Typical structure of a high-efficiency, GaAs solar cell. The p-GaAs emitter is typically doped at $N_A = 5 \times 10^{18}$ cm^{-3} and the n-GaAs base at $N_D = 2 \times 10^{17}$ cm^{-3}. (From Vernon, 1990.) (b). Equilibrium energy band diagram for the solar cell. Note the confining barrier for minority carrier holes at the n–n^+ heterojunction.

(Vernon et al., 1989). Electron–hole pairs are optically generated; holes are collected by the p-type layer (referred to as the emitter) and electrons by the n-type layer (referred to as the base). The p-type layer is heavily doped to lower its sheet resistance, which is necessary to avoid introducing lateral voltage drops as the collected holes travel to the contact grid. A heavily doped n-type layer is also placed below the active base layer. As shown in Fig. 16b, the $n-n^+$ junction produces an energy barrier for minority carrier electrons at the back of the base. This energy barrier, referred to as a back-surface field or minority-carrier mirror, confines the photogenerated carriers to the lightly doped, high-lifetime base and prevents them from diffusing into the low-lifetime substrate.

The current versus voltage characteristic of a solar cell is described by a superposition of the normal diode current (the so-called dark current) and the photogenerated current. The resulting current–voltage characteristic,

$$J(V_A) = J_D(V_A) - J_L, \qquad (89)$$

where J_D is the dark current and J_L the photogenerated current, is plotted in Fig. 17. The numbers in the figure are typical of those for a 0.5 cm by 0.5 cm high-efficiency GaAs solar cell operating under 1-sun illumination (Tobin et al., 1990). The solar cell parameters of interest are the short-circuit current, J_{sc}, the open-circuit voltage, V_{oc}, and the conversion

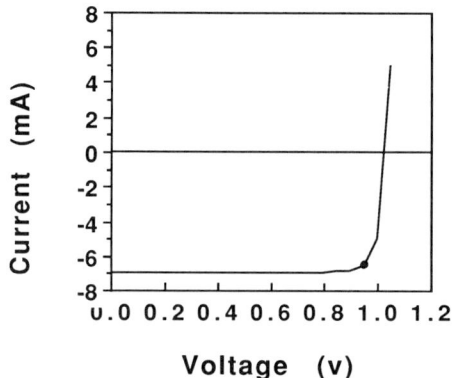

FIG. 17. Current versus voltage characteristics for a solar cell. The numbers are typical for a high-efficiency, GaAs solar cell operating under 1-sun illumination at room temperature. At the maximum power point, identified with a filled circle, the voltage is V_{mx} and the current J_{mx}.

efficiency

$$\eta = \frac{J_{sc}V_{oc}FF}{P_{inc}}, \tag{90}$$

where P_{inc} is the incident optical power density. The short-circuit current is determined by how many of the photogenerated electron–hole pairs are collected by the p–n junction. The open circuit is maximized by minimizing the diode's dark current. The fill factor,

$$FF = \frac{V_{mx}J_{mx}}{V_{oc}J_{sc}}, \tag{91}$$

is determined by the cell's series resistance and by the dark $J(V_A)$ characteristic of the junction. For high-performance cells, the series resistance is low, and cell design focuses on maximizing the short-circuit current and the open-circuit voltage.

The short-circuit current is maximized by minimizing carrier recombination losses. High-quality material with long minority-carrier lifetimes is necessary, and because the recombination velocity of a bare GaAs surface is high, the illuminated surface of the cell must be passivated with a wide energy gap $Al_{0.9}Ga_{0.1}As$ layer (Woodall, 1972). Recombination in the p^+ emitter itself is minimized by making it thin compared to a diffusion length. The moderate doping in the n-type base is used to maintain high minority-carrier lifetimes, but it is also important to minimize recombination at the back of the base. The function of the back-surface field is to produce a low minority-carrier recombination velocity at the back of the base layer.

The recombination velocity at the n–n^+ junction is determined by the height of the energy barrier for minority carriers, which is controlled by the doping ratio. The interface recombination velocity is given by (DeMoulin et al., 1987)

$$S_{nn^+} = \frac{D_p^+ N_D^-}{L_n^+ N_D^+} \frac{n_{i+}^2}{n_{i-}^2} \coth(W_n^+/L_n^+), \tag{92}$$

where the sub- and superscripts $+$ and $-$ refer to the heavily and lightly doped regions, respectively. In Eq. (92) we assume that the recombination velocity at the substrate approaches infinity. If the thickness of the heavily doped layer is much less than the diffusion length ($W_n^+ \ll L_n^+$), then Eq. (92) simplifies to

$$S_{nn^+} = \frac{D_p^+ N_D^-}{W_n^+ N_D^+} \frac{n_{i+}^2}{n_{i-}^2}. \tag{93}$$

Equation (93) shows that with a heavily doped layer beneath the base, a low interface recombination velocity should be achievable. Since carriers at the back of the base diffuse towards the junction at an effective velocity of D_p^-/W_n^-, the interface recombination velocity must be much less than D_p^-/W_n^- in order to be effective. For typical cells, we require a recombination velocity of $S \ll 10^4$ cm/s. Figure 18 shows the results of calculated recombination velocities for a typical cell. We assume the doping densities and layer thickness as shown in Fig. 16a, except that N_D^+ is varied. The results neglecting heavy doping effects suggest that the back-surface field should be an effective minority-carrier mirror. Because the heavy doping effectively widens the energy gap in n^+-GaAs, the results with heavy doping effects included show an even lower recombination velocity (Chuang et al., 1989).

If we consider the corresponding calculations for n^+–p–p^+ solar cells, we find much different effects. As shown in Fig. 18b, the results without treating heavy doping effects still suggest that the junction is an effective minority-carrier mirror, but when heavy doping effects are included, the recombination velocity becomes unacceptably high. The reason is that the energy gap effectively shrinks in p^+-GaAs, so $n_{i+} \gg n_{i-}$, which lowers the barrier height and increases S. For n^+–p–p^+ solar cells, the short-circuit can be improved by replacing the homojunction back-surface field with a heterojunction.

Since energy gap narrowing can increase minority carrier injection, it can increase a solar cell's dark current, which lowers the cell's open-circuit voltage. From Eq. (89), we find the open-circuit voltage to be

$$V_{oc} = \frac{k_B T}{q} \ln(J_{sc}/J_0), \tag{94}$$

where the saturation current density, J_0, is related to the dark current by

$$J_D(V_A) = J_0(e^{qV_A/k_B T} - 1). \tag{95}$$

To maximize V_{oc}, the cell should be designed for minimum dark current. To illustrate the effects, we consider the GaAs p/n cell that was sketched in Fig. 16a. If the recombination velocities at the front and back surfaces

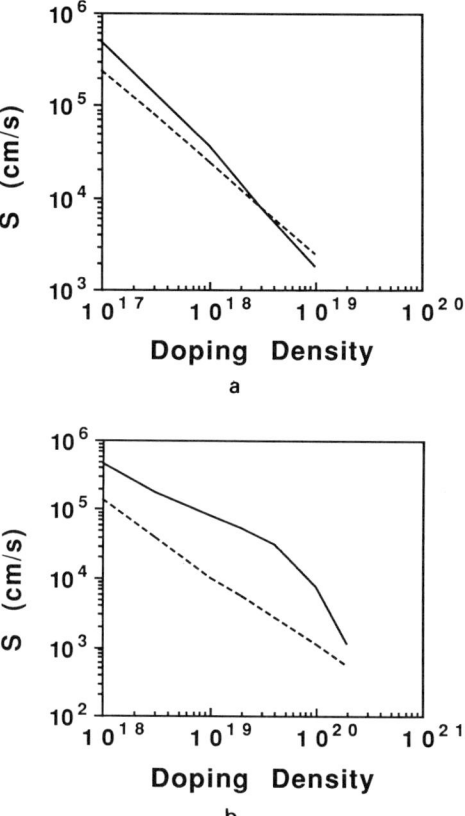

FIG. 18. (a) Computed interface recombination velocity for an n–n^+ homojunction barrier. The lightly doped layer is doped at $N_D^- = 2 \times 10^{17}$ cm^{-3}, and the doping of the n^+ layer is varied. The thickness of the heavily doped layer is 0.5 μm. The dashed line is the value computed assuming that $n_i = n_{i0}$ in both the lightly and heavily doped regions. (b) Computed interface recombination velocity for a p–p^+ homojunction barrier. The lightly doped layer is doped at $N_A^- = 2 \times 10^{17}$ cm^{-3}, and the doping of the p^+ layer is varied. The thickness of the heavily doped layer is 0.5 μm. The dashed line is the value computed assuming that $n_i = n_{i0}$ in both the lightly and heavily doped regions.

are near zero, then the injected minority carrier profiles are uniform within the quasi-neutral emitter and base regions. Within the emitter, the injected minority carrier concentration is $(n_{iE}^2/N_A) \times [\exp(qV_A/k_BT) - 1]$, and within the base it is $(n_{iB}^2/N_D) \times [\exp(qV_A/k_BT) - 1]$. The dark current consists of components due to the recombination of injected minority carriers within the quasi-neutral emitter and base regions. The emitter current component is

$$J_E = q \frac{n_{iE}^2}{N_A} \frac{W_E}{\tau_n} (e^{qV_A/k_BT} - 1), \qquad (96)$$

and the base current is

$$J_B = q \frac{n_{iB}^2}{N_D} \frac{W_B}{\tau_p} (e^{qV_A/k_BT} - 1). \qquad (97)$$

According to conventional p–n junction theory, the dark current of a p^+–n diode should be dominated by hole injection into the lightly doped base, but effective energy gap narrowing in the p^+ emitter will enhance emitter current. If we assume radiative-limit lifetimes with $\tau = 1/BN$ for both the emitter and base regions, then

$$J_{0E} = qBW_E n_{iE}^2, \qquad (98)$$

and

$$J_{0B} = qBW_B n_{iB}^2. \qquad (99)$$

It is interesting to note that for radiative limit lifetimes, the dark current is independent of the base and emitter doping densities. The ratio of emitter to base dark currents is simply

$$J_E/J_B = \frac{n_{iE}^2 W_E}{n_{iB}^2 W_B}. \qquad (100)$$

Because of effective energy gap narrowing in the emitter, $n_{iE}^2 \gg n_{iB}^2$, and, in contrast to conventional p–n junction theory, recombination within the quasi-neutral emitter is an important component of the dark current. Table II illustrates what happens in a typical GaAs p/n solar cell. For this cell, the open-circuit voltage calculated when the emitter current is neglected is 12 mV higher than the value obtained when emitter recombination is also included. For this example, $J_E/J_B = 0.56$, so the emitter current component is quite significant. Effective energy-gap narrowing in the p^+ emitter degrades the cell's performance, while effective energy-gap widening in the n^+ back-surface field enhances the performance.

TABLE II

OPEN-CIRCUIT VOLTAGE FOR THE GaAs p/n
SOLAR CELL SKETCHED IN FIG. 16a.[a]

Dark Current	V_{oc}
J_B	1.043 v
$J_E + J_B$	1.031 v

[a] Room temperature (300 K) and a 1-sun light-generated current of $J_L = 28$ mA/cm^2 are assumed.

22. QUANTUM EFFICIENCY OF GaAs PHOTODIODES

The internal quantum efficiency (QE) of a solar cell or photodetector is defined as the ratio of the electrical current generated to q times the optical flux transmitted into the semiconductor,

$$\text{QE}(\lambda) \equiv \frac{J(V_A)}{qF_{\text{inc}}}; \quad (101)$$

the applied bias, V_A, is typically set to zero or to a reverse bias. Because high-energy photons tend to be absorbed near the diode's surface, the short-wavelength internal QE provides information on the magnitude of surface recombination. For longer wavelengths, the internal QE is determined by the minority-carrier diffusion length. Measurements of the internal QE versus wavelength are used to calibrate photodetectors and as a diagnostic tool to explore recombination within a diode (Partain *et al.*, 1987). One typically fits the measured QE(λ) characteristic to a solution of the minority carrier diffusion equation in order to estimate parameters such as the surface recombination velocity S and the minority carrier diffusion length L.

Internal quantum efficiency measurements of GaAs diodes with thick emitter layers can be influenced by photon recycling. Figure 19 shows the results of some calculations using the techniques presented in Section IV. In each case, the device is a p^+–n diode with the p^+ emitter layer doped at 2×10^{18} cm^{-3} and a 1 μm thick n-type base doped at 5×10^{17} cm^{-3}. A 400 Å wide Al$_{0.2}$Ga$_{0.8}$As window was also assumed. The Shockley–Read–Hall lifetimes in the p-type layer were set to infinity and 1 ns in the n-type layer. At $N_A = 2 \times 10^{18}$ cm^{-3}, the radiative lifetime is $\tau_r = 2.5$ ns and the corresponding electron diffusion length is $L_n = \sqrt{D_n \tau_n} \simeq 3$ μm. (We have assumed that $B = 2 \times 10^{-10}$ cm^3/s and that $D_n = 35$ cm^2/s.) For

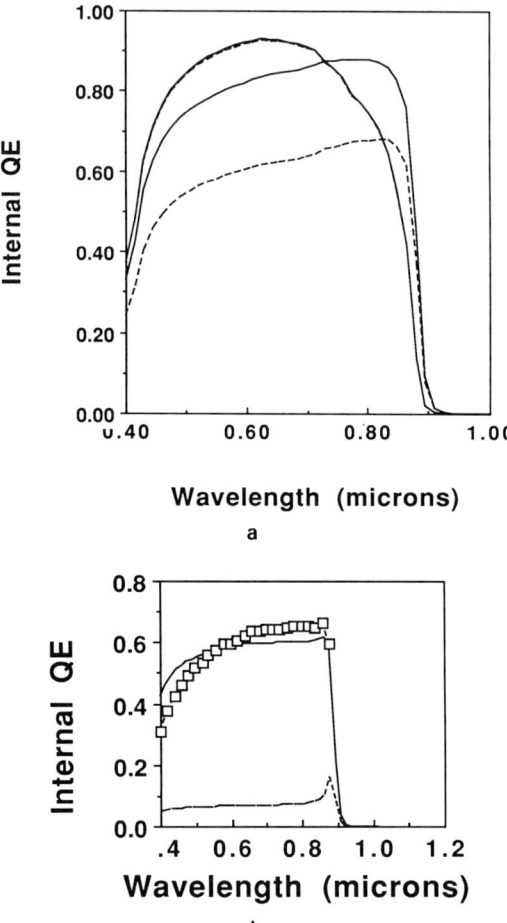

FIG. 19. (a) Computed internal quantum efficiency for a GaAs p/n diode with 0.5 and 3.0 μm thick p-layers doped at 2×10^{18} cm^{-3}. The dashed lines are theoretical calculations neglecting photon recycling, and the solid lines are numerical simulations obtained by solving Eq. (30). (b) Internal quantum efficiency for a GaAs p/n diode with a 10 μm thick p-layer doped at 2×10^{18} cm^{-3}. The dashed line is a theoretical calculation neglecting photon recycling effects, and the solid line is a numerical simulation obtained by solving Eq. (30). The data points show the measured internal quantum efficiency for this device. (Reprinted with permission from *Solid State Electron.* **33,** M. S. Lundstrom, M. E. Klausmeier-Brown, M. R. Melloch, R. K. Ahrenkiel, and B. M. Keyes, "Device-Related Materials Properties of Heavily Doped Gallium Arsenide," Copyright © 1990 Pergamon Press plc.)

4. MINORITY-CARRIER TRANSPORT IN III–V SEMICONDUCTORS

such a diffusion length, the expected internal quantum efficiencies are displayed in Fig. 19a as dashed lines. The corresponding results including photon recycling as obtained from a numerical solution to Eq. (30) are shown as solid lines. For the 0.5 μm thick layer (which is typical of the emitter layer for a high-efficiency GaAs solar cell), photon recycling has little effect, but for the 3.0 μm thick emitter, the effects are pronounced. The reason is that for the 0.5 μm thick emitter, few of the emitted photons are reabsorbed, whereas for the 3.0 μm thick emitter, a significant fraction are.

One often analyzes measured QE data by fitting it to a solution to the minority-carrier diffusion equation, ignoring the influence of photon recycling. In this case, the minority-carrier lifetime obtained exceeds the radiative lifetime. For the 3.0 μm results shown in Fig. 19a, a good fit to the calculation with photon recycled results if one assumes a lifetime of 3.8 times τ_r in the p-type layer. Since $\phi = 3.8$ is related to the probability F that emitted photons are absorbed within the p-type layer, we find that $F = 0.7$.

Figure 19b shows similar calculations for a 10 μm thick p-type layer along with the measured results for such a device (Lundstrom et al., 1990). In this case, the effects of photon recycling are pronounced. Because the layer thickness is four times the expected "radiative diffusion length," the predicted quantum efficiency is very small unless photon recycling is treated. The numerical results imply that the recycling factor for this diode is $\phi = 8.3$ or that the photon absorption probability is $F = 0.88$. The measured results suggest that the photon recycling theory presented in Section IV adequately describes photon recycling in GaAs and demonstrate that the effects on devices can be substantial.

23. MINORITY ELECTRON TRANSPORT IN InP/InGaAs HBTs

For thin-base InP/InGaAs HBT's, electron transport across the p^+ base can be very far from thermal equilibrium because electrons are launched into the base with an energy of $\sim 10\,k_B T$ and the electron scattering rate in InGaAs is low. In this section, we examine base transport in InP/InGaAs HBTs. The results show that device performance is strongly influenced by the carriers that scatter. Even a small amount of scattering can produce device characteristics that appear to be diffusive, but transport is very far from equilibrium. This section provides some additional details of a recent study by Dodd and Lundstrom (1992).

The base transit time can be deduced from either dc or ac measurements. For InGaAs transistors, the common emitter current gain is often dominated by recombination within the quasi-neutral base, so $\beta_{dc} = \tau_n/\tau_B$, where τ_n is the minority electron lifetime, and τ_B is the base transit time. The unity current gain cutoff frequency, $1/2\pi f_T = \tau_B + \tau_C$, where τ_C is the collector delay, is also influenced by the base transit time. The transit times deduced from dc and ac measurements can differ, but when the f_T is determined by extrapolating the initial fall-off in $|\beta|$ vs. f at 20 dB/decade, then the base transit time determined from the extrapolated f_T is identical to the transit time deduced from dc measurements.

It is instructive to evaluate the transit time from the base impulse response, which is evaluated by injecting a δ-function of electrons into the base and monitoring the current that exits as a function of time. The steady-state electron concentration is related to the time an electron spends in the base (Lundstrom, 1990), and the steady-state current is the integral of the impulse response. The base transit time is the base charge divided by the collector current, which gives

$$\tau_B = \frac{\int_0^\infty h(t)t\,dt}{\int_0^\infty h(t)\,dt}, \tag{102}$$

where $h(t)$ is the impulse response of the base. One can also obtain Eq. (102) by Fourier-transforming $h(t)$ to find the base transport factor $\alpha(\omega)$, from which $\beta(\omega) = \alpha/(1 - \alpha)$ can be determined. By expanding $\beta(\omega)$ for small ω and setting $|\beta| = 1$, we find the extrapolated f_T. The transit time deduced from the extrapolated f_T is precisely Eq. (102), the value deduced from steady-state analysis.

A series of Monte Carlo simulations for base widths ranging from 100 Å to 20,000 Å was performed. (Details of the Monte Carlo program are provided in Dodd and Lundstrom, 1992.) The base impulse responses were evaluated by injecting a δ-function of electrons at $t = 0$ and computing the current response by counting the number of electrons that exited the device during a sampling time ΔT. Base transit times were evaluated from the moment of the impulse response according to Eq. (102).

The base transit times obtained by Monte Carlo simulation are plotted in Fig. 20. Except for ultrathin bases, the transit times are observed to scale roughly as W_B^2, which suggests diffusive transport. Recent experimental work found similar behavior to base widths as small as 200 Å (Ritter et al., 1991). Figure 21 is a plot of the fraction of electrons that cross the base without scattering. By comparing Figs. 20 and 21, we observe that ballistic behavior (defined as τ_B scaling as W_B) occurs only when the ballistic fraction exceeds ~80%. Stated differently, the mean free path must be greater than ~$5W_B$ to observe a clear ballistic

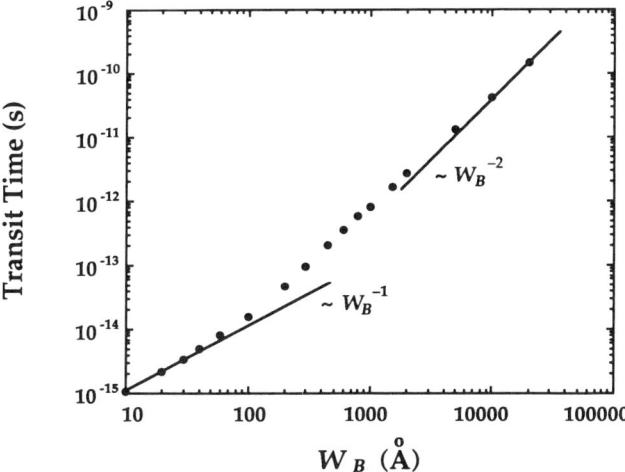

FIG. 20. Base transit times vs. base width as evaluated by Monte Carlo simulation for an $In_{0.53}Ga_{0.47}As$ p-type base doped to $7 \times 10^{19}\,cm^{-3}$. The lattice temperature is 300 K. (From Dodd and Lundstrom, 1992.)

dependence of the transit time on base width. Ballistic transport is observed only when the base is so thin that the semiclassical model of base transport itself no longer applies.

To understand why diffusive behavior is observed for such thin bases, it is instructive to examine the computed impulse responses. Figure 22 compares the impulse responses for the 300 and 2,000 Å thick bases. For

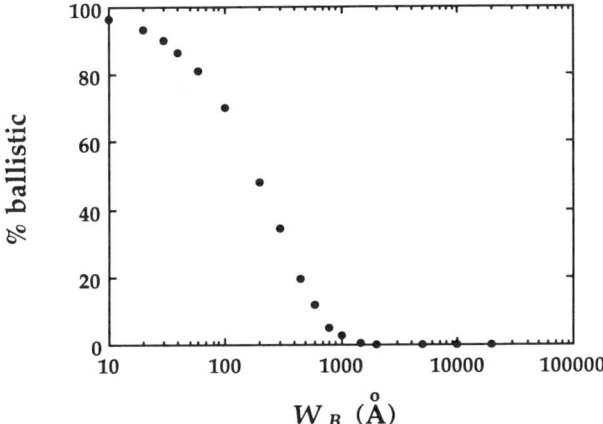

FIG. 21. Fraction of electrons injected into the p^+ InGaAs base that cross it without scattering plotted vs. base width.

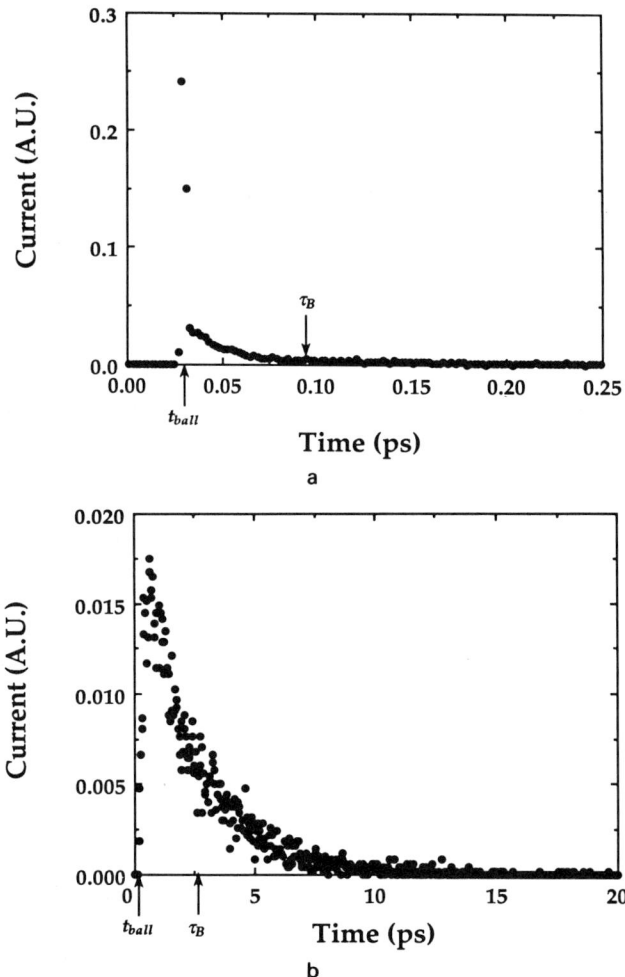

FIG. 22. (a) Current response due to a δ-function of electrons injected into the p^+ InGaAs base at $t = 0$. The base width is 300 Å, and the sampling time is $\Delta T = 2$ fs. (b) Current response due to a δ-function of electrons injected into the p^+ InGaAs base at $t = 0$. The base width is 2,000 Å, and the sampling time is $\Delta T = 40$ fs. (From Dodd and Lundstrom, 1992.)

ballistic transport, carriers would traverse the base at the injected velocity ($v_{\text{inf}} \simeq 9.0 \times 10^7$ cm/s), so the current response would be an impulse in time located at $t_{\text{ball}} = W_B/v_{\text{inj}}$. The ballistic transit time, t_{ball}, is noted in Figs. 22a and 22b, as is the actual transit time τ_B. For the 300 Å wide base, more than one-half of the carriers exit within 10 fs of t_{ball}, which suggests that transport is quasi-ballistic. The transit time, however, is determined by the moment of the impulse response, so the carriers that scatter and exit the base at long times can significantly lengthen the base transit time. The computed base delay, τ_B, is about $4 \times t_{\text{ball}}$. For the 2,000 Å wide base, the effects of scattering are much more pronounced, and the impulse response begins to appear more diffusive.

Diffusive transport implies that the steady-state distribution function within the base has a Maxwellian shape (when plotted versus longitudinal velocity) rather than being peaked about v_{inj}. The steady-state distribution functions at the middle of the 300 and 2,000 Å wide bases are displayed in Fig. 23. As expected, electrons in the 2,000 Å wide base are described by a Maxwellian distribution, but although the 300 Å wide base shows a pronounced ballistic peak, more than 80% of the electrons are located in a Maxwellian distribution. The dominance of the Maxwellian component is surprising because ~35% of the electrons crossed the base without scattering at all, and many more suffered only a few small-angle scattering events. Figure 24 is a histogram showing the number of collisions experienced by electrons as they cross the base. More than 80% percent of the carriers experience three or fewer collisions. If we regard such electrons as quasi-ballistic, then the dominance of the Maxwellian component in Fig. 23a is even more surprising.

The large Maxwellian component in the steady state arises because the steady-state distribution is especially sensitive to large-angle scattering events. Ballistic electrons cross the base at $v_z \simeq v_{\text{inj}}$, which is near the band-structure limited velocity, but the few carriers that do undergo a large-angle scattering event must diffuse out of the base. While these carriers are slowly diffusing out of the base, many more are continually being injected, so the steady-state population of scattered carriers builds up.

The steady-state populations of quasi-ballistic and diffusive electrons can be estimated with a simple argument. Assume that a flux of $n_0 v_{\text{inj}}$ electrons is injected into the base and that a fraction Γ undergo large-angle scattering. The current leaving the base is $(1-\Gamma)n_0 v_{\text{inj}} + n_{\text{mw}}(D/W_B)$, where n_{mw} is the density of scattered electrons, and (D/W_B) is roughly the velocity at which they diffuse out of the base. Under steady-state conditions, the incident and emerging fluxes are equal,

$$n_0 v_{\text{inj}} = (1 - \Gamma)n_0 v_{\text{inj}} + n_{\text{mw}}(D/W_B), \tag{103}$$

which can be solved for

$$\frac{n_{\text{mw}}}{n_0} = \Gamma \times \frac{v_{\text{inj}}}{(D/W_B)}. \tag{104}$$

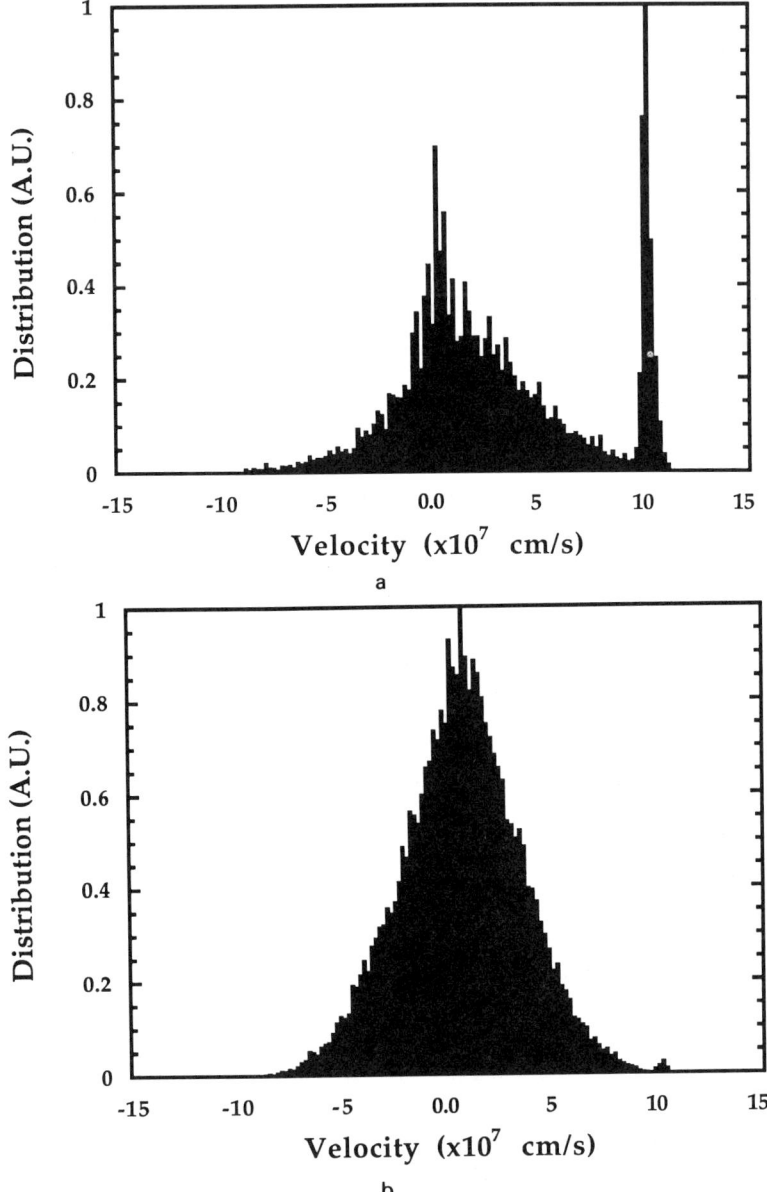

FIG. 23. (a) Steady-state distribution function for electrons in a 300 Å wide p^+, InGaAs base. The distribution function at the center of the base is shown plotted versus the longitudinal velocity. (b) Steady-state distribution function for electrons in a 2,000 Å wide p^+, InGaAs base. The distribution function at the center of the base is shown plotted versus the longitudinal velocity. (From Dodd and Lundstrom, 1992.)

4. MINORITY-CARRIER TRANSPORT IN III–V SEMICONDUCTORS

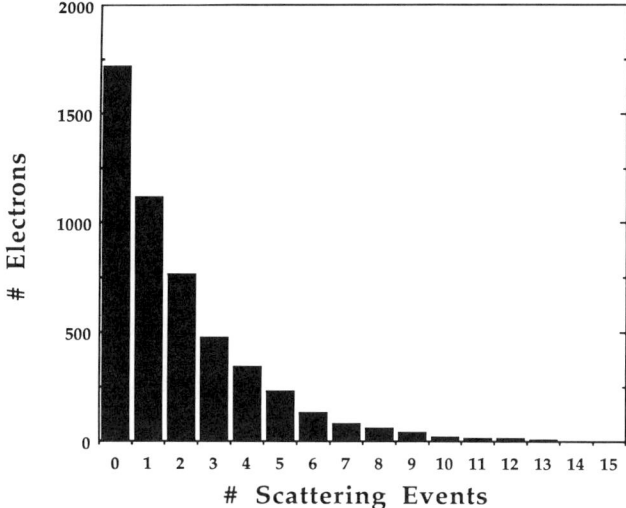

FIG. 24. Histogram showing the number of electrons experiencing collisions as they cross the 300 Å wide base. A total of 5,000 electrons were injected.

We find that the fraction of the electrons located in the Maxwellian population is not simply the fraction of carriers that undergo scattering; it is multiplied by the ratio of the ballistic to diffusive velocities. Since this ratio is very large, a small amount of scattering can produce large steady-state populations of scattered electrons. These strongly scattered electrons control the steady-state performance of the device. Because the extrapolated f_T is determined by the moment of the impulse response, it is similarly sensitive to the strongly scattering electrons.

Returning to Fig. 20, we note that over most of the range of base thicknesses of interest for HBTs, τ_B scales as $\sim W_B^{1.7}$. The linear dependence expected for ballistic transport occurs only for ultrathin bases, and the quadratic dependence expected for diffusive transport is just beginning to occur for extremely thick bases. These characteristics indicate that electron transport is off-equilibrium, even for surprisingly wide bases. Because of the InP/InGaAs launching ramp, minority electrons are hot upon entering the base. In the center of the 300 Å base structure, the average energy of electrons as computed by Monte Carlo simulation is $u = 0.114$ eV, which should be compared to $u_0 = 0.039$ eV for electrons in thermal equilibrium at 300 K. A substantial gradient in u, even for base thicknesses of a few thousand angstroms, produces thermoelectric effects that influence the transit time. If one simply equates the transit time for the 2,000 Å wide base to $W_B^2/2D_n$, one finds a diffusion coefficient of $D_n \simeq 75 \text{ cm}^2/\text{s}$. This value seems unreasonably high and suggests that base transport is off-equilibrium. Finally, note that if the scattering rate is artificially enhanced by unscreening POP scattering, then τ_B scales as W_B^2 because the enhanced scattering maintains conditions that are near-equilibrium.

To summarize, the steady-state distribution function within the base shows a Maxwellian shape even when a significant fraction of electrons cross the base ballistically. The reason is that the base electron population largely consists of the carriers that strongly scatter. The strongly scattered carriers also play a dominant role in determining the extrapolated f_T. Specific results depend on the details of injection energy, base thickness and doping, and the details of electron–hole scattering (Sadra et al., 1989), but the simple calculations presented here demonstrate how sensitive HBT transit times are to small amounts of scattering. So the rather surprising conclusion is that even for thin-base HBTs with electrons injected from a launching ramp, the concept of diffusive minority carrier transport may still apply, if the appropriate minority-carrier diffusion coefficient is used.

VII. Summary

In this chapter, we have reviewed minority carrier physics with an emphasis on relating the physics to the performance of actual devices. Bipolar devices are often controlled by the transport of minority carriers across a thin, heavily doped, quasi-neutral region. Effective energy gap narrowing associated with heavy impurity doping has a strong influence on the equilibrium np product, which controls the injection of minority carriers. Minority carrier transport is much different from the transport of majority carriers in comparably doped material. Finally, minority carriers can recombine. Recombination in III–V semiconductors is often dominated by radiative recombination, which is accompanied by the reabsorption of the emitted photons. These photon recycling effects can also produce a strong influence on devices. The results presented in this chapter show quantitatively how important these effects are for GaAs devices. Similar effects of comparable magnitude are expected in other III–V semiconductors as well.

Acknowledgement

Much of the author's own work presented in this chapter was funded by the National Renewable Energy Laboratory and by the National

Science Foundation. Special thanks to the author's collaborators, Drs. Michael Melloch and Richard Ahrenkiel, and to the students whose work was presented: Martin Klausmeier-Brown, Michael Lovejoy, Paul Dodd, Eric Harmon, and Mahesh Patkar.

References

Abram, R. A., Rees, G. J., and Wilson, B. L. H. (1978), "Heavily Doped Semiconductors and Devices," *Advances in Physics* **27**, 799.

Ahrenkiel, R. K., Dunlavy, D. J., Hamaker, H. C., Green, R. T., Lewis, C. R., Hayes, R. E., and Fardi, H. (1986). "Time-of-Flight Studies of Minority-Carrier Diffusion in $Al_xGa_{1-x}As$ Homojunctions," *Appl. Phys. Lett,* **49**, 725.

Ahrenkiel, R. K., Dunlavy, D. J., Greenberg, D., Schulpmann, J., Hamaker, H. C., and MacMillan, H. F. (1987). "Electron Mobility in p-GaAs by Time of Flight," *Appl. Phys. Lett.,* **51**, 776.

Asbeck, P. (1977). "Self-Absorption Effects on the Radiative Lifetime in GaAs–GaAlAs Double Heterostructures," *J. Appl. Phys.,* **48**, 820.

Bailbe, J. P., Marty, A., and Rey, G. (1984), "Influence of Degeneracy on Behavior of Homojunction GaAs Bipolar Transistors," *Electronics Lett.* **20**, 258.

Bardyszewski, W., and Yevick, D. (1989). "Influence of Temperature on Electron Transport in Bipolar Devices," *Appl. Phys. Lett.* **54**, 837.

Bennett, H. S. (1986). "High Dopant and Carrier Concentration Effects in Gallium Arsenide and Effective Intrinsic Carrier Concentrations," *J. Appl. Phys.* **60**, 2866.

Bennett, H. S., and Lowney, J. R. (1987). "Models for Heavy Doping Effects in Gallium Arsenide," *J. Appl. Phys.* **62**, 521.

Beyzavi, K., Lee, K., Kim, D. M., Nathan, M. I., Wrenner, K., and Wright, S. L. (1991). "Temperature Dependence of Minority-Carrier Mobility and Recombination Time in p-Type GaAs," *Appl. Phys. Lett.* **58**, 1268.

Casey, H. C., and Stern, F. (1976). "Concentration-Dependent Absorption and Spontaneous Emission of Heavily Doped GaAs," *J. Appl. Phys.* **47**, 631.

Casey, H. C., Sell, D. D., and Wecht, K. W. (1975), "Concentration Dependence of the Absorption Coefficient for n- and p-Type GaAs between 1.3 and 1.6 eV," *J. Appl. Phys.* **46**, 250.

Chuang, H. L., Klausmeier-Brown, M. E., Melloch, M. R., and Lundstrom, M. S. (1989). "Effective Minority Carrier Hole Confinement of Si-doped n^+–n GaAs Homojunction Barriers," *J. Appl. Phys.* **66**, 273.

del Alamo, J. A., and Swanson, R. M. (1987). "Modeling of Minority Carrier Transport in Heavily Doped Silicon Emitters," *Solid-State Electron.* **30**, 1127.

DeMoulin, P. D., and Lundstrom, M. S. (1989). "Projections of GaAs Solar Cell Performance Limits Based on 2-D Numerical Simulation," *IEEE Trans. Electron Dev.* **36** 897.

DeMoulin, P. D., Lundstrom, M. S., and Schwartz, R. J. (1987). "Back-Surface-Field Design for n^+p GaAs Solar Cells," *Solar Cells* **20**, 229.

Dodd, P. E., and Lundstrom, M. S. (1992). "Minority Electron Transport in InP/InGaAs HBTs," *Appl. Phys. Lett.* **61**, 465.

Dodd, P. E., Stellwag, T. B., Melloch, M. R., and Lundstrom, M. S. (1991a). "Surface and Perimeter Recombination in GaAs Diodes: An Experimental and Theoretical Investigation," *IEEE Trans. Electron Dev.* **38**, 1253.

Dodd, P. E., Melloch, M. R., and Lundstrom, M. S. (1991b). "High-Gain, Low-Leakage GaAs Pseudo-HBTs for Operation in Reduced Temperature Environments," *IEEE Electron Dev. Lett.* **12**, 629.

Dumke, W. P. (1957). "Spontaneous Radiative Recombination in Semiconductors," *Phys. Rev.* **105**, 139.

Furuta, T., and Tomizawa, M. (1990). "Velocity Electric Field Relationship for Minority Electrons in Highly Doped p-GaAs," *Appl. Phys. Lett.* **56**, 824.

Gora, T., and Williams, F. (1967). "Electronic States of Homogeneous and Inhomogeneous Mixed Semiconductors," in *II–VI Semiconducting Compounds* (D. G. Thomas, ed., p. 639. Benjamin, New York.

Harmon, E. S., Melloch, M. R., Lundstrom, M. S., and Lovejoy, M. L. (1991). "Experimental Determination of the Effects of Degenerate Fermi Statistics on Heavily p-Doped GaAs," *Appl. Phys. Lett.* **58**, 1647.

Hayes, J. R., Capasso, Malik, R. J., Gossard, A. C., and Wiegmann, W. (1983), "Optimum Emitter Grading for Heterojunction Bipolar Transistors," *Appl. Phys. Lett.* **43**, 949.

Ito, H., and Ishibashi, T. (1987). "Minority-Electron Mobility in p-Type GaAs," *J. Appl. Phys.* **65**, 5197.

Jain, S. C., and Roulston, D. J. (1991). "A Simple Expression for Band Gap Narrowing (BGN) in Heavily Doped Si, Ge, GaAs, and Ge_xSi_{1-x} Strained Layers," *Solid-State Electron.* **34**, 453.

Jain, S. C., McGregor, J. M., Roulston, D. J., and Balk, P. (1992). "Modified Simple Expression for Bandgap Narrowing in n-type GaAs," *Solid-State Electron.* **35**, 639.

Klausmeier-Brown, M. E., Lundstrom, M. S., Melloch, M. R., and Tobin, S. P. (1988). "Effects of Heavy Impurity Doping on Electron Injection in p^+–n GaAs Diodes," *Appl. Phys. Lett.* **52**, 2255.

Klausmeier-Brown, M. E., Lundstrom, M. S., and Melloch, M. R. (1989). "The Effect of Heavy Impurity Doping on AlGaAs/GaAs Bipolar Transistors," *IEEE Trans. Electron Dev.* **36**, 2146.

Klausmeier-Brown, M. E., Melloch, M. R., and Lundstrom, M. S. (1990a). "Electrical Measurements of Bandgap Shrinkage in Heavily Doped p-Type GaAs," *J. Electron. Materials* **19**, 7.

Klausmeier-Brown, M. E., Melloch, M. R., and Lundstrom, M. S. (1990b). "Transistor-Based Measurements of Electron Injection Currents in p-Type GaAs Doped from 10^{18}–10^{20} cm^{-3}," *Appl. Phys. Lett.* **56**, 160.

Kroemer, H. (1957). "Quasi-electric and Quasi-magnetic Fields in Nonuniform Semiconductors," *RCA Review*, 332.

Kroemer, H. (1982). "Heterostructure Bipolar Transistors and Integrated Circuits," *Proc. IEEE* **70**, 13.

Kuriyama, T., Kamiya, T., and Yanai, H. (1977). "Effect of Photon Recycling on Diffusion Length and Internal Quantum Efficiency in $Al_xGa_{1-x}As$–GaAs Heterostructures," *Jap. J. Appl. Phys.* **16**, 465.

Lee, S., Gopinath, A., and Pachuta, S. J. (1991). "Accurate Measurement Technique for Base Transit Time in Heterojunction Bipolar Transistors," *Electronics Lett.* **27**, 1551.

Levi, A. F. J., and Yafet, Y. (1987). "Nonequilibrium Electron Transport in Bipolar Devices," *Appl. Phys. Lett.* **51**, 42.

Lovejoy, M. L. (1992). "Minority Carrier Diffusivity Measurements in III–V Semiconductors by the Zero-Field Time-of-Flight Technique," Ph.D. thesis, Purdue University, School of Electrical Engineering, West Lafayette, Indiana.

Lovejoy, M. L., Keyes, B. M., Klausmeier-Brown, M. E., Melloch, M. R., Ahrenkiel, R. K., and Lundstrom, M. S. (1991). "Zero-Field Time-of-Flight Studies of Electron Diffusion in p^+-GaAs," *Jap. J. Appl. Phys.* **30**, L135.

Lovejoy, M. L., Melloch, M. R., Ahrenkiel, R. K., and Lundstrom, M. S. (1992a). "Measurement Considerations for Zero-Field Time-of-Flight Studies of Minority Carrier Diffusion in III–V Semiconductors," *Solid State Electron.* **35**, 251.

Lovejoy, M. L., Melloch, M. R., Lundstrom, M. S., and Ahrenkiel, R. K. (1992b). "Minority Hole Mobility in n^+-GaAs," *Appl. Phys. Lett.* **61**, 2683.

Lowney, J. R., and Bennett, H. S. (1991). "Majority and Minority Electron and Hole Mobilities in Heavily Doped GaAs," *J. Appl. Phys.* **69**, 7102.

Lundstrom, Mark (1990). *Fundamentals of Carrier Transport*. Addison-Wesley, Reading, Massachusetts.

Lundstrom, M. S., Schwartz, R. J., and Gray, J. L. (1981). "Transport Equations for the Analysis of Heavily Doped Semiconductor Devices," *Solid-State Electron.* **24**, 413.

Lundstrom, M. S., Klausmeier-Brown, M. E., Melloch, M. R., Ahrenkiel, R. K., and Keyes, B. M. (1990). "Device-Related Material Properties of Heavily Doped GaAs," *Solid-State Electron.* **33**, 693.

Marshak, A. H. (1987), "Transport Equations for Highly Doped Devices and Heterostructures," *Solid-State Electron.* **30**, 1089.

Marshak, A. H., and van Vliet, K. M. (1978a). "Electrical Currents in Solids with Position-Dependent Band Structure," *Solid-State Electron.* **21**, 417.

Marshak, A. H., and van Vliet, K. M. (1978b). "Carrier Densities and Emitter Efficiency in Degenerate Materials with Position-Dependent Band Structure," *Solid-State Electron.* **21**, 429.

Marshak, A. H., and Shrivastava, R. (1979). "Law of the Junction for Degenerate Material with Position-Dependent Band Gap and Electron Affinity," *Solid-State Electron.* **22**, 567.

Nathan, M. I., Dumke, W. P., Wrenner, K., Tiwari, S., Wright, S. L., and Jenkins, K. A. (1988). "Electron Mobility in p-GaAs," *Appl. Phys. Lett.* **52**, 654.

Nelson, R. J., and Sobers, R. G. (1978). "Minority-Carrier Lifetime and Internal Quantum Efficiency of Surface-Free GaAs," *Appl. Phys. Lett.* **49**, 6103–6108.

Pantelides, S. T., Selloni, A., and Car, R. (1985) "Energy Gap Reduction in Heavily Doped Silicon: Causes and Consequences," *Solid-State Electron.* **28**, 17.

Partain, L. D., Kuryla, M. S., Fraas, L. M., McLeod, P. S., and Cape, J. A. (1987). "A New Sequentially Etched Quantum-Yield Technique for Measuring Surface Recombination Velocity and Diffusion Lengths of Solar Cells," *J. Appl. Phys.* **61** (1987).

Patkar, M. P., Lundstrom, M. S., and Melloch, M. R. (1991). "Transistor-Based Studies of Heavy Doping Effects in n-GaAs," *Appl. Phys. Lett.* **59**, 1853.

Pritchard, R. L. (1967). *Electrical Characterization of Transistors*. McGraw-Hill, New York.

Ritter, D., Hamm, R. A., Feygenson, A., and Panish, M. B. (1991). "Diffusive Base Transport in Narrow Base InP/GaInAs Heterojunction Bipolar Transistors," *Appl. Phys. Lett.* **59**, 3431.

Sadra, K., Maziar, C. M., Streetman, B. G., and Tang, D. S. (1989). "Effects of Multiband Electron–Hole Scattering and Hole Wavefunction Symmetry on Minority-Electron Transport in GaAs," *J. Appl. Phys.* **66**, 4791.

Schuelke, R. J. and Lundstrom, M. S. (1983), "Numerical Analysis of Heteostructure Semiconductor Devices," *IEEE Trans. Electron Dev.* **30**, 1151.

Shockley, W. (1959). *Electrons and Holes in Semiconductors*. Van Nostrand, New York.

Slater, D. B. Jr., Enquist, P. M., Najjar, F. E., Chen, M. Y., Hutchby, J. A., Morris, A. S., and Trew, R. J. (1991). "Experimental Values for the Hole Diffusion Coefficient and Collector Transit Velocity in $P–n–p$ AlGaAs/GaAs HBTs," *IEEE Electron Dev. Lett.* **12**, 54.

Slotboom, J. W., and de Graaff, H. C. (1976). "Measurements of Bandgap Narrowing in Si

Bipolar Transistors," *Solid-State Electron.* **29**, 857.
Smith, A. C., Janak, J. F., and Adler, R. B. (1967). *Electronic Conduction in Solids.* McGraw-Hill, New York.
Streit, D. C., Hafizi, M. E., Umemoto, D. K., Velebir, J. R., Tran, L. T., Oki, A. K., Kim, M. E., Wang, S. K., Kim, C. W., Sadwick, L. P., and Hwu, R. J. (1991), "Effect of Exponentially Graded Base Doping on the Performance of GaAs/AlGaAs Heterojunction Bipolar Transistors," *IEEE Electron Dev. Lett.* **12**, 194.
Sutherland, J. E. and Hauser, J. R. (1977) "A Computer Analysis of Heterojunction and Graded Composition Solar Cells," *IEEE Trans. Electron Dev.* **24**, 363.
Tobin, S. P., Vernon, S. M., Bajgar, C., Wojtczuk, S. J., Melloch, M. R., Keshavarzi, A., Stellwag, T. B., Vemkatesan, S., Lundstrom, M. S., and Emery, K. A. (1990). "Assessment of MOCVD and MBE-Grown GaAs for High-Efficiency Solar Cell Applications," *IEEE Trans. Electron Dev.* **37**, 469.
Tyagi, S. D., Singh, K., Ghandhi, S. K., and Borrego, S. K. (1991). "Bandgap Narrowing in Heavily Doped N^+ Indium Phosphide," presented at the 22nd IEEE Photovoltaic Spec. Conf., Las Vegas, Nevada, October 22–25, 1991.
van Roosbroeck, W. (1950). "Theory for the Flow of Electrons and Holes in Germanium and Other Semiconductors," *Bell System Tech. J.*, pp. 560–607.
van Roosbroeck, W., and Shockley, W. (1954). "Photon-Radiative Recombination of Electrons and Holes in Germanium," *Phys. Rev.* **94**, 1558–1560.
van Vliet, C. M., and Marshak, A. H. (1982). "Wannier–Slater Theorem for Solids with Nonuniform Band Structure," *Phys. Rev. B* **26**, 6734.
Vernon, S. M., Tobin, S. P., Wojtczuk, S. J., Keavney, C. J., Bajgar, C., Sanfacon, M. M., Daly, J. T., and Dixon, T. M. (1989). *Solar Cells* **27**, 107.
von Roos, O. (1983). "Influence of Radiative Recombination on Minority-Carrier Transport in Direct Band-Gap Semiconductors," *J. Appl. Phys.* **54**, 1390.
Walukiewicz, W., Lagowski, J., Jastrzebski, L., and Gatos, H. C. (1979). "Minority Carrier Mobility in *p*-Type GaAs," *J. Appl. Phys.* **50**, 5040.
Yano, K., Nakazato, K., Miyamoto, M., Aoki, M., and Shimohigashi, K. (1989). "A High–Current-Gain Low-Temperature Pseudo-HBT Utilizing a Sidewall Base Contact Structure (SICOS)," *IEEE Electron. Dev. Lett.* **10**, 452.

CHAPTER 5

Effects of Heavy Doping and High Excitation on the Band Structure of Gallium Arsenide

Richard A. Abram

DEPARTMENT OF PHYSICS
UNIVERSITY OF DURHAM
DURHAM, UNITED KINGDOM

I. INTRODUCTION	.	259
II. MANY-BODY EFFECTS IN BULK GaAs	.	261
1. Introduction	.	261
2. Band Structure of Heavily Doped GaAs	.	263
3. Self-Energy Description of Heavy Doping Effects	.	264
4. Dielectric Function	.	266
5. Energy Shifts and Bandgap Narrowing	.	270
6. Calculations of Bandgap Narrowing in GaAs	.	272
III. MANY-BODY EFFECTS IN GaAs QUANTUM WELLS	.	279
7. Introduction	.	279
8. Screening in a Quantum Well	.	281
9. Energy Shifts and Bandgap Narrowing	.	282
10. Results of Calculations of Bandgap Narrowing	.	285
IV. EFFECTS OF THE IMPURITY CENTRES	.	288
11. Introduction	.	288
12. Semiclassical and Variational Models	.	290
13. Multiple Scattering Models	.	291
14. Quantum Wells	.	299
V. OPTICAL EXPERIMENTS	.	301
15. Introduction	.	301
16. Bulk GaAs	.	301
17. Quantum Wells	.	307
VI. ELECTRICAL EXPERIMENTS	.	309
ACKNOWLEDGMENTS	.	312
REFERENCES	.	312

I. Introduction

It is well known that the act of doping a semiconductor heavily has a significant effect on the electronic structure of the material which in general derives from the presence of a high concentration of carriers and from the random potential of the impurities. The latter influence dominates in closely compensated material; the effects of carriers alone can be seen in highly excited, lightly doped material, where a high concentration of carriers is produced by electrical injection or an external stimulus such as optical illumination. Heavily doped and highly excited semiconductors exhibit interesting optical and electrical properties which often originate from the changes induced in the electronic structure as well as from the direct effects of the carriers or impurities. The properties of these materials are of more than fundamental interest because in many devices part of the structure is either heavily doped, such as the base of a heterojunction bipolar transistor, or highly excited, as in the active region of a semiconductor laser.

In common with many areas of semiconductor physics, the study of heavily doped and highly excited semiconductors has benefited from the rapid progress in materials science and the tools of physical investigation that has occurred in recent years. For example, developments in methods of semiconductor growth, processing, and assessment have led to substantial improvements in the quality and consistency of the samples available to experimentalists. In this connection, the advent of low-dimensional structures has also provided new opportunities in the design of novel devices and fundamental experiments. Further, the refinement of established methods and the development of new techniques, particularly in the optical field, have extended the capabilities of experimental investigation. As a result, new and more detailed information on the properties of heavily doped and highly excited semiconductors has become available, which has stimulated more theoretical work and facilitated an improvement in understanding. In addition, theoreticians have been able to take advantage of the large growth in computing power and its dissemination to a wider community.

This chapter reviews the theory of the electronic structure of heavily doped and highly excited GaAs, which is chosen as being representative of the direct-gap III–V compounds and alloys. It is appropriate to concentrate on a single, well-studied material rather than to attempt coverage of all the commonly used direct-gap III–V semiconductors. GaAs is the natural choice of archetype as it has been the subject of more experimental and theoretical investigation than any other III–V material. It has also been the prototype material for most devices based on III–V semiconductors and, although it has been superseded by other materials in some applications, it continues to be widely used in its own right. The concentration on theory is far from absolute. Frequent reference is made to experimental results, and brief reviews of bandgap narrowing measurements by optical and electrical means are given in Sections V and VI, respectively. The implications of heavy doping effects for device behaviour are also mentioned in passing, but the interested reader should consult the specialist literature, including the other chapters in this volume, for detailed discussion of these matters.

The review is largely based on work reported in the last 15 years, although earlier original contributions are described where these are important for a proper discussion. The reader interested in earlier work is referred to the review of Abram *et al.* (1978), which also provides a broad introduction to the physics of heavily doped semiconductors and devices. Section II describes the many-body effects on the band structure resulting from high carrier concentrations in bulk gallium arsenide. A major feature of semiconductor physics in recent years has been the interest in quasi-low dimensional heterostructures, and Section III accordingly describes high carrier concentration effects in GaAs quantum wells. The influence on the electronic structure of the random potential due to large concentrations of ionized impurities in bulk and quantum well systems is discussed in Section IV. The chapter is concluded with a brief discussion of bandgap narrowing measurements.

The review of the theory is not intended to cover *all* the useful contributions of recent years, but rather to give a critical presentation of some of the most important developments and to provide a practical and accessible theoretical framework for describing heavy doping and many-carrier effects in GaAs. Hence, the emphasis is on useful physical concepts and calculations that provide a quantitative description of real semiconductors rather than abstract formalism.

The equations in this chapter are expressed in SI units, except where a statement to the contrary is made. The units used for the quoted values of physical quantities are more variable, with the adoption of the unit in common usage. Thus, for example, impurity and carrier densities are quoted per cubic centimeter and state energies in electron volts.

II. Many-Body Effects in Bulk GaAs

1. INTRODUCTION

The many-body interactions between carriers in heavily doped and highly excited semiconductors are known to have important effects on the physical properties of the carrier gas. This is quite apparent from optical and electrical experiments on a diverse range of systems varying from fundamental studies of electron–hole droplets through to efforts to develop better injection lasers, solar cells, and bipolar transistors. There is now a good understanding of these effects which has resulted from the early theoretical work on many-body effects in the high-density electron gas (see, for example, Gell-Mann and Brueckner, 1957; Hubbard, 1957; Nozières and Pines, 1958; Quinn and Ferrell, 1958; Hedin and Lundqvist, 1969; Mahan, 1981; Inkson, 1984) and, more recently, from the extensive range of experiments and calculations carried out on a variety of semiconductor systems. However, a *good understanding* should not be confused with a satisfactory and accessible *quantitative description*. For example, the concentrations of carriers in semiconductor systems that are studied in experiments are often such that the accuracy, or even the validity, of the many-body theory is in question. Even when the basic aspects of the many-body formalism are accepted, the calculation of numerical results requires a nontrivial computational effort, as well as the incorporation of the approximations made necessary by the realities of the semiconductor band structure, such as anisotropic, nonparabolic, multiple energy bands.

In this chapter we will concentrate on the effects of many-body interactions on the band structure and physical properties of GaAs that are relevant to the behaviour of electronic and optoelectronic devices. In particular we will be interested in the so-called one-particle properties of the carrier gas(es). It is well known that although the carriers form a strongly interacting system, many of the properties can be understood in terms of single particle-like responses or excitations (quasiparticles). This approach is particularly useful for our purposes because device behaviour is conventionally and most conveniently described in terms of single particle responses. For instance, a central property of a semiconductor laser is its optical gain, which, in the noninteracting particle picture, is seen as resulting from the stimulated emission of photons due to *individual* electrons making transitions from the conduction to the valence band of the semiconductor under the action the radiation field in the device. It is advantageous to retain the same general picture when carrier–carrier interactions are included, but to recognize that energy levels and transition matrix elements will be affected by the new circumstances.

Another example is the bipolar transistor, in which individual electrons and holes are traditionally seen as moving in the semiconductor band structure and real space under the action of a local electric field and the various scattering processes. Again the key element is the band structure — in particular, the size of the fundamental bandgap, the energy spectrum of allowed states for electrons and holes, and the band curvatures which determine the carrier effective masses and their dynamic response to applied field. These quantities can be retained in the presence of many-carrier interactions, but their values will be modified.

2. BAND STRUCTURE OF HEAVILY DOPED GaAs

There has been interest in theoretical descriptions of the many body effects in semiconductors since the early days of semiconductor research. An initial motivation for theoretical work in this area was the fact that many early experimental samples of semiconductor material were heavily doped due to poor materials technology rather than design. Later the role of heavily doped regions in electronic devices excited interest, first with the tunnel diode and later, and more importantly, with the bipolar transistors, semiconductor lasers, and solar cells. From the fundamental point of view, it was also recognized (Wolff, 1962) that heavily doped semiconductors could provide more suitable conditions for the application of many-body theory than the metals for which the theory had originally been devised. The high-density regime, which is the ideal for the operation of standard many-body theory, is defined by the condition $r_s \ll 1$, where r_s is the ratio of the interelectron spacing to the Bohr radius. This regime is relatively easy to achieve in many semiconductors because the Bohr radius is increased by a relative permittivity much greater than unity and a carrier effective mass rather less than unity, whereas $2 < r_s < 5$ is more typical of metals.

Another major impetus for the theoretical study of many-body effects in semiconductors was the discovery of electron–hole droplets in indirect-gap materials such as Si and Ge, which created a rapid expansion of interest in highly excited semiconductors in the 1970s. For reviews of the experimental and theoretical aspects of electron–hole droplets, see Rice (1977), Hensel *et al.* (1977), and Jeffries and Keldysh (1983). Although the theoretical work on electron–hole droplets has helped to form a substantial part of the general framework in which most calculations of many-body effects in semiconductors are conducted, much of it has rather less of a *direct* bearing on the particular problems discussed here than might be first thought. First, this review concerns GaAs, whilst the majority of work on droplets has been for Si and Ge; second, many of the problems in heavily doped semiconductors involve either electrons or holes but not both together; third, many calculations on droplet problems are concerned with the total energy and stability of the system at concentrations that are low compared to those of central interest here, and without direct reference to the quasiparticle properties introduced in Section II.1.

The self-energy formalism (Hedin and Lundqvist, 1969) provides a suitable method of describing the one-particle properties of heavily doped and highly excited semiconductors. In particular, it provides a very direct way of calculating the changes in the band structure of an intrinsic semiconductor that result from the introduction of electrons or holes through doping or excitation. In the next section we show how such a calculation can be done. The basic method that is described is due to Inkson (1976) and provides a formal procedure for calculating the many-body effects on an arbitrary semiconductor band structure. One of these effects, which is of particular interest for a variety of reasons, is the narrowing of the fundamental bandgap that accompanies heavy doping or high excitation. It is helpful to have simple, explicit — albeit approximate — formulae to describe the bandgap narrowing in heavily doped gallium arsenide. Following the work of Abram et al. (1984), we show how this can be achieved by adopting a simple model for the dielectric function of the semiconductor. It is instructive to present this work in some detail and, in doing so, use it as the basis for the discussion of other theories and calculations.

3. SELF-ENERGY DESCRIPTION OF HEAVY DOPING EFFECTS

For a semiconductor at zero temperature, the one-electron Green function is given by solving

$$(\hbar\omega - H)G(\mathbf{r}, \mathbf{r}', \omega) - \int \sum (\mathbf{r}, \mathbf{r}'', \omega) G(\mathbf{r}'', \mathbf{r}', \omega) \, d^3\mathbf{r}'' = \hbar \, \delta(\mathbf{r} - \mathbf{r}'), \quad (1)$$

where H represents the one-electron Hamiltonian and the average direct electron–electron interaction. The exchange and correlation effects of the electrons in the system are represented by the nonlocal, energy-dependent self-energy operator $\sum(\mathbf{r}, \mathbf{r}', \omega)$. To first order (see Hedin and Lundqvist 1969, Inkson 1984),

$$\sum (\mathbf{r}, \mathbf{r}', \omega) = (i/2\pi) \int G(\mathbf{r}, \mathbf{r}', \omega - \omega') W(\mathbf{r}, \mathbf{r}', \omega') e^{-i\omega'\delta} \, d\omega', \quad (2)$$

where δ is a positive infinitesimal and $W(\mathbf{r}, \mathbf{r}', \omega)$ is the dynamically screened Coulomb interaction whose spatial Fourier transform may be written in terms of the longitudinal dielectric function as

$$W(\mathbf{q}, \omega) = e^2/q^2 \varepsilon_0 \varepsilon(\mathbf{q}, \omega). \quad (3)$$

5. Effects of Heavy Doping and High Excitation

With, for example, the introduction of electrons into the conduction band (together with a neutralising uniform positive background charge to represent the ionised donors), the operator $\Sigma(\mathbf{r}, \mathbf{r}', \omega)$ changes, leading to shifts in the energies of the electron states. In contrast, H does not change; the direct interaction with the added electrons is exactly cancelled by the uniform positive background, and the free carrier screening provided by the added electrons is not sufficiently short-range to affect the electron–lattice interactions.

To calculate the shift of state $\mathbf{k}$ in band n due to doping, we must use G and W to obtain $\Sigma(\mathbf{r}, \mathbf{r}', \omega)$ in Eq. (2) both for the doped semiconductor and for the intrinsic semiconductor, and then evaluate

$$E_n^\Sigma(\mathbf{k}) = \operatorname{Re} \int \varphi_{n\mathbf{k}}^*(\mathbf{r}) \sum (\mathbf{r}, \mathbf{r}', \omega_n(\mathbf{k})) \varphi_{n\mathbf{k}}(\mathbf{r}') \, d^3\mathbf{r} \, d^3\mathbf{r}' \tag{4}$$

for the two cases. In Eq. (4) the $\varphi_{n\mathbf{k}}(\mathbf{r})$ are the semiconductor Bloch functions labelled by band n and wavevector $\mathbf{k}$ and with energy $\hbar\omega_n(\mathbf{k})$. The difference

$$\Delta E_n^\Sigma(\mathbf{k}) = E_n^\Sigma(\mathbf{k})|_{\text{doped}} - E_n^\Sigma(\mathbf{k})|_{\text{intrinsic}} \tag{5}$$

is then the energy shift of state $n, \mathbf{k}$.

Although our concern here is to obtain the energy shifts of certain states, it should be pointed out that the integral in Eq. (4) also has an imaginary part which gives the broadening of the energies of the quasi-particle excitations of the system resulting from their finite lifetimes (Quinn and Ferrell, 1958). To be specific, the imaginary part describes the broadening of the energy levels of holes created in states $n, \mathbf{k}$ below the Fermi surface or of electrons in states $n, \mathbf{k}$ excited above the Fermi surface. These excitations decay principally through Auger processes, of which the simplest involves the creation of an electron–hole pair through the excitation of an electron out of the Fermi sea (Landsberg, 1949; Hedin and Lundqvist, 1969). The rate of this latter process can be calculated most directly using Fermi's Golden Rule (Landsberg, 1966; Landsberg and Robbins, 1985). In general, experimental investigations of semiconductors involve creating excitations in the system, and in some cases a knowledge of the broadening of energy levels as well as their shifts is important for a satisfactory interpretation; see, for example, the discussion of optical measurements in Section V.

Returning to the evaluation of the energy shifts, the Green function has the form

$$G(\mathbf{r}, \mathbf{r}', \omega - \omega') = \sum_{n''\mathbf{k}''} \frac{\varphi_{n''\mathbf{k}''}(\mathbf{r}) \varphi_{n''\mathbf{k}''}^*(\mathbf{r}')}{\omega - \omega' - \omega_{n''}(\mathbf{k}'') + i\delta \operatorname{sgn}(\hbar\omega_{n''}(\mathbf{k}'') - \mu)}, \tag{6}$$

where δ is a positive infinitesimal and μ is the chemical potential. The Coulomb interaction may be written in terms of $W(\mathbf{q}, \omega)$ in Eq. (3) as

$$W(\mathbf{r}, \mathbf{r}', \omega') = \frac{1}{(2\pi)^3} \int W(\mathbf{q}, \omega') \exp[i\mathbf{q} \cdot (\mathbf{r} - \mathbf{r}')] \, d^3q. \qquad (7)$$

Using Eqs. (4), (2), (6), and (7), we obtain

$$E_n^\Sigma(\mathbf{k}) = \text{Re} \frac{i}{(2\pi)^4} \int \int \sum_{n''\mathbf{k}''}$$

$$\times \left| \left(\int \varphi_{n\mathbf{k}}^*(\mathbf{r}) \exp(i\mathbf{q} \cdot \mathbf{r}) \varphi_{n''\mathbf{k}''}(\mathbf{r}) \, d^3r \right) \right|^2 W(\mathbf{q}, \omega')$$

$$\times \frac{e^{-i\omega'\delta}}{\omega_n(\mathbf{k}) - \omega' - \omega_{n''}(\mathbf{k}'') + i\delta \, \text{sgn}(\hbar\omega_{n''}(\mathbf{k}'') - \mu)} \, d^3q \, d\omega'. \qquad (8)$$

The factor involving the Bloch functions can be simplified. For states normalised to the crystal volume Ω and Bloch periodic parts normalised to the unit cell volume, we have

$$\int_\Omega \varphi_{n\mathbf{k}}^*(\mathbf{r}) \exp(i\mathbf{q} \cdot \mathbf{r}) \varphi_{n''\mathbf{k}''} \, d^3r \simeq \delta_{\mathbf{k}'',\mathbf{k}-\mathbf{q}} I_{n\mathbf{k},n''\mathbf{k}''}, \qquad (9)$$

where $I_{n\mathbf{k},n''\mathbf{k}''}$ is the overlap integral over a unit cell of the Bloch function periodic parts. It follows that

$$E_n^\Sigma(\mathbf{k})$$
$$= \text{Re} \frac{i}{(2\pi)^4} \int \sum_{n''} \frac{|I_{n\mathbf{k},n''\mathbf{k}-\mathbf{q}}|^2 W(\mathbf{q}, \omega') e^{-i\omega'\delta} \, d^3q \, d\omega'}{\omega_n(\mathbf{k}) - \omega' - \omega_{n''}(\mathbf{k} - \mathbf{q}) + i\delta \, \text{sgn}(\hbar\omega_{n''}(\mathbf{k} - \mathbf{q}) - \mu)} \qquad (10)$$

Equation (3) gives the interaction $W(\mathbf{q}, \omega')$ in terms of the dielectric function of the system $\varepsilon(\mathbf{q}, \omega')$. In the next section we discuss the nature of $\varepsilon(\mathbf{q}, \omega')$ and present a simple approximation for this quantity before carrying out the integral over ω' in Eq. (10).

4. Dielectric Function

When there are electrons present in the conduction band, the dielectric function is taken to be

$$\varepsilon(\mathbf{q}, \omega) = \varepsilon_{\text{int}}(\mathbf{q}, \omega) + (\varepsilon_g(\mathbf{q}, \omega) - 1), \qquad (11)$$

where $\varepsilon_{int}(\mathbf{q}, \omega)$ is the dielectric function of the intrinsic semiconductor and $\varepsilon_g(\mathbf{q}, \omega)$ is the dielectric function of an electron gas with electron mass equal to the conduction band effective mass m_e. In effect, $(\varepsilon_g - 1)$ simply adds to the original intrinsic relative permittivity the susceptibility due to the extra electrons. There should be another term in the expression for $\varepsilon(\mathbf{q}, \omega)$ which takes account of the modification to ε_{int} resulting from the newly occupied states in the conduction band. This contribution is minor because the electrons only occupy a small part of the Brillouin zone, and it is neglected.

It is the intrinsic relative permittivity at small wavevectors and low frequencies which is relevant to the self-energy calculations, and in the elemental semiconductors this can be assumed to be constant in the relevant region of $\mathbf{q}-\omega$ space. However, in the polar semiconductors, the situation is not so straightforward. There is a lattice contribution to the intrinsic dielectric function near $\mathbf{q} = 0$ which causes it to have a frequency dependence of the form (Mahan, 1981)

$$\varepsilon_{int}(\omega) = \varepsilon_{int}(\infty) + [\varepsilon_{int}(\infty) - \varepsilon_{int}(0)] \frac{\omega_{TO}^2}{\omega^2 - (\omega_{TO} - i\delta)^2}, \qquad (12)$$

where $\varepsilon_{int}(\infty)$ and $\varepsilon_{int}(0)$ are, respectively, the optical and static relative permittivities, and ω_{TO} is the transverse optical phonon angular frequency. GaAs is only weakly polar, with a relatively small difference in $\varepsilon_{int}(\infty)$ and $\varepsilon_{int}(0)$, and under these circumstances a common approximation (the so-called ε_0 approximation — Keldysh and Silin, 1976; Kawamoto et al., 1980) is to replace $\varepsilon_{int}(\omega)$ with $\varepsilon_{int}(0)$ on the assumption that frequencies $\omega \ll \omega_{TO}$ dominate the self-energy integrals. (Another aspect of the ε_0 approximation is to replace the bare carrier mass by the polaron mass, but this has a very minor effect in GaAs.) It is perfectly feasible to carry out the self-energy calculations whilst retaining the frequency dependence of $\varepsilon_{int}(\omega)$, but in many circumstances the ε_0 approximation is accurate and has the attraction of leading to considerable simplification of the analysis. In what follows we will adopt the approximation in the interests of giving a clear illustration of the theory, and where ε_{int} appears without comment it should be interpreted as $\varepsilon_{int}(0)$.

The dielectric function of the electron gas can be written in the form

$$\varepsilon_g(\mathbf{q}, \omega) = 1 - \frac{e^2}{\varepsilon_0 q^2} P_g(\mathbf{q}, \omega), \qquad (13)$$

where P_g is the response function (see, for example, Hedin and

Lundqvist, 1969). Equation (11) can be rewritten as

$$\varepsilon(\mathbf{q}, \omega) = \varepsilon_{\text{int}}\left(1 + \frac{\varepsilon_g - 1}{\varepsilon_{\text{int}}}\right) = \varepsilon_{\text{int}}\left(1 - \frac{e^2}{\varepsilon_{\text{int}}\varepsilon_0 q^2} P_g(\mathbf{q}, \omega)\right). \quad (14)$$

The term in large parentheses in Eq. (14) is just ε_g with the permittivity of free space, ε_0, replaced by $\varepsilon_{\text{int}}\varepsilon_0$. As pointed out above, the conduction band effective mass is already used in ε_g. It is convenient to define a new quantity

$$\varepsilon_{gs}(\mathbf{q}, \omega) = 1 - \frac{e^2}{\varepsilon_{\text{int}}\varepsilon_0 q^2} P_g(\mathbf{q}, \omega) \quad (15)$$

where the subscript letter s indicates that ε_{gs} is a *semiconductor* quantity related to the behaviour of the electron gas in the conduction band. ε_{gs} is derived from the dielectric function of a free-electron gas by replacing the free-electron mass with m_e, and ε_0 with $\varepsilon_{\text{int}}\varepsilon_0$. The dielectric function of the semiconductor is then

$$\varepsilon(\mathbf{q}, \omega) = \varepsilon_{\text{int}}\varepsilon_{gs}(\mathbf{q}, \omega). \quad (16)$$

A frequently used expression for the dielectric function of a free-electron gas is the Lindhard formula (Lindhard, 1954) which is derived by using the random phase approximation (see Hedin and Lundqvist, 1969; Mahan, 1981). $\varepsilon_{gs}(q, \omega)$ can be obtained from the Lindhard formula by making the replacements described earlier. Unfortunately, the Lindhard expression has a complicated algebraic form, and although its use in the evaluation of Eq. (10) is perfectly feasible (see, for example, Berggren and Sernelius, 1981), there is some virtue in considering other approximations to $\varepsilon(\mathbf{q}, \omega)$. For example, the plasmon pole approximation (Lundqvist, 1967a, 1967b; Hedin and Lundqvist, 1969; Overhauser, 1971; Inkson, 1984; Haug and Schmitt-Rink, 1984) provides a simple algebraic expression for the inverse dielectric function which has behaviour remarkably similar to the Lindhard formula and also has the advantage of making the physics more transparent. In effect, the excitation spectrum of the electron gas is modelled by a single collective mode in such a way that the most important sum rules are satisfied and the correct behaviour is obtained in certain limits (Haug and Schmitt-Rink, 1984). The plasmon pole approximation gives for an isotropic, parabolic conduction band

$$(\varepsilon_{gs}(\mathbf{q}, \omega))^{-1} = 1 + \frac{\omega_p^2}{2\omega_1(q)}\left(\frac{1}{\omega - \omega_1(q) + i\delta} - \frac{1}{\omega + \omega_1(q) - i\delta}\right). \quad (17)$$

5. EFFECTS OF HEAVY DOPING AND HIGH EXCITATION

$\omega_1(q)$ is the *effective* plasmon dispersion relation, and a suitable choice is given by

$$\omega_1^2(q) = \omega_P^2 + \frac{4}{3}\omega_C(q)\omega_F + \omega_C^2(q) = \omega_P^2[1 + q^2/\kappa^2] + \omega_C^2(q), \quad (18)$$

in which $\hbar\omega_C(q)$ is the electron energy in the conduction band as a function of q, $\hbar\omega_F$ is the Fermi energy, κ is the Thomas–Fermi screening wavevector, and ω_p is the $q = 0$ plasmon angular frequency. For $\omega = 0$ and $q \to 0$, Eq. (17) gives the form of ε_{gs} predicted by Thomas–Fermi theory, and the second term on the right-hand side is important in this respect. The third term on the right-hand side is quartic in q and gives an approximate representation of the effects of single-particle transitions. It is important for getting the right sort of behaviour for ε_{gs}^{-1} at the larger values of q where the Thomas–Fermi theory becomes inadequate (Lundqvist, 1967b). It should be noted that the plasmon pole approximation takes no account of the damping of plasmons due to single-particle transitions. Nevertheless, many aspects of the behaviour of ε_{gs}^{-1} are very similar to those predicted by the Lindhard expression.

In n-type GaAs, the electrons normally exist in the Γ minimum, which can be assumed to be parabolic and isotropic. This case is straightforward, and Eqs. (17) and (18) can be applied without modification by using

$$\omega_C(q) = \frac{\hbar q^2}{2m_e}, \quad \omega_F = \frac{\hbar}{2m_e}(3\pi^2 n)^{2/3}, \quad \omega_p = \left(\frac{ne^2}{\varepsilon_{int}\varepsilon_0 m_e}\right)^{1/2}, \quad (19)$$

where n is the electron concentration.

For p-type GaAs, the holes normally exist in the heavy-hole and light-hole bands. The standard Lindhard expression is appropriate to an electron gas in a single band, and it is only recently that Bardyszewski (1986) has derived an expression for a hole gas in a realistic valence band structure. Bardyszewski's expression can of course be used in the self-energy calculations, but again it is worthwhile to consider the use of the plasmon pole approximation. As Bardyszewski and Yevick (1987) have pointed out, the main problem with the plasmon pole approximation in this case is that it does not include the single-particle transitions that can readily occur at small wavevectors in the valence band of p-type material because of the existence of the two bands. A phenomenological description of the effect of these interband transitions is given by the damped plasmon pole approximation (Rice, 1974), but the value to be given to the damping constant is not clear. Bardyszewski and Yevick have compared the Lindhard and the plasmon pole (damped and undamped) dielectric functions and their use in the calculation of bandgap narrowing in p-type GaAs. The three functions give rather different forms for $\varepsilon(\omega)^{-1}$ but, as reported in Section II.6, quite similar values of bandgap narrowing for a hole concentration $p < 10^{20}$ cm^{-2}. In view of this result, in some applications it is worth persisting with the simple plasmon pole approximation, its shortcomings being offset by its facility of use and apparent accuracy. Abram et al. (1984) have discussed the changes that can be made to Eqs. (17) and (18) to take account of the more complex nature of the valence band.

5. Energy Shifts and Bandgap Narrowing

Now that we have an explicit expression for $\varepsilon(\mathbf{q}, \omega')$, it is possible to carry out the integral over ω' in Eq. (10). For the doped semiconductor,

$$E_n^{\Sigma d}(\mathbf{k}) = \text{Re} \frac{i}{(2\pi)^4} \int \sum_{n''} |I_{n\mathbf{k},n''\mathbf{k}-\mathbf{q}}|^2 \frac{e^2}{q^2 \varepsilon_0 \varepsilon_{\text{int}} \varepsilon_{\text{gs}}(\mathbf{q}, \omega')}$$
$$\times \frac{e^{-i\omega'\delta}}{\omega_n(\mathbf{k}) - \omega' - \omega_{n''}(\mathbf{k}-\mathbf{q}) + i\delta \, \text{sgn}(\hbar\omega_{n''}(\mathbf{k}-\mathbf{q}) - \mu)} d^3q \, d\omega', \tag{20}$$

where the superscript d is used to denote "doped."

The integral over ω' can be accomplished by contour integration. Two sets of poles contribute to the integral: those in the Green function corresponding to occupied states and those in $\varepsilon_{\text{gs}}(q, \omega')$. The result is

$$E_n^{\Sigma d}(\mathbf{k}) = \frac{1}{(2\pi)^3} \int_{\text{all } \mathbf{q}} \sum_{\text{all } n''} |I_{n\mathbf{k},n''\mathbf{k}-\mathbf{q}}|^2 \frac{e^2}{q^2 \varepsilon_0 \varepsilon_{\text{int}}} \frac{[\omega_p^2/2\omega_1(q)] d^3\mathbf{q}}{(\omega_n(\mathbf{k}) - \omega_1(q) - \omega_{n''}(\mathbf{k}-\mathbf{q}))}$$
$$- \frac{1}{(2\pi)^3} \int_{n'', \mathbf{k}-\mathbf{q} \text{ occupied}} \sum_{n''} |I_{n\mathbf{k},n''\mathbf{k}-\mathbf{q}}|^2$$
$$\times \frac{e^2}{q^2 \varepsilon_0 \varepsilon_{\text{int}}} \frac{d^3\mathbf{q}}{\varepsilon_{\text{gs}}(q, \omega_n(\mathbf{k}) - \omega_{n''}(\mathbf{k}-\mathbf{q}))}. \tag{21}$$

The first term comes from the poles in the interaction ε_{gs} and is called the *Coulomb hole term*; the second comes from the poles in the Green function and is called the *screened exchange term* (Hedin, 1965). The division into two terms is a result of evaluating the integral in Eq. (20) in a particular way that is suggested by the mathematical structure of the integrand, and is not unique.

In fact, the traditional division has been into a *Hartree–Fock exchange term* (essentially the second term of Eq. (21) but without the free carrier screening of the Coulomb interaction) and a *correlation term* (defined as the difference between the Hartree–Fock exchange energy and the complete energy). Such a division is attractive because the correlation energy is often small and the Hartree–Fock energy, which is straightforward to evaluate because the free carrier screening does not enter the calculation, can be a good approximation to the complete result (see, for example, Mahan 1980).

The virtue of the alternative approach is that the screened exchange term has a clear physical significance: The electron in the state in question has an exchange interaction with the other electrons of the system, and the resultant energy is calculated using the screened electron–electron interaction that pertains. Furthermore, the Coulomb hole energy is quite insensitive to the state considered. This is because the integral for the Coulomb hole energy is over *all* states, and the electron–electron interaction is quite short-range, giving a factor in the integrand which is slowly varying in wavevector space.

However, it is important to emphasise that, in general, both the screened exchange and Coulomb hole terms make an important contribution to the energy shifts of states. Some calculations (see, for example, Wolff, 1962; Bonch-Bruevich, 1966a, 1966b) have ignored the Coulomb hole term, a procedure that can lead to serious errors in certain applications.

For the intrinsic semiconductor,

$$E_n^{\Sigma i} = \operatorname{Re} \frac{i}{(2\pi)^4} \int \sum_{n''} |I_{n\mathbf{k},n''\mathbf{k}-\mathbf{q}}|^2 \frac{e^2}{q^2 \varepsilon_0 \varepsilon_{\text{int}}}$$

$$\times \frac{e^{-i\omega'\delta}}{\omega_n(\mathbf{k}) - \omega' - \omega_{n''}(\mathbf{k}-\mathbf{q}) + i\delta \operatorname{sgn}(\hbar\omega_{n''}(\mathbf{k}-\mathbf{q}) - \mu)} d^3\mathbf{q}\, d\omega'. \quad (22)$$

Again the integral can be carried out by contour integration, but this time ε_{gs} is not present in the integrand and there are no poles in the interaction (ε_{int} is assumed to be a constant). Hence,

$$E_n^{\Sigma i}(\mathbf{k}) = -\frac{1}{(2\pi)^3} \int_{\text{all } \mathbf{q}} \sum_{n''=\text{valence bands}} |I_{n\mathbf{k},n''\mathbf{k}-\mathbf{q}}|^2 \frac{e^2}{q^2 \varepsilon_0 \varepsilon_{\text{int}}} d^3\mathbf{q}. \quad (23)$$

The energy shift of the state at wavevector **k** in band n due to many-body effects is

$$\Delta E_n^\Sigma(\mathbf{k}) = E_n^{\Sigma d}(\mathbf{k}) - E_n^{\Sigma i}(\mathbf{k}). \tag{24}$$

Equations (21), (23), and (24) constitute the basic formulae used to calculate energy shifts and bandgap narrowing in doped or excited semiconductors. In physical terms they describe how the added carriers change the energies of all states in the system by (1) increasing the number of electron–electron interactions in the system and (2) changing the screening of the old and new interactions.

6. Calculations of Bandgap Narrowing in GaAs

a. n-Type GaAs

To evaluate the shift of the conduction band edge it is necessary to calculate $\Delta E_n^\Sigma(\mathbf{k})$ with n corresponding to the conduction band ($n = c$) and $\mathbf{k} = 0$. For the intrinsic semiconductor, there are no electrons in the conduction band; for the doped semiconductor, the conduction band is filled to some Fermi level. It follows from Eqs. (21), (23), and (24) (Abram et al., 1984) that for an isotropic conduction band,

$$\Delta E_c^\Sigma(0) = -\frac{1}{(2\pi)^3} \frac{e^2}{\varepsilon_0 \varepsilon_{\text{int}}} \int_{\text{all } \mathbf{q}} \frac{1}{q^2} \frac{\omega_p^2}{2\omega_1(q)} \frac{1}{(\omega_1(q) + \omega_c(q) - \omega_c(0))} d^3\mathbf{q}$$

$$-\frac{1}{(2\pi)^3} \frac{e^2}{\varepsilon_0 \varepsilon_{\text{int}}} \int_{\text{occ } \mathbf{q}} \frac{1}{q^2} \frac{1}{\varepsilon_{\text{gs}}(q, \omega_c(q) - \omega_c(0))} d^3\mathbf{q}, \tag{25}$$

where the second integral is over the occupied conduction band states.

The calculation of the valence band shift is carried out in a similar manner. However, an additional complication is the existence of two constituents of the valence band: the heavy-hole and light-hole bands which form the degenerate valence band edge. If the valence bands are taken as isotropic, the total change in the heavy-hole valence band edge is (Abram et al., 1984)

$$\Delta E_{\text{HH}}^\Sigma(0) = \frac{1}{(2\pi)^3} \frac{e^2}{\varepsilon_0 \varepsilon_{\text{int}}} \int_{\text{all } \mathbf{q}} \frac{1}{q^2} \frac{\omega_p^2}{2\omega_1(q)}$$

$$\times \frac{\frac{1}{4}(1 + 3\cos^2\theta_q)}{[\omega_1(q) + \omega_{\text{HH}}(0) - \omega_{\text{HH}}(q)]} d^3\mathbf{q} + \frac{1}{(2\pi)^3} \frac{e^2}{\varepsilon_0 \varepsilon_{\text{int}}} \int_{\text{all } \mathbf{q}} \frac{1}{q^2} \frac{\omega_p^2}{2\omega_1(q)}$$

$$\times \frac{\frac{3}{4}\sin^2\theta_q}{[\omega_1(q) + \omega_{\text{HH}}(0) - \omega_{\text{LH}}(q)]} d^3\mathbf{q}. \tag{26}$$

It is clear that ΔE_{HH}^{Σ} is positive, representing a rise of the heavy-hole band edge. The corresponding expression for the light-hole band edge shift ΔE_{LH}^{Σ} is obtained by the interchange of the subscripts HH and LH in Eq. (26), and it follows that $\Delta E_{LH}^{\Sigma}(0) = \Delta E_{HH}^{\Sigma}(0)$. Hence, the valence band edge rises by $\Delta E_{HH}^{\Sigma}(0)$, as given in Eq. (26).

The change in the GaAs bandgap is therefore

$$\Delta E_g = \Delta E_c^{\Sigma}(0) - \Delta E_{HH}^{\Sigma}(0). \qquad (27)$$

It is apparent that the bandgap narrowing in n-type gallium arsenide is caused by a significant movement of the valence band as well as the conduction band, and we will see in the next section that the obverse holds for p-type material. This highlights the important fact that the theory considers the effect of the carriers on the *whole* band structure. Prior to Inkson's 1976 paper, theories of band gap narrowing due to heavy doping (see, for example, Haas, 1962; Wolff, 1962; Bonch-Bruevich, 1966a, 1966b) had only considered the effect on the band in which the carriers resided. In fact, consideration of the effect on the whole band structure was implicit in the theory of electron–hole droplets that had been developed a little earlier than Inkson's work (see, for example, Rice, 1977), although this was obscured to some extent by the coexistence of carriers in both the conduction and valence bands.

The integral expressions for the self-energies of the states of doped GaAs have been written in terms of unspecified band dispersions $\omega_n(\mathbf{k})$, where n corresponds to the band labels C, HH, or LH. In fact almost all calculations of this kind have been carried out by using the approximation of isotropic, parabolic bands, despite the ease with which realistic band structure can be incorporated in numerical evaluations of the integrals, and it is not clear what errors are incurred when this is done. However, the approximation of isotropic, parabolic bands is also embedded rather more deeply in most calculations. The Fermi surface, for instance, is determined by the band structure for a given carrier concentration, and it has a direct influence on the self-energy integrals. The plasmon pole approximation for the dielectric function is based on isotropic, parabolic bands, and so are the other frequently used standard forms such as that due to Lindhard (1954). In common with most recent calculations intended to provide accurate numerical results, the heavy-hole/light-hole structure of the valence band has been explicitly included in the self-energy integrals derived earlier; ignoring the structure leads to substantial error.

Figure 1 shows the shifts in the conduction and valence bands at their edges and at the Fermi wavevector as a function of electron concentra-

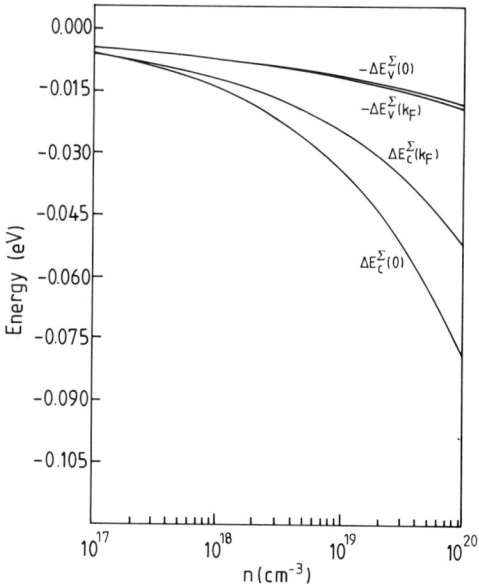

FIG. 1. Electron concentration dependence of the many-body shifts in the energies of the conduction and valence bands of n-type GaAs at their edges and at the Fermi wavevector. (Adapted from Sernelius, 1986a, Fig. 3.)

tion obtained by Sernelius (1986a) using the Lindhard dielectric function. The predicted shifts in the band edges are very similar to those obtained using the plasmon pole approximation. The total narrowing of the fundamental gap varies approximately as $n^{1/3}$, and the major contribution comes from a downward shift in the conduction band. The upward movement of the valence band is quite small over the concentration range considered, and this is largely due to the small change in the screened exchange contribution to the self-energy of the valence band states that results from the introduction of the electron gas. It is small because the screening effect of the electron gas is relatively weak — very much weaker than in silicon, for example, where the large conduction band effective mass and multiple valleys give a large effect.

The energy shifts of *all* the states near the band edges are needed to interpret or model optical absorption and luminescence spectra, and this information can be obtained by using the theory outlined. A reasonable understanding of optical spectra can be achieved with a knowledge of the energy shifts in the conduction and valence bands at $\mathbf{k} = 0$ and at the Fermi wavevector, as given in Fig. 1. The valence band states are moved by essentially the same small amount, implying a rigid shift in the band.

On the other hand, the movement of the conduction band is far from rigid, with a 30 meV greater fall at the band edge than the Fermi level for $n = 10^{20}$ cm^{-3}, which results in a stretching of the band.

b. *p-Type* GaAs

For *p*-type GaAs, holes are present in the valence band, and the ionised acceptor centres are considered as being smeared out into a neutralising uniform negative background charge. The basic principles of the bandgap-narrowing calculation are the same as for *n*-type GaAs, and it is convenient to continue to deal with *electron* states and energies. The energies that must be calculated are again given by Eqs. (21) and (23), but now ε_{gs} is derived from the behaviour of a hole gas, and ω_p^2 and $\omega_1(q)$ must be redefined accordingly. Abram *et al.* (1984) have given the relevant formulae.

Just as in *n*-type GaAs, the conduction band moves down and the valence band moves up, causing bandgap narrowing. The heavy-hole/light-hole structure of the valence band is again explicitly included in the self-energy integrals. However, it is rather more difficult to take satisfactory account of the valence band structure in the plasmon pole approximation for the dielectric function of a *p*-type semiconductor, as explained in Section II.4. Nevertheless, it turns out that the plasmon pole approximation works quite well. Figure 2 shows the results of Bardyszewski and Yevick (1987) for the bandgap narrowing and valence-band shifts predicted by calculations based on three different models of the dielectric function.

The Lindhard, plasmon pole, and damped plasmon pole dielectric functions all give similar results for $p < 10^{20}$ cm^{-3}, and the energy changes are close to proportional to the cube root of the carrier concentration, as in the *n*-type case. There is a marked minimum in the bandgap narrowing obtained from the undamped plasmon pole approximation around $p = 10^{21}$ cm^{-3}. This feature is not a genuine physical effect, but an artefact of the *undamped* plasmon pole approximation which results when the light-hole Fermi surface passes through a singularity in the integrand of the screened exchange term. It is smeared out when a *damped* plasmon pole is used. It is apparent from Fig. 2 that the largest contribution to the bandgap narrowing now comes from the upward shift of the valence band. Hence, as in the *n*-type case the shift in band containing the carriers dominates the bandgap narrowing, but the dominance is not so marked in the *p*-type case. This is a manifestation of the rather stronger screening in *p*-type material. Similar results have been

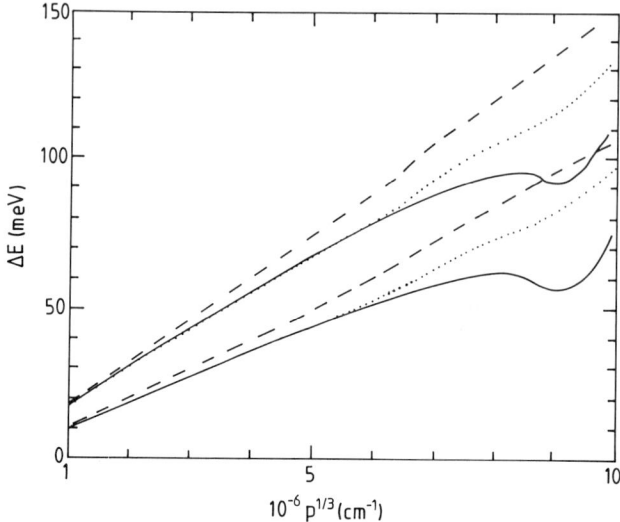

FIG. 2. Hole concentration dependence of the many-body shift in the energy of the valence band edge (lower curves) and the bandgap narrowing (upper curves) in p-type GaAs. Solid curves: plasmon pole approximation; dashed curves: damped plasmon pole approximation; dotted curves: use of the Lindhard dielectric function. (From Bardyszewski and Yevick, 1987, Fig. 3.)

obtained by Sernelius (1986b) and Abram *et al.* (1984) using the Lindhard and plasmon pole dielectric functions, respectively.

c. *The Electron–Hole Plasma in* GaAs

It is possible to produce large, equal concentrations of electrons and holes in intrinsic GaAs by optical excitation. The foregoing theory can also be applied to the description of the electron–hole plasma. Figure 3 shows the concentration dependences of the shifts in the valence and conduction bands, as well as the total bandgap narrowing, as obtained by Abram *et al.* (1984).

Much of the theory relevant to the electron–hole plasma in GaAs is embedded in the enormous literature on the theory of electron–hole droplets and plasmas in the common elemental and compound semiconductors, and it would be inappropriate to attempt a full review here. Therefore, the discussion is confined to a mention of some of the more apposite original papers and reviews. A fine review of the early work on the theory of electron–hole liquids in *indirect-gap* semiconductors has been given by Rice (1977). Although the article concentrates on Si and

5. EFFECTS OF HEAVY DOPING AND HIGH EXCITATION

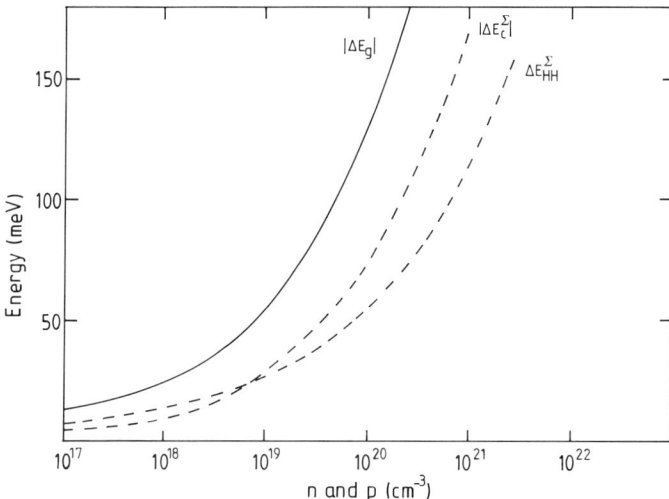

FIG. 3. Electron–hole pair concentration dependence of the band edge and bandgap shifts due to many-body effects in GaAs with an electron–hole plasma. (From Abram *et al.*, 1984, Fig. 10.)

Ge, much of the theory discussed — including, for example, the important contributions of Combescot and Nozières (1972), Brinkman *et al.* (1972), Brinkman and Rice (1973), and Vashista *et al.* (1973, 1974) — has more general application. In view of the discussion of the plasmon pole approximation in this review, the use of the approximation in the description of electron–hole droplets in germanium by Rice (1974) and Rösler and Zimmerman (1975) is also of interest.

Beni and Rice (1978) have carried out calculations on the ground-state energies of electron–hole systems in a wide range of semiconductors, including GaAs. Vashista and Kalia (1982) have shown that when the exchange-correlation energy is expressed in reduced units it is, to a good approximation, independent of the details of the band structure of the semiconductor. Simple universal formulae for the exchange-correlation energy have been presented by Vashista and Kalia and by Thuselt (1983). Vashista and Kalia give

$$E_{xc}(r_s) = -\frac{4.8316 + 5.0879 r_s}{0.0152 + 3.0426 r_s + r_s^2}, \qquad (28)$$

where r_s is determined using the excitonic Bohr radius and E_{xc} is in units of the excitonic binding energy. If it is assumed that the many-body effects cause a rigid movement of the bands, then the change in bandgap at electron–hole pair density n is given by (Rice, 1977; Mahan, 1981; Seitz, 1940) as

$$\Delta E_g = E_{xc} + n \frac{\partial E_{xc}}{\partial n}. \qquad (29)$$

Optical techniques, particularly gain and luminescence measurements, have been used extensively to study highly excited gallium arsenide, as discussed in Section V. GaAs and similar direct-gap materials form the active regions of semiconductor lasers and light-emitting diodes, and as a consequence there has been considerable theoretical interest in the optical properties of the electron–hole plasma in GaAs and related semiconductors. A detailed review of the general theory has been given by Haug and Schmitt-Rink (1984). Examples of other contributions are Brinkman and Lee (1973) and Haug and Tran Thoai (1980) on the gain spectrum, and Selloni et al. (1984) on luminescence line shape analysis.

d. Finite Temperatures

The theory of earlier sections refers to GaAs at zero temperature. Experimental measurements on heavily doped and highly excited semiconductors can be carried out over a wide range of temperature extending from that of liquid helium to above room temperature, and the majority of electronic devices operate at room temperature or above. It is therefore important to know how the self-energy theory is affected by finite temperatures. Inspection of Eq. (10) shows that the effects of temperature will enter the self-energy integrals through the dielectric function and the carrier occupancy of the band states. Lowney (1986a, 1989) and Bennett and Lowney (1987) have used the plasmon pole approximation to calculate the effect of large concentrations of electrons and/or holes on the band structure of GaAs at 300 K. The change in the dielectric function is described by using the appropriate finite temperature value of κ in Eqs. (17) and (18). Childs (1987) has gone further in making a change to the functional form of the dielectric function whilst retaining the plasmon pole approximation. Following the same approach as proposed by Hedin and Lundqvist (1969) for calculating the self-energy due to electron–phonon interactions, Childs has generalized Eq. (17) to finite temperatures by writing

$$(\varepsilon_{gs}(q,\omega))^{-1} = 1 + \omega_p^2 \left(\frac{1+N_q}{\omega^2 - (\omega_1(q) - i\delta)^2} - \frac{N_q}{\omega^2 - (\omega_1(q) + i\delta)^2} \right) \qquad (30)$$

where N_q is the Bose–Einstein distribution function for plasmons. The plasmon dispersion relation $\omega_1(q)$ has the same form as Eq. (18), but κ is given by the relevant finite temperature value. Plasmon damping is not included by Bennett and Lowney or Childs, but this will occur in the nondegenerate gases of both carrier types. The effect of plasmon damping on the results is difficult to quantify, as indeed is the general accuracy of a theory which is being applied outside its intended domain of the high-density, *degenerate* electron gas. These uncertainties have led others such as Balslev and Stahl (1986) (see also Bányai and Koch, 1986; Lanyon and Tuft, 1979; and Landsberg et al., 1985) to develop simple electrostatic models with the principal aim of producing agreement with experimental results rather than providing a full description of all the physical mechanisms. Bardyszewski and Yevick (1987) have claimed that Balslev and Stahl's calculation can be regarded as an *unreliable* simplification of the self-energy approach.

Childs (1987) has shown how, with the dielectric function of Eq. (30), an integral expression for the bandgap narrowing can be derived by using essentially the same procedure as for the zero-temperature case discussed in Section II.5. Thuselt (1983) has also reported a similar analysis. Figure 4 shows the results obtained by using Child's formulae for the conduction and valence band edge shifts and the bandgap narrowing as a function of doping concentration in n- and p-type gallium arsenide at 300 K. The zero-temperature bandgap narrowing results are also shown for comparison. At high carrier concentrations, the room- and zero-temperature carrier distributions are both degenerate, and consequently the bandgap narrowing results are very similar except for the minimum in the p-type material around $p = 10^{21}$ cm^{-3} at 0 K, which was also seen in Fig. 2. We saw in Section II.6.b that this spurious feature is removed by the use of a *damped* plasmon pole model, and it is apparent here that the blurring of the light-hole Fermi surface by finite temperature has the same effect. At the lower end of the carrier concentration range shown in Fig. 4, the bandgap narrowing at room temperature is noticeably smaller than at 0 K. This is a result of the change that occurs in the carrier distribution with increasing temperature, but the absolute difference in energy shifts remains small simply because all shifts are small in this regime.

III. Many-Body Effects in GaAs Quantum Wells

7. INTRODUCTION

The self-energy formalism can be used to study many-body effects in quantum wells by using essentially the same approach as for bulk

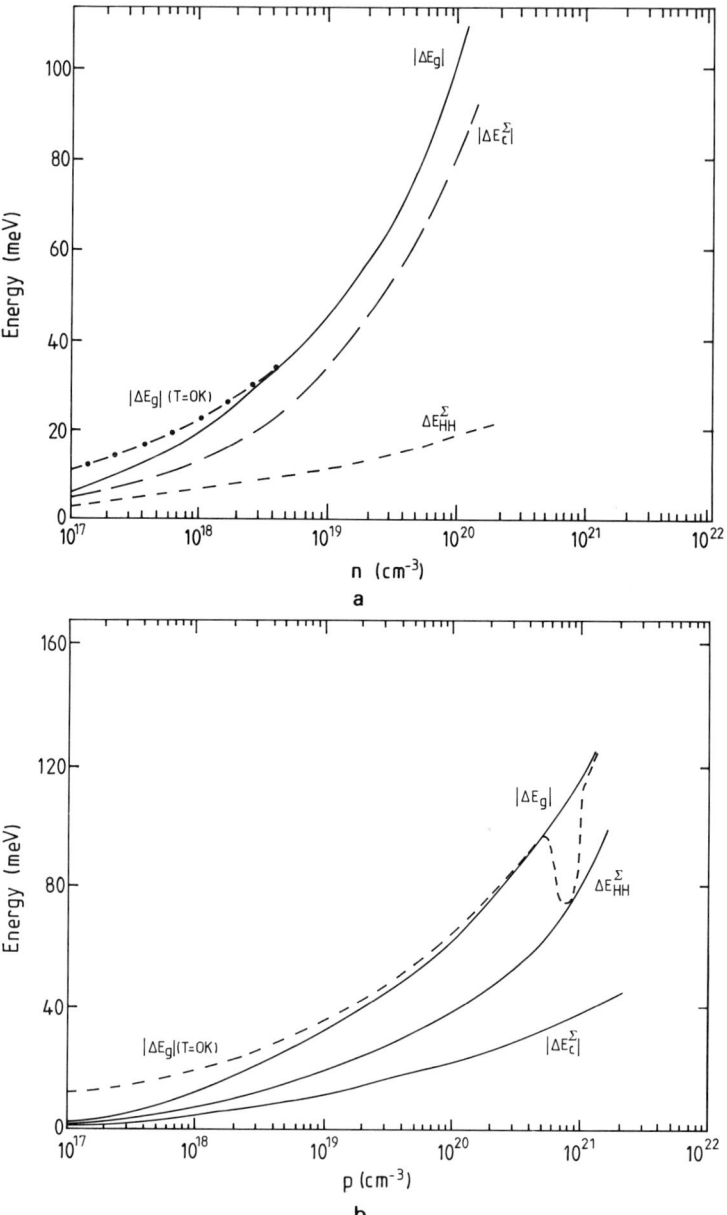

FIG. 4. (a) Electron concentration dependence of the rise in the valence band edge, fall in the conduction band edge, and bandgap narrowing due to many-body effects in n-type GaAs at 300 K. The bandgap narrowing at zero temperature is also shown. (b) Hole concentration dependence of the same quantities as in (a) for p-type GaAs (Childs, private communication).

material. However, the quantum-well heterostructure does present some new problems. First, the band structure is quite different from the bulk, particularly in respect of the subbands that result from the bound states of the well. Anticrossing features commonly occur in the valence subbands and can lead to highly nonparabolic band dispersion. The precise forms of the conduction and valence subbands depend on the well width and on the well and barrier materials and must be calculated afresh for each case. Second, the essential inhomogeneity of the structure and the carrier distributions complicates the description of the screening of the carrier–carrier interaction. To illustrate the theory, we consider an electron gas in a type I quantum well formed by a GaAs layer in the x–y plane sandwiched between layers of a suitable barrier material such as AlGaAs.

8. SCREENING IN A QUANTUM WELL

To calculate self-energies it is necessary to have a description of the electron–electron interactions in the quantum well. The screening of the electron–electron interaction is rather more difficult to treat than in the bulk, because the well and barrier layers have different dielectric properties and their geometrical form has a summetry different from the isotropic bare Coulomb potential produced by an electron. The screened potential of an electron in the quantum well therefore has an anisotropy which reflects the form of the heterostructure. In the plane of the well, the screening mechanism is the two-dimensional analogue of bulk screening; the electrons are completely free to move in the plane, and the electron gas provides screening by adjusting its local density. However, the electron density can only assume a different dependence on z if some electrons make real or virtual transitions to other subbands, and this suppresses the strength of the screening in the direction normal to the well layer. Furthermore, the electron gas normally only exists in the well layer, and its ability to screen in the z direction is restricted by the small range of z over which changes of electron density can be induced. These considerations have led most workers to adopt approximate screening models which concentrate on the in-plane response of the electron gas.

For example, a common approach is to consider that the screening of a point charge at (x, y, z_0) *inside* the well is in effect due to the electron gas concentrated into a sheet of infinitesimal thickness in the plane $z = z_0$. Now if the difference in the relative permittivities of the intrinsic semiconductors constituting the well and barriers is ignored, it is straightforward to show (see, for example, Vinter, 1976) that the dynamically screened electron–electron interaction is

$$W(\mathbf{q}, z, \omega) = \frac{e^2}{2q\varepsilon_0\varepsilon(\mathbf{q}, \omega)} e^{-q|z-z_0|}. \qquad (31)$$

The dependence of the interaction on z is given explicitly in Eq. (31), but the Fourier transform has been taken with respect to the dependence on the in-plane coordinates x and y; q is the magnitude of the *in-plane* wavevector $\mathbf{q}$. The dielectric function $\varepsilon(\mathbf{q}, \omega)$ can be expressed in terms of the dielectric function of the electron gas in the same way as the bulk, viz.,

$$\varepsilon(\mathbf{q}, \omega) = \varepsilon_{\text{int}}\varepsilon_{\text{gs}}(\mathbf{q}, \omega). \qquad (32)$$

ε_{gs} is derived from the dielectric function of a free electron gas by replacing the free electron mass with m_e, and ε_0 with $\varepsilon_{\text{int}}\varepsilon_0$.

The plasmon pole approximation for the dielectric function has proved to be valuable in bulk problems and can also be used to effect in the case of a quantum well. The plasmon pole approximation for ε_{gs} is the same as Eq. (17), except that $\omega_1(q)$ is now chosen to be appropriate to the quasi–two-dimensional system. Stern (1967) has derived the static dielectric function of a purely two-dimensional gas, and that result provides a suitable way of choosing $\omega_1(q)$. When

$$\omega_1^2(q) = \omega_p^2[1 + q/s(q)] + \omega_c^2(q)$$

where

$$s(q) = (e^2 m_e/2\pi\varepsilon_{\text{int}})\{1 - \theta(q - 2k_F)[1 - (2k_F/q)^2]^{1/2}\}, \qquad (33)$$

$\varepsilon_{\text{gs}}(q, \omega)^{-1}$ reduces to the form derived by Stern in the limit of $\omega = 0$ and $q \to 0$. As in the bulk, $\omega_1(q)^2$ also contains a term $\omega_c(q)^2$ which gives an approximate description of the effect of single-particle transitions at large q.

9. Energy Shifts and Bandgap Narrowing

To illustrate the many-body effects on the band structure of a quantum well, we consider the application of the self-energy formalism to the case of a well which is heavily doped n-type. The quantum well wavefunctions can generally be written as

$$\phi_{n\mathbf{\kappa}}(\mathbf{r}) = \chi_{n\mathbf{\kappa}}(\mathbf{\rho}, z)e^{i\mathbf{\kappa}\cdot\mathbf{\rho}}, \qquad (34)$$

5. EFFECTS OF HEAVY DOPING AND HIGH EXCITATION

where κ is the in-plane wavevector and ρ the in-plane position vector. The contribution to the energy of any electronic state from many-body effects in a doped quantum well can then be derived in a similar manner to the bulk (compare Eq. (20)) and is

$$E_n^{\Sigma d}(\kappa) = \operatorname{Re} \frac{i}{(2\pi)^3} \int \sum_{n''} J_{n\kappa,n''\kappa-\mathbf{q}} \frac{e^2}{2q\varepsilon_0 \varepsilon_{\text{int}} \varepsilon_{\text{gs}}(\mathbf{q}, \omega')}$$
$$\times \frac{e^{-i\omega'\delta}}{\omega_n(\kappa) - \omega' - \omega_{n''}(\kappa - \mathbf{q}) + i\delta \operatorname{sgn}(\hbar \omega_{n''}(\kappa - \mathbf{q}) - \mu)} d^2\mathbf{q}\, d\omega', \tag{35}$$

where

$$J_{n\kappa,n''\kappa-\mathbf{q}}$$
$$= \int \chi_{n\kappa}^*(\rho, z)\chi_{n''\kappa-\mathbf{q}}(\rho, z)e^{-q|z-z'|}\chi_{n\kappa}(\rho', z')\chi_{n''\kappa-\mathbf{q}}^*(\rho', z')\, d^3\mathbf{r}\, d^3\mathbf{r}'. \tag{36}$$

It is instructive to consider the evaluation of J by using an envelope function representation of the wavefunctions. In each of the layers of the heterostructure, the function $\chi_{n\kappa}$ can be written as a sum over the zone centre Bloch function periodic parts $u_j(r)$ of the bulk material:

$$\chi_{n\kappa}(\rho, z) = \sum_j f_{jn\kappa}(z)u_j(\mathbf{r}), \tag{37}$$

with the envelope function $f_{jn\kappa}(z)$ taking on appropriate forms in the well and the barriers. For a general in-plane wavevector κ, the χ can have significant contributions from the conduction, heavy-hole, light-hole, and spin split-off Bloch states, and under those circumstances the integrand in Eq. (36) has a rather complicated form.

However, the situation is simpler when the wavevectors are close to zero. For example, close to the edge of the lowest conduction band subband and with a reasonably deep well, the function χ inside the well $(0 < z < L)$ is essentially $u_c(\mathbf{r})$, the conduction band Bloch periodic part (of either spin configuration), multiplied by an envelope function which is approximately $\sin(\pi z/L)$. A similar situation exists for states at the valence band edge which, in the normal unstrained case, will be dominated by the heavy-hole contribution. These considerations suggest that a one-band, infinite-well approximation will be quite accurate in the evaluation of $J_{n\kappa,n''\kappa''}$ when the wavevectors are small, and when n and n''

correspond to subbands of the same type (e.g., both heavy-hole–like) which are strongly bound by the well. The integral defining J in Eq. (36) occurs in the theory of Auger recombination in quantum wells and has been evaluated by Smith (1985) in the approximation just given for an arbitrary pair of subbands of the same type. Das Sarma et al. (1990) have also derived J in the same approximation for the case where n and n'' both denote the same ground subband, and they obtain

$$J(q) = \frac{8}{(q^2L^2 + 4\pi^2)} \left(\frac{3}{8}qL + \frac{\pi^2}{qL} - \frac{4\pi^4(1 - e^{-qL})}{q^2L^2(q^2L^2 + 4\pi^2)} \right). \tag{38}$$

At the same level of approximation, J vanishes if n and n'' correspond to different types of subbands because of the orthogonality of the Bloch periodic parts that occur in the wavefunctions of the two states.

Equation (35) gives the contribution to the energy of a state from the many-body interactions with the electrons which have been added by doping *and* those that fill the valence subbands. The shift in the energy of a state *due to doping* is obtained by subtracting from that contribution the part of the energy of the same state in an undoped well which is derived from the interaction with the electrons in the valence subbands. If only the first conduction subband is occupied, then the shift in the conduction subband edge is given by

$$\Delta E_c^\Sigma(0) = -\frac{1}{(2\pi)^2} \frac{e^2}{\varepsilon_0 \varepsilon_{int}} \int \sum_{n''} J_{c10n''\mathbf{q}} \frac{1}{2q} \frac{[\omega_p^2/2\omega_1(q)]\, d^2\mathbf{q}}{(\omega_1(q) + \omega_{cn''}(q) - \omega_{c1}(0))}$$

$$-\frac{1}{(2\pi)^2} \frac{e^2}{\varepsilon_0 \varepsilon_{int}} \int_{occ\,\mathbf{q}} \frac{1}{2q} \frac{d^2\mathbf{q}}{\varepsilon_{gs}(q, \omega_{c1}(q) - \omega_{c1}(0))}, \tag{39}$$

and the shift in the valence band edge is given by

$$\Delta E_v^\Sigma(0) = \frac{1}{(2\pi)^2} \frac{e^2}{\varepsilon_0 \varepsilon_{int}} \int \sum_{n''} J_{v10n''\mathbf{q}} \frac{1}{2q} \frac{[\omega_p^2/2\omega_1(q)]\, d^2\mathbf{q}}{[\omega_1(q) - (\omega_{v1}(0) - \omega_{vn''}(q))]}. \tag{40}$$

As in the bulk, the conduction band edge falls, the valence band edge rises, and the bandgap is reduced. Similar formulae can be derived to calculate energy changes in states away from the zone centre and in other subbands, and the method is applicable to an excited system with an electron–hole plasma as well as to the cases of n- and p-type doping.

10. Results of Calculations of Bandgap Narrowing

Vinter (1976) used the self-energy method to study many-body effects in the n-type inversion layer of a silicon metal–oxide–semiconductor (MOS) structure. The MOS structure is of course rather different from the GaAs quantum well, and it has the added complication that it is necessary to take account of the significantly different relative permittivities of the silicon and silicon oxide. Nevertheless, the essential physics of the two structures is the same, and Vinter's paper is frequently cited by workers concerned with the theory of many-body effects in III–V heterostructures. The paper is entirely concerned with the many-body effects in the lowest conduction subband, for which the energy shift and the change in effective mass are calculated by using an approach which is essentially the same as that described in Section III.9. It is assumed that only the ground subband is occupied with carriers, and no attempt is made to calculate the shifts of higher subbands or the narrowing of the fundamental gap. The effect on the excited subbands is, however, addressed in other publications (Vinter and Stern, 1976; Vinter, 1977).

Das Sarma and co-workers (1989, 1990) have used the self-energy method to calculate bandgap narrowing in a quantum well due to the presence of an electron–hole plasma. The quantum wells for the electrons and holes are taken to be infinite square-well potentials, and only the lowest conduction and valence subbands are assumed to be occupied by carriers. The subbands are each obtained by a simple one-band effective mass calculation and are therefore parabolic with effective masses m_e and m_h. Das Sarma *et al.* find that, when the carrier density and the well width are expressed in terms of r_s for the material, the bandgap narrowing, in units of the effective Rydberg, shows an approximate universality, dependent on r_s and the dimensionless well width but roughly independent of the band structure details of the material. Figure 5, taken from Das Sarma *et al.* (1990), shows the universal results for the bandgap narrowing as a function of r_s for a number of well widths obtained by using the random phase approximation. The individual theoretical results for GaAs, InAs, and GaSb all fall within 10% of the curves. Also shown in Fig. 5 are some experimental points, which will be discussed in Section V, and the bandgap narrowing predicted for bulk material. Das Sarma *et al.* have also calculated the corresponding results by using the plasmon pole approximation to describe the dielectric response, although no details of the approximation are given. There is semiquantitative agreement between the two sets of

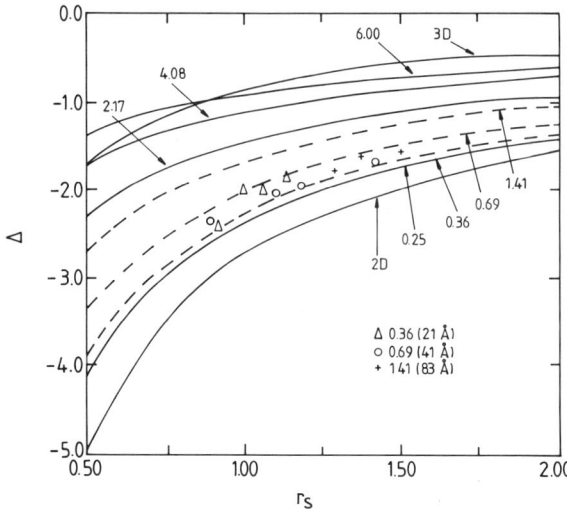

FIG. 5. Bandgap narrowing Δ in a quantum well containing an electron–hole plasma, calculated using the random phase and ε_0 approximations. Δ is expressed in units of the effective two-dimensional excitonic Rydberg, and the carrier density is given in terms of r_s. The points are some experimental measurements of Tränkle et al. (1987a, 1987b), and the dashed curves are the theoretical values for the well widths used in the experiments. The results for the two- and three-dimensional limits are also shown. (From Das Sarma et al., 1990, Fig. 1a.)

results which is poorest (about 15% discrepancy) at small well widths and r_s values.

By modulation-doping the heterostructure, it is possible to have an electron gas in the well without holes or ionized donors in close proximity. The exchange and correlation effects can be described in the same way as in the simple doped case but, as in the silicon inversion layer, there is the complication that the well is not charge-neutral. This case has been studied by Bauer and Ando (1985, 1986a) using two different approaches. One calculation uses the plasmon pole approximation in the self-energy method. The electron and hole envelope functions are represented by Gaussians in which the width parameters are determined variationally. The other calculation is based on density functional theory (Kohn and Sham, 1965), which has been used successfully in other electronic structure problems in space charge layers (Ando, 1976; Ando et al., 1982). The basic method is to account for the exchange and correlation effects with an exchange-correlation potential which is added to the Hartree potential resulting from the space charge. Bauer and Ando use a local approximation to the exchange-correlation potential which again is based on the plasmon pole approximation (Gunnarson and Lundqvist, 1976).

Bauer and Ando (1986a) have calculated electron and hole subband energies as a function of electron concentration for a modulation doped 250 Å wide GaAs/AlGaAs quantum well containing an electron gas. As expected, the Hartree potential (including the potential representing the confinement effect of the heterostructure) acts to raise the conduction subbands, whereas exchange and correlation have the opposite effect. Both influences increase with electron concentration, but the change in the Hartree energy dominates strongly at the higher concentrations, resulting in a substantial rise in the energies of the subbands. The first valence subband is raised (that is, the energy of a hole at the top of the subband is reduced) by the Hartree and many-body effects acting in the same direction. The dependence of the magnitude of the Hartree energy on electron concentration is stronger for holes than for electrons. The additional many-body contribution to the hole energy is approximately independent of electron density. The results of calculations of the bandgap of the quantum well relative to the bulk value are shown as a

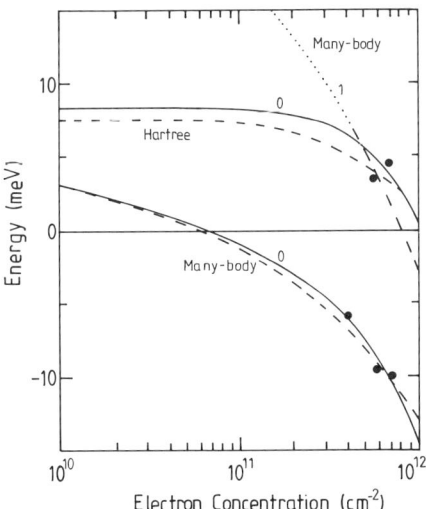

FIG. 6. Calculated values of the energy shifts relative to bulk GaAs of the luminescence spectrum of an n-type modulation-doped 250 Å GaAs/AlGaAs quantum well in the Hartree approximation, and with exchange and correlation effects included (many-body). Full curve: variational (Gaussian) wavefunctions; dashed curve: numerical wavefunctions; 0: transition between the first subbands; 1: transition between second conduction subband and the first valence subband. The points are the experimental results of Pinczuk et al. (1984). (From Bauer and Ando, 1986a, Fig. 6.)

function of electron concentration by the curves labelled 0 in Fig. 6. The full and dashed curves refer to the two different methods of calculation described earlier, both in the Hartree approximation and after the inclusion of exchange and correlation (many-body) effects. Also shown by the curve labelled 1 is the predicted energy separation of the second conduction subband and the first valence subband. At low concentrations, the quantum confinement produces a larger gap than the bulk, but above an electron density of 7×10^{10} cm^{-2}, there is a reduction compared to the intrinsic bulk value.

Bauer and Ando (1986b) have also used density functional theory to describe the electronic structure of GaAs quantum wells which are p-type modulation-doped or occupied by an electron–hole plasma. Kleinman and Miller (1985) and Kleinman (1983) have also considered similar systems, as well as the simple electron gas, but their approach to calculating the bandgap narrowing is rather different. They choose to calculate the exchange-correlation contribution to the total energy following the basic procedure used by Brinkman and Rice (1973) for three-dimensional problems. The effect of carrier confinement on the carrier–carrier interactions is described not by the function J given by Eq. (38), but by a model potential of the form $v(r) = (1 - e^{-\gamma r})/r$, where γ is a function of the well width. One aspect of the calculation which appears to be unsatisfactory is the assumption that the shifts in electron and hole energies are identical. Bauer and Ando (1986b), in their study of p-type modulation-doped wells, comment that this assumption is a crude approximation, but they do nevertheless obtain similar results for the total bandgap reduction. They quote a reduction of 12 meV at a hole density of 5.3×10^{10} cm^{-2} in a 107 Å wide quantum well.

IV. Effects of the Impurity Centres

11. Introduction

The act of doping introduces impurity centres into an array of sites in the semiconductor. The impurities can replace one of the native atoms or take up positions in the interstices of the material. In this review our essential concern is with electrically active substitutional impurities, although the basic ideas discussed are applicable more generally. The established techniques of semiconductor technology make possible rapid spatial variations of dopant type and/or density, but in the vicinity of any given point it is normal to assume that the impurity centres are distributed randomly. Because the average impurity separation is generally several lattice constants, it is a good approximation to ignore the discrete nature of the lattice in describing the impurity distribution. Of course, some correlation of the positions of impurity can occur by accident or design, but that will not be considered here.

A single donor centre in gallium arsenide will produce a hydrogen-like state which, in the simple effective mass model, has a Bohr radius of

$$a = 4\pi\hbar^2\varepsilon_{int}\varepsilon_0/m_c e^2 \quad [= 0.53\varepsilon_{int}(m_0/m_c) \text{ Å}] \quad (41)$$

and an energy below the band edge of

$$E_d = m_c e^4/32\pi^2\hbar^2\varepsilon_{int}^2\varepsilon_0^2 \quad [= (13.6/\varepsilon_{int}^2)(m_c/m_0) \text{ eV}], \quad (42)$$

where ε_{int} is the static relative permittivity of the intrinsic semiconductor and m_c is the electron effective mass. Taking $\varepsilon_{int} = 13.0$ and $m_c = 0.067 m_0$ for GaAs gives $a = 103$ Å and $E_d = 5.4$ meV.

Doped GaAs will contain many impurity centres, and a useful measure of the level of doping is the dimensionless quantity $N^{1/3}a$, where N is the doping density. When $N^{1/3}a \ll 1$, the impurity separation greatly exceeds the Bohr radius of the individual donor states and is classified as the lightly doped regime. In lightly doped GaAs at zero temperature, the combined effects of the random arrangement of the impurities and electron–electron repulsion cause the electrons to be localized on the individual donors with state wavefunctions that closely resemble those of isolated donors. As the doping level is increased, a simple one-electron tight-binding model would suggest the occurence of state mixing, but the theoretical description is complicated by the strong influence of the disorder and the electron–electron interactions. An elegant physical picture of the ground state and excitations of this system and the metal–insulator transition that occurs as the donor density is increased is provided by the Hubbard band model (Hubbard, 1964) discussed in some detail by Mott (1974) and others. One-electron models of the impurity band are valid when the electron–electron interactions have been substantially reduced by adding compensating acceptors to the semiconductor to depopulate the donor impurity band. Depopulation also occurs at room temperature because the electrons are readily excited to the conduction band. One-electron tight-binding models of the impurity band have been reviewed by Abram *et al.* (1978).

The subject of this review is *heavily doped* gallium arsenide for which $N^{1/3}a \geq 0.5$ or $N \geq 10^{17}$ cm^{-3}. In this regime the impurity band has broadened and merged with the conduction band, and the large majority of the states occupied by electrons in uncompensated material (at any temperature) are extended rather than localized. Therefore, it is more appropriate to take free conduction band states as the starting point and to consider how these are modified by the presence of the impurity centres, instead of attempting a tight-binding calculation for strongly coupled impurity states with the added complication of a parent conduction band. Most scattering theories and related approaches are based on the approximation of linear screening, in which it is assumed that the potential at any point is given by the superposition of the *individually* screened potentials of the impurity centres. This approximation is valid when the potential variations are small compared to the Fermi level (or $k_B T$ in the case of classical statistics) (Halperin and Lax, 1966; Hwang, 1970). Hence, the linear screening model gives the impurity potential at **r** as

$$V(\mathbf{r}) = \sum_i \frac{-e^2}{4\pi\varepsilon_{\text{int}}\varepsilon_0 |\mathbf{r} - \mathbf{R}_i|} \exp(-\kappa |\mathbf{r} - \mathbf{R}_i|) \tag{43}$$

where $\mathbf{R}_i$ are the impurity locations. The reciprocal screening length κ depends on the electron distribution and hence on the density of states and so should be obtained self-consistently. However, a common approximation is to use the value of κ given by the unperturbed conduction band, which can give acceptable accuracy if the Fermi level is well into the band.

12. Semiclassical and Variational Models

The density of states $\rho(E)$ in the conduction band of a doped semiconductor is a quantity of interest for many applications, and a number of calculations have been directed at obtaining $\rho(E)$ with rather less concern for describing the nature of the states themselves. A simple calculation of this type is due to Kane (1963) and has the same physical basis as the Thomas–Fermi theory of screening. It is assumed that the potential $V(\mathbf{r})$ is sufficiently slowly varying in space that a local density of states at energy E can be assigned at each point **r** which is given by the unperturbed conduction band density of states at energy $E - V(\mathbf{r})$ above the band edge. $\rho(E)$ is then obtained by averaging over all **r** or, equivalently, over the distribution of the local potential $V(\mathbf{r})$. The model predicts the formation of a bandtail and the loss of some states from the body of the conduction band. However, the semiclassical nature of the model is a fundamental limitation on its accuracy. Fluctuations in the potential tend to confine the lower energy states, and the neglect of the kinetic energy associated with this (an inherently quantum-mechanical effect) causes the theory to exaggerate the tail.

Halperin and Lax (1966) developed a quantum-mechanical variational theory of the tail states. The model has been reviewed in detail by Abram *et al.* (1978) and will not be discussed again at length here. Suffice it to say that, although Halperin and Lax's theory overcomes the deficiencies of Kane's model, it is only valid for deep tail states. Lloyd and Best (1975) have reported a variational theory which is of more general validity. They have shown that in a variational model for the density of states, the best choice is the one which maximizes the quantity

$$P(E) = \int_{-\infty}^{E} \int_{-\infty}^{E'} \rho(E'') \, dE'' \, dE'. \tag{44}$$

This is a rigorously derived maximum principle for the density of states which in the deep tail reduces to the variational method used by Halperin and Lax. However, the main difficulty with the Lloyd–Best theory is in devising an accurate unified model of the electronic states which embraces states both in the deep tail and in the body of the band and which can be subjected to the variational principle. In recent years a number of papers have appeared reporting refinements or applications of these variational methods, prominent examples being Sa-yakanit and Glyde (1980) (and references therein) and Soukoulis *et al.* (1984).

Mahan (1980) has described the effect of ionized donors on the bandgap of heavily doped Si and Ge by carrying out a variational calculation on a *regular array* of donor centres. However, although this approach allows a well-defined and straightforward variational calculation, the arrangement of impurity centres in the model is fundamentally different from the real situation, as discussed by Berggren and Sernelius (1984).

13. Multiple Scattering Model

a. Basic Theory

Several workers have considered the effects of impurity scattering within the framework of a self-energy description. The usual starting point is the equation that gives the one-electron Green's function G in $\omega, \mathbf{k}$ space for the envelope part of the wavefunction of the conduction band:

$$(\hbar\omega - E(\mathbf{k}) + i\delta)G(\mathbf{k}, \mathbf{k}', \omega)$$
$$- \frac{1}{(2\pi)^3} \int V(\mathbf{k}, \mathbf{k}'')G(\mathbf{k}'', \mathbf{k}', \omega) \, d^3\mathbf{k}'' = \delta_{\mathbf{k}\mathbf{k}'}, \quad (45)$$

where $E(\mathbf{k})$ is the conduction band dispersion relation and $V(\mathbf{k} - \mathbf{k}')$ is the matrix element of the impurity potential of Eq. (43) between plane wave states normalized to the volume of the system. In the absence of the impurity potential, the solution of Eq. (45) is quite straightforward, *viz.*,

$$G_0(\mathbf{k}, \mathbf{k}', \omega) = \frac{\delta_{\mathbf{k}\mathbf{k}'}}{(\hbar\omega - E(\mathbf{k}) + i\delta)}. \quad (46)$$

When the impurities are randomly located in the system, the Green's function depends on their spatial distribution, and each particular arrangement of impurities will give a different function $G(\mathbf{k}, \mathbf{k}', \omega)$. However, it has been shown that in a large system, the ensemble average of G is a good representation of the Green's function of a particular system, except for a number of special cases which constitute an infinitesimal fraction of all the possible arrangements (Kohn and Luttinger, 1957). A perturbation expansion solution of Eq. (45) may be written starting with G_0 as the zero-order term:

$$G(\mathbf{k}, \mathbf{k}', \omega) = G_0(\mathbf{k})\delta_{\mathbf{k}\mathbf{k}'} + G_0(\mathbf{k})V(\mathbf{k} - \mathbf{k}')G_0(\mathbf{k}')$$
$$+ \sum_{\mathbf{k}_1} G_0(\mathbf{k})V(\mathbf{k} - \mathbf{k}_1)G_0(\mathbf{k}_1)V(\mathbf{k}_1 - \mathbf{k}')G_0(\mathbf{k}') + \ldots, \quad (47)$$

where the second wavevector is not shown in the G_0 because it is always the same as the first, and ω has also been suppressed in the expansion because it is the same in all terms. It is now possible to take the ensemble average of $G(\mathbf{k}, \mathbf{k}', \omega)$ and to sum the terms in the expansion at various levels of approximation. The simplest nontrivial approximation is to retain only those terms corresponding to single or double scattering off the *same* impurity, and sum over all impurities. It is then possible to write the ensemble-averaged Green's function as

$$\langle G(\mathbf{k}, \mathbf{k}', \omega) \rangle = \frac{\delta_{\mathbf{k}\mathbf{k}'}}{(\hbar\omega - E(\mathbf{k}) - \Sigma(\mathbf{k}, \omega))}, \quad (48)$$

5. Effects of Heavy Doping and High Excitation

where the self-energy is given by

$$\Sigma(\mathbf{k}, \omega) = Nv(0) + \frac{N}{(2\pi)^3} \int \frac{|v(\mathbf{k} - \mathbf{k}')|^2 \, d^3\mathbf{k}'}{\hbar\omega - E(\mathbf{k}') + i\delta}. \tag{49}$$

In Eq. (49), N is the ionized impurity concentration and $v(\mathbf{k} - \mathbf{k}')$ is the Fourier transform of a *single* impurity potential (or, equivalently, the matrix element between plane wave states $\mathbf{k}$ and $\mathbf{k}'$ multiplied by the system volume Ω).

The first term on the right-hand side of Eq. (49) is simply the shift in the average electron potential energy due to the interaction with all the impurities in the system. In fact, as pointed out by Wolff (1962), this shift is exactly cancelled by the *direct* screened Coulomb interaction that each electron has with every other electron, and the cancellation reflects the charge neutrality of the system. The same effect was noted in the discussion of many-body effects in Section II.3, where the donor charge was assumed to be distributed in a *uniform* positive background. The second term in the self-energy expression of Eq. (49) has real and imaginary parts and as such describes both a shift and a broadening of the states due to scattering. The *principal part* of the integral gives the shift of the state at energy $\hbar\omega$ and wavevector $\mathbf{k}$, and the imaginary part reduces to the scattering rate expression that would be deduced from Fermi's Golden Rule (Mattuck, 1976). The density of states can be obtained by the standard procedure (Economou, 1983) of calculating

$$\rho(\hbar\omega) = -\frac{1}{\pi\Omega} \operatorname{Im} \sum_{\mathbf{k}} \langle G(\mathbf{k}, \mathbf{k}, \omega) \rangle. \tag{50}$$

The derivation of Eqs. (48) and (49) is essentially equivalent to second-order perturbation theory, and as such the results are prone to the limitations associated with that approximation. Apart from any concern about the quantitative accuracy of the energy shifts predicted for the body of the conduction band in certain circumstances, the description of the low kinetic energy states close to band edges is generally unreliable. Nevertheless, the theory provides a simple expression for the electron self-energy resulting from the electron–impurity interaction and has frequently been used in calculations on gallium arsenide and other semiconductors (see, for example, Parmenter, 1955; Wolff, 1962; Berggren and Sernelius, 1981; Sernelius 1986a, 1986b).

b. Self-consistent Calculations

A major advance in the description of the impurity effects was provided by the work of Ghazali and Serre (1982) and Serre and Ghazali (1983), who made use of some multiple scattering theory that had been published considerably earlier by Klauder (1961). Klauder had developed a formalism at a number levels of approximation for the electron self-energy due to impurity scattering in solids. However, the better approximations involved coupled integral equations, and Klauder only carried out explicit calculations for the algebraically tractable case of a one-dimensional disordered array of delta function potentials. Serre and Ghazali's contribution was to take Klauder's so called fifth approximation and develop an efficient numerical solution to the coupled equations for the case of a doped semiconductor: a three-dimensional system of randomly located, screened Coulomb potentials.

The basis of Klauder's fifth approximation is the following. Equation (49) is a self-energy expression that results when only single and double scattering processes at a single impurity are considered, and the approximation can be improved by including higher levels of multiple scattering. Following the comments earlier, it should also be noted that the single scattering term will be cancelled by the direct electron–electron interaction and need not be considered explicitly in any enhanced calculation. If double and all higher levels of multiple scattering at a single impurity are considered, the self-energy is given by the infinite series obtained by iterating the following relation for the function $K(\mathbf{k}, \mathbf{q}; E)$ (the so-called vertex function in field theory):

$$K(\mathbf{k}, \mathbf{q}; E) = \frac{1}{(2\pi)^3} \int v(\mathbf{q}' - \mathbf{q}) \langle G(\mathbf{k} + \mathbf{q}'; E) \rangle [Nv(-\mathbf{q}) + K(\mathbf{k}, \mathbf{q}'; E)] \, d^3\mathbf{q} \tag{51}$$

and then using

$$\sum(\mathbf{k}, E) = K(\mathbf{k}, 0; E). \tag{52}$$

In the spirit of the perturbation theory described above, $\langle G(\mathbf{k}, E) \rangle$ is the zero-order Green function ($= G_0(\mathbf{k}, \mathbf{k}; E)$ of Eq. (46)). However, the formalism we have described takes no account of scattering processes in which two or more impurities are involved. To attempt to correct this omission, Klauder (1961) proposed the replacement of the zero-order Green function in Eq. (51) by its modified form when the effects of

5. Effects of Heavy Doping and High Excitation

scattering have been included:

$$\langle G(\mathbf{k}, E)\rangle = [E - E(\mathbf{k}) - \sum(\mathbf{k}, E)]^{-1}. \tag{53}$$

In physical terms this means that the scattering by all other impurities is taken into account in some sense, albeit approximate, during the propagation of the electron between the scatterings by any single impurity. The formalism includes at least some of the important processes to all orders in the impurity interaction and to all levels of multiple impurity scattering, and is therefore expected to be suited to strong and weak impurity potentials and to high and low impurity densities. The penalty to be paid for the improvement in the physical description is that the set of *coupled* Eqs. (51)–(53) now has to be solved self-consistently, and this is generally only possible by numerical means.

Figure 7 shows the numerical results of Ghazali and Serre (1982) for the conduction band density of states as a function of energy for a range of donor concentrations. A single, isotropic parabolic band is assumed, and the electron–donor interaction is taken to be a Thomas–Fermi screened Coulomb potential. The energies, lengths, densities of states, and impurity concentrations are given in units of the effective Rydberg R,

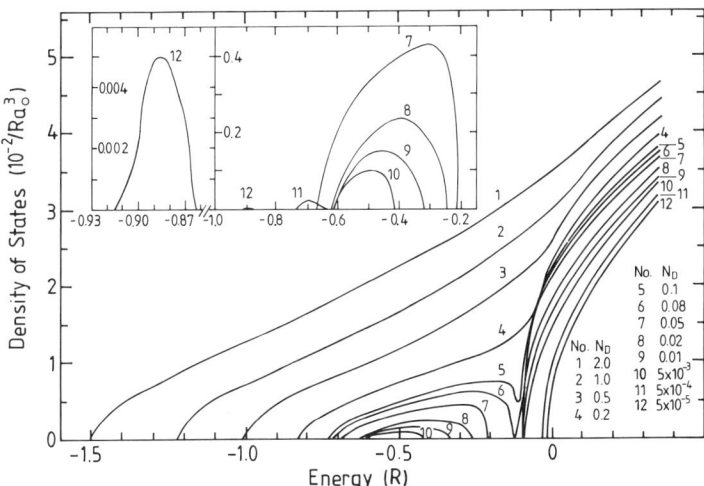

FIG. 7. Conduction band density of states as a function of energy for different donor concentrations obtained by a self-consistent multiple scattering calculation. The units of energy, density of states, and impurity concentration are defined in the text. (From Ghazali and Serre, 1982.)

effective Bohr radius a_0, $(Ra_0^3)^{-1}$, and $\pi/3(4a_0)^3$ respectively. The bands have been shifted rigidly by the Hartree–Fock exchange energy to take account of many-body effects. The highest concentration considered (curve 1) corresponds to $N = 3 \times 10^{16}$ cm^{-3} in GaAs and is therefore below the levels of main interest here. Nevertheless, it is worthwhile to briefly examine the results. In curve 1 the density of states shows a distinct tail, but the remarkable capability of the theory is demonstrated at the lower concentrations (curves 7–12), where an impurity band, separated from the conduction band, is apparent. At the lowest concentrations, the impurity band is narrow and approaching the isolated donor energy of $-E_d$. As the donor concentration is increased, the impurity band broadens and exhibits the asymmetry predicted by other methods (Matsubara and Toyozawa, 1961; Yonezawa, 1964; Gaspard and Cyrot-Lackmann, 1973, 1974). A further increase in donor density causes the impurity band to merge with the host band and eventually to disappear into the conduction band tail.

Although the theory appears to work remarkably well at low impurity concentrations, it must be recognized that there are shortcomings in the detailed physical model adopted by Serre and Ghazali in this regime. In particular, no account is taken of electron–electron correlation, which is known to be an important influence on the behaviour of electrons in narrow impurity bands (Mott, 1974) of uncompensated material. Also, the screening is based on the assumption of free electrons in a simple conduction band and will not be properly described for dilute impurity concentrations at low temperatures because the carriers will have frozen out. Nevertheless, the general approach is well founded and is expected to give qualitatively sensible results for the conduction band at low impurity concentrations and a good quantitative description of heavily doped semiconductors.

Serre and Ghazali (1983) have also shown how the method can be used to describe both the valence and conduction bands in n- and p-type and compensated semiconductors. Their results demonstrate how, in contrast to the simple double scattering theory (Eqs. (48) and (49)), the multiple scattering method is sensitive to the sign of the impurity potential, and is so in accord with expectations. For example, the method predicts that a low concentration of donors will produce a distinct impurity band close to the conduction band, whilst little change occurs in the density of states of the valence band. At higher donor densities a tail does grow on the valence band, and its characteristics relative to the conduction band tail depend on the values of the hole and electron effective masses.

In the limit of high impurity densities, a simplification to the method can also be considered, which produces a rather less computationally demanding calculation. This involves ignoring $K(\mathbf{k}, \mathbf{q}'; E)$ in the integral on the right-hand side of Eq. (51). The expression for the self-energy is similar to that of the double scattering approximation, but now the propagation between the scatterings is described by the modified Green function. This is Klauder's third approximation (Klauder, 1961), essentially a self-consistent Born approximation, and an examination of the terms included in the perturbation series suggests that it is appropriate for high impurity densities and weak scattering (Serre and Ghazali, 1983).

Lowney (1986a, 1986b), Bennett (1986), and Bennett and Lowney (1987) have employed Klauder's third and fifth approximations in a series of calculations of impurity and many-body effects in Si and GaAs. Bennett and Lowney (1987) give results for the densities of states of the conduction and valence bands of n- and p-type GaAs for impurity concentrations ranging from 10^{17} cm^{-3} to 10^{20} cm^{-3}. An example of the work on impurity effects in n-type GaAs is shown in Fig. 8. A significant feature is a much smaller shift in the top of the valence band than that given by the second-order perturbation theory results of Sernelius (1986a).

Haufe and Unger (1985) and Schwabe et al. (1986) have pointed out that the application of Klauder's fifth approximation is considerably simplified if the screened electron–impurity interaction in **k**-space,

$$v(\mathbf{k} - \mathbf{q}) = \frac{e^2}{\varepsilon_0 \varepsilon_{\text{int}}} \frac{1}{|\mathbf{k} - \mathbf{q}|^2 + \kappa^2}, \tag{54}$$

is approximated by

$$v(\mathbf{k} - \mathbf{q}) = -\frac{e^2}{2\varepsilon_0 \varepsilon_{\text{int}}} \frac{1}{(k^2 + \kappa^2)^{1/2}(q^2 + \kappa^2)^{1/2}}. \tag{55}$$

The factorized form of the approximate $v(\mathbf{k} - \mathbf{q})$ facilitates the *algebraic* solution of the set of self-consistent Eqs. (51)–(53). The only numerical tasks that remain are the determination of the zeros of a simple algebraic expression and the evaluation of the density of states function. Schwabe et al. demonstrate good agreement between the conduction band densities of states obtained by the full and approximate methods for donor concentrations of 0.02, 0.2, and 2.0 in the units of Ghazali and Serre. Schwabe et al. then go on to report the use of the theory in the interpretation of the results of photoluminescence experiments on heavily doped n-type InP.

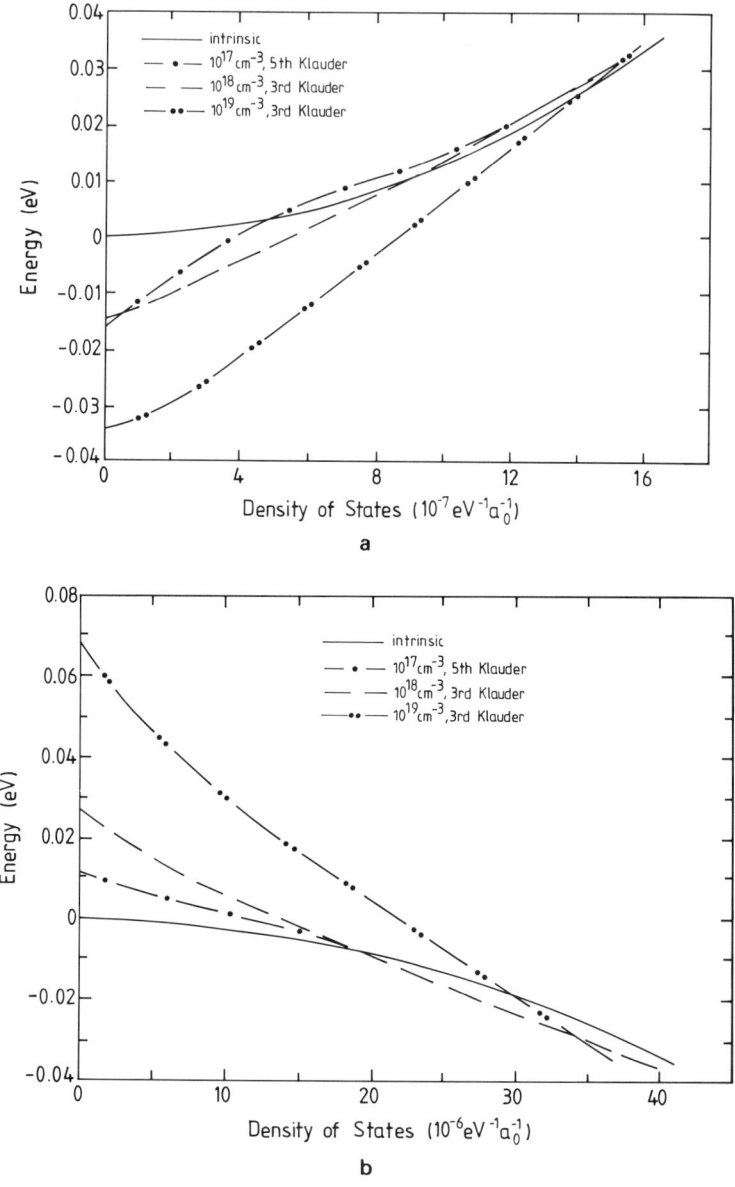

FIG. 8. (a) The conduction-band density of states of GaAs at 300 K for intrinsic material and for three different donor concentrations, obtained in multiple scattering calculations and illustrating the effect of the electron–impurity interaction. (b) The valence-band density of states for the materials in (a). From Bennett and Lowney, 1987, Figs. 6 and 7.)

14. Quantum Wells

Just as with bulk material, to describe the effects of impurities on the band structure of quantum wells and other quasi–two-dimensional structures, it is necessary to have a model for the random potential created by the ionized dopants. However, heterostructures have extra complications because of the introduction of structural features with different symmetry from the carrier–impurity interaction and because of the flexibility in geometrical and material parameters of the system. For example, in addition to the presence of the basic heterostructure, quantum wells are often doped in limited spatial regions. For example, it is possible for only the well itself to be doped or for doping to be confined to a layer or layers of doped material remote from the quantum well.

Whatever configuration of impurities exists, the potential in the well (and other parts of the system) is due to the ionized impurities screened by the dielectric response of the system. It is normal to assume that the linear screening model discussed in Section IV.11 is valid and its use requires a knowledge of the potential due to an individual impurity centre. In many cases, the large majority of carriers reside in the well and are then the only significant source of free carrier screening. Under these circumstances, a simple model that is often used to obtain the potential due to an individual impurity is to approximate the mobile charge in the quantum well by an infinitesimally thin sheet of charge of the same density per unit area and to use the Thomas–Fermi model of carrier response (see, for example, Hess, 1979). Consider a well layer parallel to the x–y plane with centre at $z = 0$. It then follows (see, for example, Stern, 1967; Vinter, 1976; Hess, 1979) that a singly ionized donor located at $(0, 0, z_0)$ and screened by a sheet of degenerate electrons at $z = 0$ produces a potential

$$v_{[2]}(\mathbf{\rho}, z) = \frac{1}{(2\pi)^2} \int \phi_{\mathbf{q}}(z) e^{i\mathbf{q} \cdot \mathbf{\rho}} \, d^2\mathbf{q}, \tag{56}$$

where

$$\phi_{\mathbf{q}}(z) = \frac{-e^2}{2\varepsilon_0 \varepsilon_{\text{int}} q} \left\{ e^{-q|z-z_0|} - \frac{s}{q+s} e^{-q(|z|+|z_0|)} \right\}. \tag{57}$$

In Eqs. (56) and (57), $\mathbf{\rho} = (x, y)$ and $s = ne^2/2\varepsilon_0 \varepsilon_{\text{int}} E_F$ is the degenerate screening constant for an electron density of n *per unit area*. (For the nondegenerate case, E_F should be replaced by $k_B T$).

More complete models of the screening of the impurity potential, which take account of the spatial extent of the carrier states and other relevant features of the system, are possible and have been used in band-structure and transport calculations as appropriate. The details of these models depend to some extent on the particular structure under consideration, and most attention has been directed to the accumulation layer at a heterointerface rather than the quantum well, prompted mainly by the extensive experimental studies of the silicon–silicon dioxide structures. For detailed discussion of the more sophisticated models, the reader is referred to the original papers and recent reviews (Stern and Howard, 1967; Vinter, 1976; Ando et al., 1982).

In fact, there have been relatively few theoretical studies of the effects of impurities on the band structure of GaAs quantum wells and related structures. An early calculation was by Ekenberg (1984) and concerned the deep tail region of uniformly doped n-type GaAs/Al$_x$Ga$_{1-x}$As quantum wells and superlattices. Ekenberg shows how the method of Halperin and Lax (1966) can be adapted to the quasi–two-dimensional context by the use of a suitable trial wavefunction and other appropriate modifications. However, a serious shortcoming of the calculation is the use of a *bulk-like* screened Coulomb potential for the electron–impurity interaction. The same approximation is made by Takeshima (1986a), who considers a single GaAs well in which only the well layer is doped. His concern is to provide a description of the density of states throughout the lowest subband, and he therefore employs a modified form of the semiclassical approach of Bonch-Bruevich (1966a) in an infinite square-well model. Takeshima's results for a donor density of 10^{19} cm^{-3} in a 60 Å wide well show a substantial bandtail and an associated loss of states in the body of the subband. This effect is much stronger than in three dimensions, and the author explains that it is a direct result of the steplike nature of the quantum-well density of states. The results also show a general upward energy shift of the band (of the order of 100 meV), which the author appears to ascribe to the direct electron–electron interaction or Hartree potential. It is not clear why this effect is so strong in view of the compensating charge in the well provided by the ionized donors. In a later paper, Takeshima (1986b) improves the description of the screening, adopts an finite square-well potential model, and includes the exchange energy. Rorison et al. (1988) have also used a similar semiclassical model in the interpretation of carrier mobility and photoluminescence lineshapes in InGaAs/InP quantum wells.

Before ending the discussion of this topic it is worth pointing out that Gold et al. (1988) have applied Serre and Ghazali's multiple scattering method to calculating the density of electronic states in the accumulation

5. EFFECTS OF HEAVY DOPING AND HIGH EXCITATION

layer of a sodium doped metal–oxide–silicon field-effect transistor (MOSFET). The results show a band tail at high impurity concentrations and the formation of a distinct impurity band as the doping is decreased. Gold *et al.* also examine the effectiveness of the factorized potential approximation in the case of two dimensions. Clearly application of the multiple scattering method to the quantum well problem would be very welcome.

V. Optical Experiments

15. INTRODUCTION

The most direct method of measuring the band structure parameters of a semiconductor is through optical experiments. A simple introduction to the theory of the optical properties of heavily doped semiconductors has been given by Abram *et al.* (1978) and will not be repeated here. Suffice it to say that the interpretation of the results of optical experiments requires consideration of, amongst other things, the changes to the densities of states in the bands including band tailing; the filling of those bands by degenerate or near-degenerate carriers; and the modifications to the optical matrix elements as a result of disorder, including the possible loss of wavevector conservation in optical transitions.

16. BULK GaAs

The optical absorption and emission experiments of Casey *et al.* (1975) and their subsequent interpretation by Casey and Stern (1976) constitute a thorough and comprehensive study of the optical properties of heavily doped GaAs at room temperature. The latter paper is also a useful source of references of earlier optical studies. Casey and Stern used a hybrid Kane/Halperin and Lax model and suitably modified optical matrix elements to calculate the absorption and emission spectra of n- and p-type gallium arsenide. The bandgap narrowing due to many-body effects was then deduced by comparing the theoretical and experimental absorption spectra. For p-type samples with hole concentrations varying between 1.2×10^{18} cm^{-3} and 1.6×10^{19} cm^{-3}, Casey and Stern find a simple relation between the bandgap narrowing *due to many-body effects alone*, $\Delta E_{g\,ee}$, and the hole concentration p:

$$\Delta E_{g\,ee} = -1.6 \times 10^{-8} p^{1/3}, \tag{58}$$

where $\Delta E_{\text{g ee}}$ is measured in electron volts and p in inverse cubic centimeters. No comparable result is quoted for n-type material.

There are problems in measuring the change in bandgap in a heavily doped semiconductor by using optical absorption. The high density of carriers that exist in either the valence or the conduction band precludes absorption transitions involving states close to the edge of the occupied band. The result is the so called Moss–Burstein shift (Moss, 1954; Burstein, 1954) of the absorption edge away from the energy of the fundamental gap. To deduce the gap it is *essential* to have a theoretical model which involves a description of the local electronic structure and the carrier distribution (such as attempted by Casey and Stern). A further complication in absorption measurements is that intraband absorption can obscure the interband absorption edge. As a result of these problems, photoluminescence has proved to be a much more popular technique for making bandgap-narrowing measurements. Here, photo-generated minority carriers recombine radiatively with the majority carriers, and the transitions involve the band states of particular interest. The photoluminescence signal can therefore give a rather more direct measure of the bandgap and band filling, as well as information on the majority-carrier distribution. Another technique, photoluminescence excitation spectroscopy, monitors the intensity of the luminescence as a function of the excitation photon energy. In contrast to conventional absorption measurements, it is only sensitive to interband excitations and does not suffer from the problems caused by intraband absorption (Wagner, 1985).

Whatever optical technique is employed, it is important to recognise that a *thorough* interpretation of the results should include not only consideration of the shifts in the band states (real part of the self-energy), but also the broadening of levels (imaginary part of the self-energy) due to final-state lifetime effects (see, for example, Landsberg, 1966; Landsberg and Robbins, 1985; Haug and Tran Thoai, 1980; Selloni *et al.*, 1984; Sernelius, 1986b; Childs and Abram, 1989). A semiconductor laser operates with a high concentration of carriers in its active region, and it is important to include lifetime broadening effects in models of these devices. For example, Asada *et al.* (1984) and Kucharska and Robbins (1990) have shown how quantum-well laser properties, including the gain and spontaneous emission spectra, are substantially modified by lifetime broadening effects.

Olego and Cardona (1980), Titkov *et al.* (1981), Saito *et al.* (1989), and Borghs *et al.* (1989) have carried out low-temperature photoluminescence experiments on p-type GaAs during the last decade. They have each deduced bandgap narrowing as a function of hole concentration from the

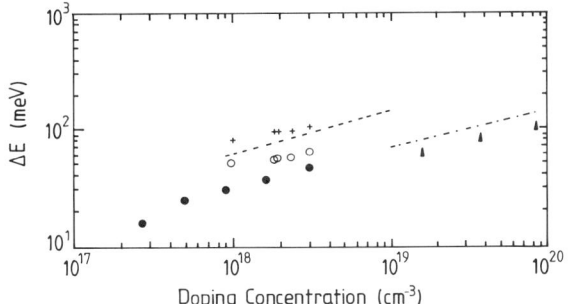

FIG. 9. Fermi level shifts and bandgap narrowing as a function of doping concentration of n- and p-type GaAs deduced from low-temperature optical measurements. △: bandgap narrowing results of Olego and Cardona (1980) for p-type; ○: Fermi level shift for n-type; ●: Fermi level shift for p-type; +: bandgap narrowing for n-type — all results due to Borghs et al. The broken lines show the theoretical bandgap-narrowing results (many-body *and* impurity effects) of Bennett and Lowney (1987) for GaAs at 300 K. (From Borghs et al., 1989, Fig. 6.)

spectra, and their results are in general agreement. The measurements of Olego and Cardona and Borghs et al. are shown in Fig. 9. Titkov et al. and Saito et al. find the gap change at the higher hole concentrations is described by a relation of the form

$$\Delta E_g = -2.2 \times 10^{-18} p^{1/3}. \qquad (59)$$

Note that Eqs. (58) and (59) describe different quantities. Equation (58) gives the change due to the many-body effects *only* (or at least any effects not explicitly included in the Casey and Stern calculation), whilst Eq. (59) gives the experimentally observed change in bandgap due to all the effects of doping. Of course, if *exponential* band tails are present, there are no definite band edges or band gaps. However, if the bands have the appearance of those in Fig. 8, then there is no such problem. Any additional exponential band tailing due to density fluctuations is likely to make only a small modification to Fig. 8 and have only a marginal effect on the measured spectra. The experimental measurements of total bandgap narrowing are in good agreement with the theoretical results of Bennett and Lowney (1987) (as described in Section IV.13.b), but it should be remembered that the latter were calculated for room temperature and will only be *directly* relevant as long as the hole gas remains degenerate at that temperature.

Borghs et al. (1989) and Yao and Compaan (1990) have reported results on bandgap narrowing in n-type gallium arsenide, again derived

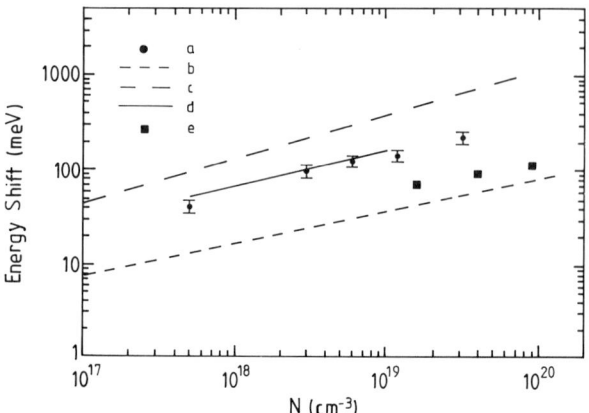

FIG. 10. Bandgap narrowing as a function of doping concentration for n-type GaAs measured by low-temperature optical experiments — points (a). Also shown are the results of Olego and Cardona (1980) for p-type material — points (e); Eq. (58) due to Casey and Stern (1976) — curve (b); the theoretical results of Bennett and Lowney (1987) — curve (d); and the theoretical results of Sernelius (1986a) — curve (c). (From Yao and Compaan, 1990, Fig. 3.)

from low-temperature photoluminescence experiments. These results are shown in Figs. 9 and 10, together with the theoretical predictions of Bennett and Lowney (1987) and Sernelius (1986a). The latter's calculations give a much larger gap narrowing but, as with p-type material, Bennett and Lowney's results are in close accord with the experimental data. Borghs et al. have fitted an expression similar to Eq. (59) to their results and obtain

$$\Delta E_g = -7.3 \times 10^{-8} n^{1/3}. \tag{60}$$

Jain and Roulston (1991) have taken simplified formulae from the theories of impurity and many-body effects, along with the results of optical experiments, to derive simple closed-form expressions for bandgap narrowing as a function of carrier concentration. In fact, the expression for n-type GaAs does not agree well with experiment. Jain et al. (1992) have attributed the discrepancy to the description of the impurity scattering which is based on the results of Sernelius (1986a). Those results were obtained using the second-order perturbation theory approach, the limitations of which were discussed in Section IV.13.a. When the bandgap change due to impurity scattering is described by an expression fitted to the results of self-consistent multiple scattering calculations (see Section IV.13.b), Jain et al. (1992) obtain much better agreement with experiment. For n-type GaAs, they give

$$\Delta E_g = 62.00(n\ 10^{-18})^{1/3} + 7.47(n\ 10^{-18})^{1/4}\ \text{meV}, \tag{61}$$

where n is in cm^{-3}. For p-type GaAs, the equivalent formula given by Jain and Roulston (1991) is

$$\Delta E_g = 9.71(p\ 10^{-18})^{1/3} + 12.19(p\ 10^{-18})^{1/4} + 3.88(p\ 10^{-18})^{1/2}\ \text{meV}. \tag{62}$$

Jain et al. (1990) have also given similar formulae for the important III–V compounds and some of their alloys.

The electron–hole plasma in GaAs has been investigated expermentally by several workers. For example, Göbel et al. (1973), Hildebrand et al. (1978), and Tanaka et al. (1980) have measured gain and luminescence spectra and interpreted them by assuming that the **k**-selection rule for optical transitions had broken down. Their results showed a discrepancy between theory and experiment, with the measured bandgap more than an excitonic binding energy below the theoretical value. However, the problem appears to have been resolved by Capizzi et al. (1984), who have reported the results of low-temperature photoluminescence experiments on Al$_x$Ga$_{1-x}$As for a range of alloy compositions including $x = 0$, viz., GaAs. In this work the samples were in the form of mesa heterostructures matched to the incident laser spot dimension and excitation depth so as to produce a nearly uniform plasma density and temperature. In contrast to the papers just described, Capizzi et al. analysed their results by retaining **k**-conservation whilst introducing a calculated broadening of the one-particle density of states according to the theory of Haug and Tran Thoai (1980). The experimental results for the bandgap change due to the plasma are shown in Fig. 11 for GaAs and Al$_x$Ga$_{1-x}$As. The new method of interpreting the spectra leads to much better agreement with established theory, as demonstrated by the closeness of most of the experimental results to the universal function of Vashista and Kalia (1982) presented in Section II.6.c. The deviation of theory and experiment for the samples with aluminium fractions of 0.40 and 0.42 at the higher carrier concentrations can be explained in terms of the electron occupancy of the X-valleys in addition to the Γ-valley.

A high-density electron–hole plasma is a basic feature of the semiconductor injection laser, and the device provides a convenient vehicle for observing the properties of the plasma and the effect it has on band structure. A recent example of this type of experiment is the work of Tarucha et al. (1984), who have measured the carrier-concentration dependence of the bandgap in multi–quantum-well and conventional

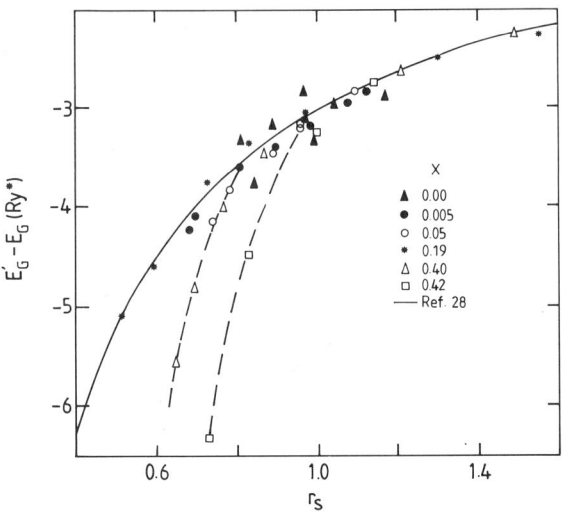

FIG. 11. Bandgap narrowing as a function of r_s for $Al_xGa_{1-x}As$ (for various x) with an electron–hole plasma. The solid curve is the theoretical result derived from the universal function of Vashishta and Kalia (1982) (see Eqs. (28) and (29) of this chapter). The dashed curves are guides to the eye. (From Capizzi *et al.*, 1984, Fig. 5.)

double-heterostructure GaAs/AlGaAs lasers by observing the emission spectrum as a function of injection current. The conventional laser studied had an undoped GaAs active layer, and the values of bandgap narrowing deduced for this device are in general agreement with those in Fig. 11.

It was emphasized in Section II.6.a that the many-body effects have an influence on the whole band structure, and not just on the band or bands occupied by carriers. An experimental demonstration of this in $Al_xGa_{1-x}As$ has been given by Bohnert *et al.* (1988) and Kalt *et al.* (1989). They have carried out time-resolved photoluminescence studies of the alloy for compositions when the bandgap is direct-gap ($x < 0.43$) and indirect-gap ($x > 0.43$). In the indirect samples, optical excitation products an electron–hole plasma with nearly all (99%) of the electrons in the X valleys. Luminescence is seen because of recombination of electrons in the Γ and X valleys and facilitates measurement of both the direct and indirect gaps.

The results show that the two gaps are reduced by different amounts, and this can be explained by many-body theory. Bohnert *et al.* use the theory of Vashista and Kalia (1982) to interpret the results, but here it is convenient to refer to the formalism of Section II.5. For the state at the bottom of an X valley, there will be significant Coulomb hole and screened exchange contributions to the energy shift resulting from the introduction of the optically generated carriers. For the state at the bottom of the Γ valley, there will also be a Coulomb hole energy shift, but the screened exchange contribution will be very small because of the low density of carriers in that valley. Of course, there is also a contribution to the bandgap narrowing from the upward movement of the valence band, but that produces an identical shift in the Γ and X luminescence spectra. An interesting feature of the experiment is that it is able to provide some information on the *separate* contributions to the energy shifts in the conduction band.

17. QUANTUM WELLS

Optical studies of quantum wells have concentrated on modulation-doped systems and the electron–hole plasma in nominally undoped structures. An exception is the work of Harris *et al.* (1990), who have carried out a photoluminescence study of the transition from nondegenerate to degenerate doping in *n*-type GaAs/AlGaAs quantum wells. At the lower doping levels, they find evidence of electrons remaining bound to donors up to room temperature, and for the degenerate samples the spectra indicate relaxation of **k** conservation in the optical transitions. Tränkle *et al.* (1987a, 1987b) have carried out high excitation photoluminescence experiments on GaAs/AlGaAs, InGaAs/InP, and GaSb/AlSb multiple quantum-well structures. Their bandgap narrowing results for GaAs are seen in Fig. 5 to be quite close to the theoretical predictions of Das Sarma *et al.* (1990). However, in contrast to the theory, the experiments do not show any systematic dependence of bandgap narrowing on well width. Bongiovanni and Staehli (1989) have carried out similar experiments but have adopted a similar procedure to Capizzi *et al.* (1984) to interpret the results. They find that the bandgap narrowing *does* become larger as the well width is made smaller. This behaviour is in qualitative agreement with the trend in Fig. 5, although the gap-narrowing values are substantially smaller than the results shown there. Bongiovanni and Staehli have also used a theory based on the plasmon pole approximation to describe their experimental results. They are able to achieve satisfactory agreement between theory and experiment and predict that the bandgap narrowing for a particular well width *expressed in units of exciton binding energy for that well width* will be independent of the well width.

The measurements of Tarucha *et al.* (1984) on multi–quantum-well lasers, which were mentioned earlier, give values that are slightly larger than those of Bongiovanni and Staehli, but again smaller than those of Tränkle et al. However, the rather thin barriers ($\simeq 30$ Å) in the lasers rather complicate direct comparison with the other work. A nonspectroscopic method of measuring bandgap narrowing in quantum-well lasers has been reported by Böttcher et al. (1990). This technique involves measuring the junction potential to obtain the chemical potential, and the small signal carrier lifetime to deduce the carrier density from the device current. Although the method avoids the need to interpret broadened spectra to obtain the change in bandgap, it is necessary to calculate the chemical potential of a noninteracting plasma, and this leads to some uncertainty. The bandgap change deduced for a 70 Å GaAs/AlGaAs well is approximately 30 meV at $n = 7 \times 10^{12}$ cm^{-2}, and therefore somewhat below the values obtained by Tränkle et al. at $n = 10^{12}$ cm^{-2}.

Gain measurements have also been used to determine bandgap narrowing in quantum wells. For example, Weber et al. (1988) have used a pump and probe technique with excitation pulses of several nanoseconds to measure the optical properties of 100 Å GaAs/AlGaAs quantum wells. They present results for the chemical potential and bandgap for a carrier concentration range similar to the photoluminescence experiments discussed earlier. The values of bandgap narrowing obtained are similar to the results of Tränkle et al. for the one well width studied. Weber et al. were also able to observe energy shifts in the excited subbands even when these are essentially unoccupied. This is another example where carriers in one band affect the energy of other bands, as discussed earlier for bulk AlGaAs. Varying subband shifts have also been seen in photoluminescence studies of InGaAs/InP quantum wells by Reinecke et al. (1990).

Pinczuk et al. (1984) have studied n-type modulation-doped GaAs/AlGaAs quantum wells by photoluminescence. Their results have been analysed by Bauer and Ando (1986a), who deduce the bandgap changes *relative to bulk gallium arsenide* and find they agree well with their own theoretical results, as illustrated in Fig. 6. As pointed out in Section III.10, Bauer and Ando (1986b) also report good agreement between their theory and the experimental gap narrowing of 12 meV obtained by Kleinman and Miller (1985) for a p-type ($p = 5.3 \times 10^{10}$ cm^{-3}) modulation-doped 107 Å quantum well. Delande et al. (1986, 1987) have used both illumination and a Schottky gate to vary the carrier concentration in n-type modulation-doped quantum wells, and they measured bandgap narrowing by photoluminescence. In the Schottky gate experiments, the gap narrowing varies from approximately 10 meV at $n = 10^{11}$ cm^{-2} to 18 meV at $n = 6 \times 10^{11}$ cm^{-2}, which is generally in accord with the other results on modulation-doped systems.

VI. Electrical Experiments

Measurements of the electrical characteristics of transistors and related devices have been used extensively to investigate the effects of heavy doping on the fundamental bandgap. Compared to optical experiments, electrical measurements have the virtue of being straightforward to perform, but they do not give such direct or detailed information on the band structure. Not surprisingly, silicon was the first material to have the effects of heavy doping studied by the use of transistor measurements, and it remains the semiconductor most investigated by this approach. A review of the early experiments on silicon transistors and the associated theoretical concepts has been given by Abram *et al.* (1978). More recently, Wagner and del Alamo (1988) have compared the results of electrical and optical studies of silicon. The electronic effects of heavy doping on III–V bipolar transistors are discussed in detail elsewhere in this volume, but it is appropriate to give a brief summary here to place the work in the context of the theory discussed in the previous sections.

Slotboom and de Graaf (1976) were the first to show how measurements of the collector current as a function of emitter-base voltage in silicon *npn* transistors could give information on the bandgap of the heavily doped base. More recently, similar experiments have been carried out on gallium arsenide transistors. For the simplest device with a uniformly doped base in which there is negligible minority-carrier recombination, the collector current density is given by

$$J_n = J_{n0}[\exp(eV_{be}/k_B T) - 1], \quad (63)$$

where V_{be} is the emitter-base voltage and, for base width W_b and electron diffusion coefficient D_{bn},

$$J_{n0} = (eD_{bn}n_b)/W_b = (eD_{bn}n_{ie}^2)/p_b W_b. \quad (64)$$

n_{ie}^2 is the product of the electron and hole densities in the base and for a lightly doped semiconductor is given by

$$n_{ie}^2 = n_{i0}^2 = N_c N_v e^{-E_g/k_B T} \quad (65)$$

in the usual notation. However, for a heavily doped semiconductor in which the bands are assumed to simply move rigidly due to the many-body and impurity effects to produce a change in bandgap ΔE_g,

$$n_{ie}^2 = N_v N_c F_{1/2}(E_F/k_B T) \exp[-(E_g + \Delta E_g + E_F)/k_B T]$$
$$= n_{i0}^2 \exp\{-(\Delta E_g + E_F - k_B T \ln[F_{1/2}(E_F/k_B T)])/k_B T\}, \quad (66)$$

where E_F is the depth of the Fermi level below the valence band edge. ΔE_g is negative, but the other energy terms in the exponential, which derive from the degeneracy of the holes, provide a positive contribution (Abram et al., 1978). In the nondegenerate limit, only ΔE_g remains, and n_{ie}^2 provides a *direct* measure of bandgap change. With degeneracy, an *effective* bandgap change can be defined by

$$\Delta E_{g\,\text{eff}} = \Delta E_g + E_F - k_B T \ln[F_{1/2}(E_F/k_B T)]. \tag{67}$$

The effective band gap narrowing is less than the *actual* narrowing because of the effects of degeneracy. As Klausmeier-Brown et al. (1989) have observed, this difference has some similarity with the Moss–Burstein shift in the optical absorption spectrum. Of course, in general the bands do not move rigidly, and the densities of states near the band edges are in fact modified. This has the effect of introducing an additional contribution to the effective bandgap change which can be formally represented by an extra term on the right-hand side of Eq. (67).

Klausmeier-Brown et al. (1989, 1990), and Harmon et al. (1991a, 1991b) have reported measurements of collector current versus emitter-base bias in gallium arsenide homojunction transistors to study bandgap narrowing in n- and p-type material. The same group has also carried out measurements of electron injection in p^+n diodes which can be interpreted in a similar fashion (Klausmeier-Brown et al., 1988, 1989). Harmon et al. (1991a) have studied *npn* transistors with heavily doped bases and have deduced $D_{bn}n_{ie}^2$ from Eq. (65) by using the current measurements of J_{n0} and Hall measurements of $p_b W_b$. They find that, with increasing base doping, $D_{bn}n_{ie}^2$ first rises because of bandgap narrowing in the base and then falls as a result of degeneracy. By taking a reasonable value for D_{bn}, the authors find the results are quantitatively described by the bandgap narrowing formula of Jain et al. (1990), although the experimental values are a little higher. A fit of a $p^{1/3}$ law to the bandgap narrowing versus hole concentration gives

$$\Delta E_g = -2.85 \times 10^{-8} p^{1/3}, \tag{68}$$

in which the prefactor of 2.85 is rather larger than the 2.2 obtained from the optical experiments discussed in Section V.16.

Harmon et al. (1991b) report the results of experiments on both *npn* and *pnp* transistors to obtain $D_{bn}n_{ie}^2$ for p-type and $D_{bp}n_{ie}^2$ for n-type material. The results for n-type gallium arsenide show a monotonic fall in $D_{bp}n_{ie}^2$ for $n > 10^{17}$ cm^{-3} which the authors ascribe to the earlier onset of degeneracy in n-type material and a decrease of D_{bp} with doping. The theoretical bandgap-narrowing results of Bennett and Lowney (1987) and Jain *et al.* (1992) give close agreement with the experiments, whereas the results of Jain *et al.* (1990) predict that $D_{bp}n_{ie}^2$ increases rather than decreases with doping. This is further evidence that the second-order perturbation theory treatment of impurity scattering overestimates the gap narrowing in n-type gallium arsenide.

Heterojunction bipolar transistors are the subject of intense interest because of their advantageous high-frequency properties. These devices have a heavily doped base and at first sight would appear to provide a convenient means of studying bandgap narrowing. There has been some work using these transistors (see, for example, the measurement of effective bandgap narrowing in p-type gallium arsenide by Tiwari and Wright, 1990) but Klausmeier-Brown *et al.* (1989) have pointed out that there are uncertainties over the properties of the heterojunction which make the homojunction transistor the more favoured device for the studies considered here.

The extraction of bandgap narrowing values from the transistor experiments requires a value for the minority-carrier diffusion coefficient, of which there is some uncertainty, as well as descriptions of majority-carrier degeneracy and the modified densities of states at the band edges. The perturbations of the band edges are normally neglected, with some justification in the majority-carrier band where tail states are a small fraction of the total occupied. However, a high proportion of the minority carriers are expected to exist in the tail states, and a correct description of these would seem to be important. Despite these difficulties, transistor measurements have made and continue to make a useful contribution to the study of bandgap narrowing.

In an effort to avoid some of the difficulties in experimental interpretation described earlier, Van Mieghem *et al.* (1990) have attempted a different kind of electrical measurement to determine bandgap narrowing. The method was first proposed by Van Overstraeten *et al.* (1973) and is based on a capacitance measurement of an abrupt symmetrically doped diode. The inverse capacitance squared versus applied voltage plot for an abrupt symmetrical $p-n$ junction has an intercept on the voltage axis which is a function of the semiconductor band gap and the doping concentration. Van Mieghem *et al.* show how capacitance measurements on a gallium arsenide diode can be analysed to obtain the *average* of the bandgap narrowing in the n- and p-type materials. For samples with doping up to 3×10^{18} cm^{-3}, the average band gap narrowing dependence on doping density obeys the $\frac{1}{3}$ power law, and its magnitude is in agreement with the optical results of the same group (Borghs *et al.*, 1989) reported in Section V.16.

A virtue of the capacitance technique is that the capacitance is essentially determined by the response of the majority carriers, whose behaviour is more easily described than that of the minority carriers, but the method is not without its problems. The n- and p-type bandgap changes are not determined separately and, as pointed out by Lowney (1985, 1992), the result is not representative of the neutral bulk. This latter observation follows from the fact that the apparent bandgap narrowing obtained from capacitance measurements is determined by the conditions at the edges of the depletion region where the carrier density varies rapidly. In the depletion region, the carrier concentrations are smaller than in the bulk and, as a result, the free carrier screening is reduced. Hence, there is a higher impurity contribution and a lower many-body contribution to the bandgap narrowing. In general, the bandgap narrowing is different from that in the bulk, although accidental near-cancellation of the two effects can occur. Lowney and Bennett (1989) have also discussed the influence of rapid spatial variations of doping or carrier concentration on bandgap narrowing.

Acknowledgments

It is a pleasure to acknowledge the contributions of G. N. Childs, J. Gilman, D. T. Hughes, J. C. Inkson, S. C. Jain, R. W. Kelsall, P. T. Landsberg, J. R. Lowney, N. F. Mott, G. J. Rees, D. J. Robbins, P. A. Saunderson, and B. L. H. Wilson, with whom fruitful collaboration or discussion has taken place on various aspects of semiconductors relevant to this article. I would also like to thank those colleagues who have communicated results prior to publication and the authors and publishers who have allowed the reproduction of figures from the literature.

References

Abram, R. A., Rees, G. J., and Wilson, B. L. H. (1978). *Adv. Phys.* **27,** 799.
Abram, R. A., Childs, G. N., and Saunderson, P. A. (1984). *J. Phys. C: Solid State Phys.* **17,** 6105.
Ando, T. (1976). *Surf. Sci.* **68,** 128.
Ando, T., Fowler, A. B., and Stern, F. (1982). *Rev. Mod. Phys.* **54,** 437.
Asada, M., Ama, A. K., and Suematsu, Y. (1984). *IEEE J. Quant. Electron.* **20,** 745.
Balslev, I., and Stahl, A. (1986). *Solid State Commun.* **59,** 371.
Bányai, L., and Koch, S. W. (1986). *Z. Phys. B—Condensed Matter* **63,** 283.

Bauer, G. E. W., and Ando, T. (1985). *Phys. Rev. B* **31**, 8321.
Bauer, G. E. W., and Ando, T. (1986a). *J. Phys. C: Solid State Phys.* **19**, 1537.
Bauer, G. E. W., and Ando, T. (1986b). *Phys. Rev. B* **34**, 1300.
Bardyszewski, W. (1986). *Solid State Commun.* **57**, 873.
Bardyszewski, W., and Yevick, D. (1987). *Phys. Rev. B* **35**, 619.
Beni, G., and Rice, T. M. (1978). *Phys. Rev. B* **18**, 768.
Bennett, H. S. (1986). *J. Appl. Phys.* **60**, 2866.
Bennett, H. S., and Lowney, J. R. (1987). *J. Appl. Phys.* **62**, 521.
Berggren, K. F., and Sernelius, B. E. (1981). *Phys. Rev. B* **24**, 1971.
Berggren, K. F., and Sernelius, B. E. (1984). *Phys. Rev. B* **29**, 5575.
Bonch-Bruevich, V. L. (1966a). *Semiconds. Semimetals* **1**, 101.
Bonch-Bruevich, V. L. (1966b). *The Electronic Theory of Heavily Doped Semiconductors*. Elsevier, Amsterdam.
Bohnert, K., Kalt, H., Smirl, A. L., Norwood, D. P., Boggess, T. F., and D'Haenens, I. J. (1988). *Phys. Rev. Lett.* **60**, 37.
Bongiovanni, G., and Staehli, J. L. (1989). *Phys. Rev. B* **39**, 8359.
Borghs, G., Bhattacharyya, K., Deneffe, K., Van Mieghem, P., and Mertens, R. (1989). *J. Appl. Phys.* **66**, 4381.
Böttcher, E. H., Kirstaedter, N., Grundmann, M., Bimberg, D., Harder, C., and Meier, H. (1990). In *Proceedings of the 20th International Conference on the Physics of Semiconductors, Thessaloniki, Greece, 1990* (E. M. Anastassakis and J. D. Joannopoulos, eds.), p. 371. World Scientific, Singapore.
Brinkman, W. F., and Lee, P. A. (1973). *Phys. Rev. Lett.* **31**, 237.
Brinkmann, W. F., and Rice, T. M. (1973). *Phys. Rev. B* **7**, 1508.
Brinkman, W. F., Rice, T. M., Anderson, P. W., and Chui, S. T. (1972). *Phys. Rev. Lett.* **28**, 961.
Burstein, E. (1954). *Phys. Rev.* **93**, 632.
Capizzi, M., Modesti, S., Frova, A., Staehli, J. L., Guzzi, M., and Logan, R. A. (1984). *Phys. Rev. B* **29**, 2028.
Casey, H. C., and Stern, F. (1976). *J. Appl. Phys.* **47**, 631.
Casey, H. C., Sell, D. D., and Wecht, K. W. (1975). *J. Appl. Phys.* **46**, 250.
Childs, G. N. (1987). Ph.D. Thesis, University of Durham, Durham, U.K.
Childs, G. N., and Abram, R. A. (1989). *Solar Energy Materials* **18**, 399.
Combescot, M., and Nozières, P. (1972). *J. Phys. C: Solid State Phys.* **5**, 2369.
Das Sarma, S., Jalabert, R., and Yang, S. R. E. (1989). *Phys. Rev. B* **39**, 5516.
Das Sarma, S., Jalabert, R., and Yang, S. R. E. (1990). *Phys. Rev. B* **41**, 8288.
Delande, C., Orgonasi, J., Meynadier, M. H., Brum, J. A., Bastard, G., Weimann, G., and Schlapp, W. (1986). *Solid State Commun.* **59**, 613.
Delande, C., Bastard, G., Orgonasi, J., Brum, J. A., Lui, H. W., Voos, M., Weimann, G., and Schlapp, W. (1987). *Phys. Rev. Lett.* **59**, 2690.
Economou, E. N. (1983). *Green's Functions in Quantum Physics*. Springer-Verlag, Berlin.
Ekenberg, U. (1984). *Phys. Rev. B* **30**, 3367.
Gaspard, J. P., and Cyrot-Lackmann, F. (1973). *J. Phys. C: Solid State Phys.* **6**, 3077.
Gaspard, J. P., and Cyrot-Lackmann, F. (1974). *J. Phys. C: Solid State Phys.* **7**, 1829.
Gell-Mann, M., and Brueckner, K. (1957). *Phys. Rev.* **106**, 364.
Ghazali, A., and Serre, J. (1982). *Phys. Rev. Lett.* **48**, 886.
Göbel, E. O., Herzog, H., Pilkuhn, M. H., and Zschauer, K. H. (1973). *Solid State Commun.* **13**, 719.
Gold, A., Serre, J., and Ghazali, A. (1988). *Phys. Rev. B* **37**, 4589.
Gunnarson, O., and Lundqvist, B. I. (1976). *Phys. Rev. B* **13**, 4274.

Haas, C. (1962). *Phys. Rev.* **125**, 1965.
Halperin, B. I., and Lax, M. (1966). *Phys. Rev.* **148**, 722.
Harmon, E. S., Melloch, M. R., Lundstrom, M. S., and Lovejoy, M. L. (1991a). *Appl. Phys. Lett.* **58**, 1647.
Harmon, E. S., Patkar, M. P., Lovejoy, M. L., Lundstrom, M. S., and Melloch, M. R. (1991b). Paper presented at the Electronic Materials Conference, Boulder, Colorado, 1991.
Harris, C., Monemar, B., Kalt, H., Köhler, K., and Schweizer, T. (1990). In *Proceedings of the 20th International Conference on the Physics of Semiconductors, Thessaloniki, Greece, 1990* (E. M. Anastassakis and J. D. Joannopoulos, eds.), p. 1377. World Scientific, Singapore.
Haufe, A., and Unger, K. (1985). unpublished paper presented at the EPS Condensed Matter Division Conference, West Berlin, 1985.
Haug, H., and Schmitt-Rink, S. (1984). *Prog. Quant. Electr.* **9**, 3.
Haug, H., and Tran Thoai, D. B. (1980). *Phys. Stat. Sol.* (*b*) **98**, 581.
Hedin, L. (1965). *Phys. Rev.* **139**, A796.
Hedin, L., and Lundqvist, S. (1969). *Solid State Physics* **23**, 1.
Hess, K. (1979). *Appl. Phys. Lett.* **35**, 484.
Hensel, J. C., Phillips, T. G., and Thomas, G. A. (1977). *Solid State Physics* **32**, 88.
Hildebrand, O., Göbel, E. O., Romanek, K. M., Weber, H., and Mahler, G. (1978). *Phys. Rev. B* **17**, 4775.
Hubbard, J. (1957). *Proc. Roy. Soc. A* **240**, 539.
Hubbard, J. (1964). *Proc. Roy. Soc. A* **277**, 237.
Hwang, C. J. (1970). *J. Appl. Phys.* **41**, 2668.
Inkson, J. C. (1976). *J. Phys. C: Solid State Phys.* **9**, 1177.
Inkson, J. C. (1984). *Many-Body Theory of Solids*. Plenum, New York.
Jain, S. C., and Roulston, D. J. (1991). *Solid State Electron.* **34**, 453.
Jain, S. C., McGregor, J. M., and Roulston, D. J. (1990). *J. Appl. Phys.* **68**, 3747.
Jain, S. C., McGregor, J. M., Roulston, D. J., and Balk, P. (1992). *Solid State Electron.* **35**, 639.
Jeffries, C. D., and Keldysh, L. V. eds (1983). *Electron–Hole Droplets in Semiconductors*. North-Holland, Amsterdam.
Kalt, H., Bohnert, K., Smirl, A. L., and Boggess, T. F. (1989). In *Proceedings of the 19th International Conference on the Physics of Semiconductors, Warsaw 1988* (W. Zawadzki, ed.). Polish Academy of Sciences, Warsaw.
Kane, E. O. (1963). *Phys. Rev.* **131**, 79.
Kawamoto, G., Kalia, R., and Quinn, J. J. (1980). *Surf. Sci.* **98**, 589.
Keldysh, L. V., and Silin, A. P. (1976). *Sov. Phys. JETP* **42**, 535.
Klauder, J. R. (1961). *Ann. Phys.* **14**, 43.
Klausmeier-Brown, M. E., Lundstrom, M. S., Melloch, M. R., and Tobin, P. (1988). *Appl. Phys. Lett.* **52**, 2255.
Klausmeier-Brown, M. E., Lundstrom, M. S., and Melloch, M. R. (1989). *IEEE Trans. Electron Devices* **36**, 2146.
Klausmeier-Brown, M. E., Melloch, M. R., and Lundstrom, M. S. (1990). *Appl. Phys. Lett.* **56**, 160.
Kleinman, D. A. (1983). *Phys. Rev. B* **28**, 871.
Kleinman, D. A., and Miller, R. C. (1985). *Phys. Rev. B* **32**, 2266.
Kohn, W., and Luttinger, J. M. (1957). *Phys. Rev.* **108**, 590.
Kohn, W., and Sham, L. J. (1965). *Phys. Rev.* **140**, A1133.
Kucharska, A. I., and Robbins, D. J. (1990). *IEEE J. Quant. Electron.* **26**, 443.

Landsberg, P. T. (1949). *Proc. Phys. Soc.* **A62**, 806.
Landsberg, P. T. (1966). *Phys. Stat. Sol.* **15**, 623.
Landsberg, P. T., and Robbins, D. J. (1985). *Solid State Electron.* **28**, 137.
Landsberg, P. T., Neugroschel, A., Lindholm, F. A., and Sah, C. T. (1985). *Phys. Stat. Sol.* (b) **130**, 255.
Lanyon, H. P. D., and Tuft, R. A. (1979). *IEEE Trans. Electron Devices* **26**, 1014.
Lindhard, J. (1954). *Dan. Math. Phys. Medd.* **28**, No. 8.
Lloyd, P., and Best, P. R. (1975). *J. Phys. C: Solid State Phys.* **8**, 3752.
Lowney, J. R. (1985). *Solid State Electron.* **28**, 187.
Lowney, J. R. (1986a). *J. Appl. Phys.* **60**, 2854.
Lowney, J. R. (1986b). *J. Appl. Phys.* **59**, 2048.
Lowney, J. R. (1989). *J. Appl. Phys.* **66**, 4279.
Lowney, J. R. (1992). Private communication.
Lowney, J. R., and Bennett, H. S. (1989). *J. Appl. Phys.* **65**, 4823.
Lundqvist, S. (1967a). *Phys. Kondens. Mater.* **6**, 193.
Lundqvist, S. (1967b). *Phys. Kondens. Mater.* **6**, 206.
Mahan, G. D. (1980). *J. Appl. Phys.* **51**, 2634.
Mahan, G. D. (1981). *Many-Particle Physics.* Plenum, New York.
Matsubara, T., and Toyozawa, Y. (1961). *Prog. Theor. Phys.* **26**, 739.
Mattuck, R. D. (1976). *A Guide to Feynman Diagrams in the Many-Body Problem.* McGraw-Hill, New York.
Moss, T. S. (1954). *Proc. Phys. Soc. B* **67**, 775.
Mott, N. F. (1974). *Metal–Insulator Transitions.* Taylor and Francis, London.
Nozières, P., and Pines, D. (1958). *Phys. Rev.* **109**, 1009.
Olego, D., and Cardona, M. (1980). *Phys. Rev. B* **22**, 886.
Overhauser, A. W. (1971). *Phys. Rev. B* **3**, 1888.
Parmenter, R. H. (1955). *Phys. Rev.* **97**, 587.
Pinczuk, A., Shah, J., Störmer, H. L., Miller, R. C., Gossard, A. C., and Wiegmann, W. (1984). *Surf. Sci.* **142**, 492.
Quinn, J. J., and Ferrell, R. A. (1958). *Phys. Rev.* **112**, 812.
Reinecke, T. L., Broido, D. A., Lach, E., Forchel, A., Weimann, G., Schlapp, W., and Gruetzmacher, D. (1990). In *Proceedings of the 20th International Conference on the Physics of Semiconductors, Thessaloniki, Greece, 1990* (E. M. Anastassakis and J. D. Joannopoulos, eds.) p. 1113. World Scientific, Singapore.
Rice, T. M. (1974). *Nuovo Cim.* **23B**, 226.
Rice, T. M. (1977). *Solid State Physics* **32**, 1.
Rorison, J., Kane, M. J., Herbert, D. C., Skolnick, M. S., Taylor, L. L., and Bass, S. J. (1988). *Semicond. Sci. Technol.* **3**, 12.
Rösler, M., and Zimmermann, R. (1975). *Phys. Stat. Sol.* (b) **67**, 525.
Saito, K., Yamada, T., Akatsuka, T., Fukamachi, T., Tokimitsu, E., Konagai, M., and Takahashi, K. (1989). *Jap. J. Appl. Phys.* **28**, L2081.
Sa-yakanit, V., and Glyde, H. R. (1980). *Phys. Rev. B* **22**, 6222.
Schwabe, R., Haufe, A., Gottschalch, V. and Unger, K. (1986). *Solid State Commun.* **58**, 485.
Seitz, F. (1940). *Modern Theory of Solids.* McGraw-Hill, New York.
Selloni, A., Modesti, S., and Capizzi, M. (1984). *Phys. Rev. B* **30**, 821.
Sernelius, B. E. (1986a). *Phys. Rev. B* **33**, 8582.
Sernelius, B. E. (1986b). *Phys. Rev. B* **34**, 5610.
Serre, J., and Ghazali, A. (1983). *Phys. Rev. B* **28**, 4704.
Slotboom, J. W., and de Graaf, H. C. (1976). *Solid State Electron.* **19**, 857.

Smith, C. (1985). Ph.D. Thesis, University of Durham, Durham, U.K.
Soukoulis, C. M., Cohen, M. H., and Economou, E. N. (1984). *Phys. Rev. Lett.* **53,** 616.
Stern, F. (1967). *Phys. Rev. Lett.* **18,** 546.
Stern, F., and Howard, W. E. (1967). *Phys. Rev.* **163,** 816.
Takeshima, M. (1986a). *Phys. Rev. B* **33,** 4054.
Takeshima, M. (1986b). *Phys. Rev. B* **34,** 1041.
Tanaka, S., Kobayashi, H., Saito, S., and Shionoya, S. (1980). *J. Phys. Soc. Jap.* **49,** 1051.
Tarucha, S., Kobayashi, H., Horikoshi, Y., and Okamoto, H. (1984). *Jap. J. Appl. Phys.* **23,** 874.
Thuselt, F. (1983). *Phys. Lett.* **94A,** 93.
Titkov, A. N., Chaĭkina, E. I., Komova, É. M., and Ermakova, N. G. (1981). *Sov. Phys. Semicond.* **15,** 198.
Tiwari, S., and Wright, S. L. (1990). *Appl. Phys. Lett.* **56,** 563.
Tränkle, G., Leier, H., Forchel, A., Haug, H., Ell, C., and Weimann, G. (1987a). *Phys. Rev. Lett.* **58,** 419.
Tränkle, G., Lach, E., Forchel, A., Scholz, F., Ell, C., Haug, H., Weimann, G., Griffiths, G., Kroemer, H., and Subbanna, S. (1987b). *Phys. Rev. B* **36,** 6712.
Van Mieghem, P., Mertens, R. P., Borghs, G., and Van Overstraeten, R. J. (1990). *Phys. Rev. B* **41,** 5952.
Van Overstraeten, R. J., de Man, H. J., and Mertens, R. P. (1973). *IEEE Trans. Electron Devices* **20,** 290.
Vashista, P., and Kalia, R. K. (1982). *Phys. Rev. B* **25,** 6492.
Vashista, P., Bhattacharyya, P., and Singwi, K. S. (1973). *Phys. Rev. Lett.* **30,** 1248.
Vashishta, P., Das, S. G., and Singwi, K. S. (1974). *Phys. Rev. Lett.* **33,** 911.
Vinter, B. (1976). *Phys. Rev. B* **13,** 4447.
Vinter, B. (1977). *Phys. Rev. B* **15,** 3947.
Vinter, B., and Stern, F. (1976). *Surf. Sci.* **58,** 141.
Wagner, J. (1985). *Solid State Electron.* **28,** 25.
Wagner, J., and del Alamo, J. A. (1988). *J. Appl. Phys.* **63,** 425.
Weber, C., Klingshirn, C., Chemla, D. S., Miller, D. A. B., Cunningham, J. E., and Ell, C. (1988). *Phys. Rev. B* **38,** 12748.
Wolff, P. A. (1962). *Phys. Rev.* **126,** 405.
Yao, H., and Compaan, A. (1990). *Appl. Phys. Lett.* **57,** 147.
Yonezawa, F. (1964). *Prog. Theor. Phys.* **30,** 357.

CHAPTER 6

An Introduction to Nonequilibrium Many-Body Analyses of Optical and Electronic Processes in III–V Semiconductors

David Yevick*
Witold Bardyszewski†

DEPARTMENT OF ELECTRICAL ENGINEERING
QUEEN'S UNIVERSITY
KINGSTON, ONTARIO, CANADA

I.	INTRODUCTION	318
II.	$k \cdot p$ BAND STRUCTURE	319
III.	SECOND QUANTIZATION	322
	1. *Long-Range Electron–Phonon Potential*	324
	2. *Short-Range Electron–Phonon Potential*	325
	3. *Electron–Photon Interaction*	328
IV.	ENSEMBLE PROPERTIES	329
	4. *Wigner Distribution Function*	329
	5. *Nonequilibrium Green's Functions*	333
V.	ELECTRON SELF-ENERGY	343
	6. *Imaginary Part of the Self-Energy and Inelastic Losses*	343
	7. *Real Part of the Self-Energy and Bandgap Renormalization*	348
VI.	DIELECTRIC FUNCTION MODELS	350
VII.	INTRABAND PROCESSES AND TRANSPORT	358
	8. *Hot Electron Transport*	358
	9. *Low-Field Electron Mobility*	360
VIII.	INTERBAND PROCESSES	364
	10. *Auger Processes*	364
	11. *Gain and Stimulated Recombination*	370
IX.	CONCLUSIONS	380
	APPENDIX A	380
	APPENDIX B	385
	APPENDIX C	385
	REFERENCES	386

* Supported by the National Sciences and Research Council of Canada, Bell Northern Research, Corning Glass, the Ontario Centre for Materials Research and the Ontario Laser and Lightwave Research Center.

† Permanent address: Department of Theoretical Physics, Warsaw University, Hoza 69, 00 681 Warsaw, Poland.

I. Introduction

This chapter is intended as an overview of nonequilibrium quantum-mechanical transport theory for experimental physicists and engineers with a background in advanced quantum mechanics. We examine both optical and transport processes of semiconductor materials and devices. Although these were originally considered as separate fields, we will present a unified description of both sets of processes that is particularly suited to the study of general optoelectronic devices. Despite the widespread use of simple one-particle models, it should be realized that electrons and holes in a semiconductor are in fact quasi-particle excitations with energies and lifetimes that are determined by many-body effects. Hence, the manner in which a full description of the carrier system can be reduced to a one-particle model for both equilibrium and nonequilibrium configurations, as well as the corrections to such a model, is of great importance to future device analysis. We will therefore concentrate on describing important relaxation effects in semiconductor components and devices within an approximate but mathematically justifiable many-body theory framework. Indeed, while present *ab initio* treatments of the full electronic structure of solids only yield the correct properties of the ground state, semi-empirical descriptions of electron bands in most undoped or slightly doped materials correctly reflect the results of numerous experimental measurements and phenomenological theories. Accordingly, to justify or improve the predictions of simple models for optical and transport properties of practical semiconductor optoelectronic devices at high doping or excitation levels, we here attempt to describe free carriers in quasi-equilibrium under conditions of high doping or excitation. Since the number of free carriers remains far smaller than the total number of electrons in the solid, we limit our discussion to the valence and conduction electrons, which are treated as quasi-particles. We then examine the details of the interactions of the quasi-particles, such as collective excitations and screening, from both a simplified effective mass picture and a more detailed analysis including the properties of the Bloch electron wavefunctions and the valence band structure. A particularly appealing technique for analyzing such interactions at finite temperature is provided by the nonequilibrium Green's function procedure introduced in the 1960s by Kadanoff and Baym (Kadanoff and Baym, 1962; Baym and Kadanoff, 1961), Keldysh (1965), Craig (1968), and others. This technique is not only conceptually simpler than the standard Matsubara formalism (Fetter and Walecka, 1971) when applied to equilibrium finite-temperature calculations, but also has additional practical advantages that lead to significantly reduced theoreti-

6. INTRODUCTION TO NONEQUILIBRIUM MANY-BODY ANALYSES 319

cal and calculational effort. We will accordingly here seek to provide the essential information necessary to understand and, if required, perform calculations of optical and transport processes within this framework. We will attempt to present the underlying physical concepts of many-body theory on the level of basic quantum mechanics, referring the reader to the literature for a more rigorous mathematical treatment.

Our chapter is organized as follows. We first summarize the $k \cdot p$ band-structure method, which yields simple algebraic formulas for the band structure in the vicinity of the Γ point. These expressions permit analytic or numerical evaluation of the multi-dimensional phase-space integrals that arise in many-body calculations. Next, we introduce the electron, phonon, and photon wave functions in second-quantized form. The $k \cdot p$ hamiltonian, H', describing the interactions among these fields is introduced and its physical content summarized. We then discuss the quantum-mechanical generalization of the particle density distribution functions, namely the finite-temperature nonequilibrium Green's functions. After deriving appropriate equations of motions, we demonstrate that the standard Boltzmann formalism can be recovered by application of the gradient expansion. The quantum analogue of the classical collision term incorporates the effects of screening, which we describe by a finite-temperature RPA dielectric function. Subsequently, we adapt the formalism to minority carrier transport, carrier-induced energy bandshifts, minority-carrier lifetimes, Auger recombination rates, and gain spectra and conclude with a discussion of the relative accuracy of many-body and one-particle theories in device contexts.

II. $k \cdot p$ Band Structure

Because of the great complexity of many-body analyses, only a small subset of existing band-structure models can in practice be readily employed. Fortunately, in most electronic and optoelectronic devices, the principal contributions to physically interesting quantities such as the gain spectra, carrier lifetimes, and bandgap shifts are generated by carriers with energies close to those of the band extrema. Therefore, in direct bandgap semiconductors the $E(k)$ dispersion relations are only required near the Γ point, where the computationally efficient $k \cdot p$ band structure formalism is highly accurate. The $k \cdot p$ model proceeds from the known symmetries of the conduction and valence band wave functions at $k = 0$. The energies of the conduction and valence band edges in the absence of the spin–orbit interaction are assigned values E_c and E_v, while the potential acting on the electrons is the sum of an periodic effective crystal potential, $V(\vec{r})$, and the spin–orbit interaction, $(\hbar/4m_0^2c^2)[\vec{\nabla} V \times (-i\hbar\vec{\nabla})]\cdot\vec{\sigma}$. The symbols $\vec{\sigma}$, m_0, and c denote the spin operator, the unrenormalized electron mass, and the speed of light in free space, respectively. Inserting the Bloch representation, $\phi_{n,\vec{k}}(\vec{r}) = \frac{1}{\sqrt{\Omega}}e^{i\vec{k}\cdot\vec{r}}u_{n\vec{k}}(\vec{r})$, in which $u_{n,\vec{k}}(\vec{r})$ are periodic with respect to the lattice spacing, for the electronic wave function into the resulting Schrödinger equation and setting $E'_{n\vec{k}} = E_{n\vec{k}} - \hbar^2k^2/2m_0$ yields (Kane, 1957; 1966; Bir and Pikus, 1974)

$$\left\{\frac{p^2}{2m_0} + V(\vec{r}) + \frac{\hbar}{m_0}\vec{k}\cdot\vec{p} + \frac{\hbar}{4m_0^2c^2}[\vec{\nabla} V \times \vec{p}]\cdot\vec{\sigma} \right.$$
$$\left. + \frac{\hbar^2}{4m_0^2c^2}[\vec{\nabla} V \times \vec{k}]\cdot\vec{\sigma}\right\}u_{n\vec{k}}(r) = E'_{n\vec{k}}u_{n\vec{k}}(r). \quad (1)$$

In the following equations, we will often suppress the band index n in our notation $E_{n\vec{k}}$ for the energy of a state of wavevector $\vec{k}$ and will also where convenient set the Planck constant, $\hbar$, and the system volume, Ω, equal to unity. We now observe that for small lattice wavevectors, $\vec{k}$, the electronic wave functions, $u_{n,\vec{k}}(r)$, are approximately unchanged from their $\vec{k} = 0$ form, and we can apply standard time-independent perturbation theory to express these as a linear superposition of the solutions $u_{n,\vec{k}=0}(r)$ to the unperturbed equation

$$\left\{\frac{p^2}{2m_0} + V\right\}u_{n,0}(r) = E'_{n,0}u_{n,0}(\vec{r}). \quad (2)$$

Since the change in a state wave function generated by an admixture of a second state through the perturbing potential varies inversely with the energy difference, if we are interested in the dispersion of a given band we can to a good approximation restrict our basis set to a small number of closely lying energy levels. In a typical III–V compound, we therefore consider only the conduction, light- and heavy-hole, and spin–orbit split-off bands, the latter three of which are degenerate in the absence of spin–orbit coupling. We denote the spin × space solutions of Eq. (2) corresponding to the s-symmetric conduction band by $S\uparrow$ and $S\downarrow$, while in terms of the linear combinations of space wave functions $R_\pm = (1/\sqrt{2})(X \pm iY)$, the p-symmetric valence band wave functions are instead labeled $R_\pm\uparrow$, $R_\pm\downarrow$, $Z\uparrow$, and $Z\downarrow$. Assuming first for simplicity that $\vec{k}$ is directed along the z direction and applying the symmetries of the $k\cdot p$ and the spin–orbit terms, we obtain immediately the following matrix representation of the Hamiltonian operator in terms of the basis functions $\phi_1 = iS\uparrow$, $\phi_2 = -R_+\downarrow$, $\phi_3 = Z\uparrow$ and $\phi_4 = R_-\downarrow$ and their Kramers (space × spin) conjugates $\phi_5, \ldots \phi_8$:

$$\mathcal{H} = \begin{pmatrix} H & 0 \\ 0 & H \end{pmatrix}, \tag{3}$$

where

$$H = \begin{pmatrix} E_c & 0 & kP & 0 \\ 0 & E_v - \dfrac{\Delta}{3} & \dfrac{\sqrt{2}\Delta}{3} & 0 \\ kP & \dfrac{\sqrt{2}\Delta}{3} & E_v & 0 \\ 0 & 0 & 0 & E_v + \dfrac{\Delta}{3} \end{pmatrix}, \tag{4}$$

$$P = -i\frac{\hbar}{m_0}\langle S|p_z|Z\rangle, \tag{5}$$

and

$$\Delta = i\frac{3\hbar}{4m_0^2 c^2}\left\langle X \left| \frac{\partial V}{\partial x}p_y - \frac{\partial V}{\partial y}p_x \right| Y \right\rangle. \tag{6}$$

We next express the dispersion relations and wave functions for nonzero k in terms of the parameters E_c, E_v, Δ, and P by solving the eigenvalue problem $\mathcal{H}Y = EY$. If the bandgap at $k = 0$ is represented by E_g, the spin–orbit split-off, light-hole, and conduction band energies are given by the three solutions of the characteristic equation

$$E'(E' - E_g)(E' + \Delta) - k^2 P^2 \left(E' + \frac{2\Delta}{3}\right) = 0. \tag{7}$$

The energy-wavevector dispersion relations represented by Eq. (7) may be approximated in the neighborhood of $k = 0$ by quadratic expressions. The spin–orbit split-off band is then

$$E_{so}(k) = -\Delta + \frac{\hbar^2 k^2}{2m_0} - \frac{P^2 k^2}{3(E_g + \Delta)}, \tag{8}$$

while for the conduction band

$$E_c = E_g + \frac{\hbar^2 k^2}{2m_0} + \frac{P^2 k^2}{3}\left(\frac{2}{E_g} + \frac{1}{E_g + \Delta}\right). \tag{9}$$

Since the wave functions are in general not known, Δ and P are determined from the measured spin–orbit split-off band-edge energy and the light hole and spin–orbit split-off masses. Further, the heavy-hole band-edge effective mass, which is positive unless additional bands are included in the theory, is equated to its experimental value. The associated eigenvectors of H for $\vec{k} = k\hat{e}_z$ are given by $R_- \downarrow$ and $R_+ \uparrow$ for the heavy-hole valence band and

$$a_n(iS)\downarrow + b_n R_- \uparrow + c_n Z \downarrow \tag{10}$$

and

$$a_n(iS)\uparrow - b_n R_+ \downarrow + c_n Z \uparrow \tag{11}$$

for the remaining bands. The band label $n \in \{c, l, s\}$ determines which solution E'_n is inserted into the expressions below for the constants a_n, b_n, and c_n:

$$a_n = kP(E'_n + 2\Delta/3)\frac{1}{N},$$

$$b_n = \frac{\sqrt{2}\Delta}{3}(E'_n - E_g)\frac{1}{N}, \tag{12}$$

$$c_n = (E'_n - E_g)(E'_n + 2\Delta/3)\frac{1}{N}.$$

The normalizing constant N is further chosen so that $a^2 + b^2 + c^2 = 1$.

III. Second Quantization

While the previous section concerned single-particle energy states, we here consider the behavior of interacting electrons, which is most conveniently modeled in a second-quantized formalism. Our reference system is the vacuum configuration in which the valence and conduction band states are filled and empty, respectively. The creation of an additional quasi-particle at a specified position $\vec{r}$ is then represented by the application of an operator $\Phi^\dagger(\vec{r})$ to this vacuum state. (Bir and Pikus, 1974, p. 293). In analogy with the Fourier decomposition of a delta function, an electron localized at a given point is described by a sum over the complete set of normalized electronic wavefunctions satisfying the given boundary conditions. Therefore, if we define $u_{n\vec{k}}(\vec{r})$ as the periodic part of the Bloch wave function for a particle in band n and introduce an annihilation operator $a_{n,\vec{k}}$ that, acting on a state containing a single electron in momentum $\vec{k}$ in band n, yields the vacuum state, we find that the field operator that annihilates a particle at position $\vec{r}$ is

$$\Phi(\vec{r}) = \frac{1}{\sqrt{\Omega}} \sum_{n,\vec{k}} a_{n,\vec{k}} e^{i\vec{k}\cdot\vec{r}} u_{n,\vec{k}}(\vec{r}). \tag{13}$$

The sum over $\vec{k}$ is restricted to the Brillouin zone; we omit spin indices for convenience. Similarly, if the operator $a^\dagger_{n,\vec{k}}$ creates a particle in the specified electronic state, to generate an additional particle at $\vec{r}$, we apply

$$\Phi^\dagger(\vec{r}) = \frac{1}{\sqrt{\Omega}} \sum_{n,\vec{k}} a^\dagger_{n,\vec{k}} e^{-i\vec{k}\cdot\vec{r}} u^*_{n,\vec{k}}(\vec{r}). \tag{14}$$

Neglecting spin variables, Fermi–Dirac statistics are implemented through the anticommutation relation $[a_{n,\vec{k}'}, a^\dagger_{m,\vec{k}}]_+ = \delta_{\vec{k}',\vec{k}} \delta_{n,m}$ or, equivalently,

$$[\Phi(\vec{r}), \Phi^\dagger(\vec{r}')]_+ = \delta(\vec{r} - \vec{r}'), \tag{15}$$

while $[\Phi(\vec{r}), \Phi(\vec{r}')]_+ = [\Phi^\dagger(\vec{r}), \Phi^\dagger(\vec{r}')]_+ = 0$.

Having represented the electron field in terms of second-quantized operators, we wish to examine the electron–electron, electron–photon, and electron–phonon interactions. The Hamiltonian of the noninteracting electronic field and the electron density operator are, respectively, in terms of the one-particle Hamiltonian $\mathcal{H}$ and the state energies $E_n(\vec{k})$,

$$\begin{aligned} H &= \int d^3r\, \Phi^\dagger(\vec{r}) \mathcal{H} \Phi(\vec{r}) \\ &= \frac{1}{\Omega} \sum_{n,\vec{k},n',\vec{k}'} a^\dagger_{n,\vec{k}} a_{n',\vec{k}'} \int d^3r\, e^{-i\vec{k}\cdot\vec{r}} u^*_{n\vec{k}}(\vec{r}) \mathcal{H} e^{i\vec{k}'\cdot\vec{r}} u_{n'\vec{k}'}(\vec{r}) \\ &= \sum_{n,\vec{k}} E_n(\vec{k}) a^\dagger_{n,\vec{k}} a_{n,\vec{k}} \end{aligned} \tag{16}$$

and

$$\rho(\vec{r}) = \Phi^\dagger(\vec{r})\Phi(\vec{r}) = \frac{1}{\Omega}\sum_{n,\vec{k},n',\vec{k}'} a^\dagger_{n,\vec{k}} a_{n',\vec{k}'} e^{-i\vec{k}\cdot\vec{r}} u^*_{n\vec{k}}(\vec{r}) e^{i\vec{k}'\cdot\vec{r}} u_{n'\vec{k}'}(\vec{r}). \quad (17)$$

From these two equations, we conclude that the various components of the interaction Hamiltonian can be obtained by inserting appropriate potentials $V(\vec{r})$ into

$$H_{\text{int}} = \int d^3r \rho(\vec{r}) V(\vec{r}). \quad (18)$$

Generally, $V(\vec{r})$ for the electron–phonon coupling is calculated in the linear approximation with respect to the lattice displacements. In terms of phonon creation and annihilation operators $b^\dagger_{\vec{q},\lambda}$ and $b_{\vec{q},\lambda}$ and the position vector, $\vec{R}_l$, of the lth cell, the displacement of the ath basis atom in the lth primitive cell resulting from the phonon field is (Bir and Pikus, 1974)

$$\vec{u}_{l,a} = \sum_{\vec{q},\lambda} \left(\frac{\hbar}{2\Omega r \omega_{\vec{q},\lambda}}\right)^{1/2} e^{i\vec{q}\cdot\vec{R}_l} (\vec{v}^\lambda_a(\vec{q}) b_{\vec{q},\lambda} + \{\vec{v}^\lambda_a(-\vec{q})\}^* b^\dagger_{-\vec{q},\lambda}). \quad (19)$$

In the preceding equation, r, λ, $\omega_{\vec{q},\lambda}$ and $\vec{v}$ designate crystal density, phonon branch, phonon frequency, and a normalized polarization vector.

Since the electron density can be represented as a lattice periodic term multiplied with a slowly varying envelope function, we divide the integral in Eq. (18) into a sum of contributions from the wavelength components of the potential $V(\vec{r})$ that are large compared to the lattice constant and from the remaining, rapidly varying quasi-periodic part of $V(\vec{r})$. The long- and short-range parts of the electron–phonon interaction originate in different physical processes.

1. Long-Range Electron–Phonon Potential

The main contributions to the long-range interactions with phonons are associated with the polarization of the lattice by out-of-phase oscillation of the different sublattices induced by optical phonons. Further, piezoelectric electron–phonon coupling in III–V crystals results primarily from the strain-dependent ionicity of the crystal (Rode, 1975, p. 34). In such cases the effective Hamiltonian matrix element between two electronic

6. INTRODUCTION TO NONEQUILIBRIUM MANY-BODY ANALYSES

states in bands n and n' assumes the general form

$$\langle n'\alpha'\vec{k}'| H_{\text{e-ph}}^{\text{long range}} |n, \alpha, \vec{k}\rangle$$
$$= \sum_{\vec{q},\lambda} (M_{\vec{q},\lambda}b_{\vec{q},\lambda} + M^*_{-\vec{q},\lambda}b^{\dagger}_{-\vec{q},\lambda})\langle n'\alpha'\vec{k}' + \vec{q} |n\alpha\vec{k}\rangle_B \delta_{\vec{k}',\vec{k}+\vec{q}}. \quad (20)$$

We will denote the overlap integrals between periodic parts of Bloch waves divided by the unit cell volume Ω_0 as

$$B^{n'n}_{\alpha'\alpha}(\vec{k}+\vec{q},\vec{k}) \equiv \langle n'\alpha'\vec{k}+\vec{q} | n\alpha\vec{k}\rangle_B \equiv \frac{1}{\Omega_0}\int_{\Omega_0} d^3r u^*_{n'\alpha'\vec{k}+\vec{q}}(\vec{r})u_{n\alpha\vec{k}}(\vec{r}), \quad (21)$$

which should be distinguished from the overlap of full Bloch waves denoted by brackets without the subscript B. For the polar coupling induced by longitudinal optical phonons,

$$M_{\vec{q}}^{\text{polar optical}} = \frac{ie}{q}\left(\frac{2\pi\hbar\omega_{\text{LO}}}{\Omega}\right)^{1/2}\left(\frac{1}{\varepsilon_\infty}-\frac{1}{\varepsilon_0}\right), \quad (22)$$

where ε_0 and ε_∞ are the low- and high-frequency limits of the dielectric constant, e is the bare electron charge, and the longitudinal optical phonon frequency is ω_{LO}. In terms of the piezoelectric constant B, the piezoelectric electron–phonon coupling is (Bir and Pikus, 1974)

$$M_{\vec{q}}^{\text{piezo}} = \frac{4\pi e}{\varepsilon_0}B\left(\frac{\hbar}{2r\Omega\omega_{\vec{q}}}\right)^{1/2}f(\hat{q}). \quad (23)$$

The function

$$f(\hat{q}) = \frac{2}{q^2}(v_x q_y q_z + v_y q_z q_x + v_z q_x q_y) \quad (24)$$

depends only on the direction of $\vec{q}$. Often, as in calculations involving M^2, f is replaced by its angular average

$$\frac{1}{4\pi}\int_0^\pi \sin\theta\, d\theta \int_0^{2\pi} d\phi\, |f(\hat{q})|^2 = \frac{4}{15}. \quad (25)$$

2. SHORT-RANGE ELECTRON–PHONON POTENTIAL

The short-range interaction with phonons is instead described by deformation potentials $V_{ij}(\vec{r})$. This quantity can be computed in the case of homogeneous strain described by the strain tensor $\hat{\mathscr{E}}$ with components $\mathscr{E}_{ij} = \frac{1}{2}[(\partial u_i/\partial x_j) + (\partial u_j/\partial x_i)]$ from the difference,

$$V_{ij}(\vec{r}) = \frac{1}{2 - \delta_{ij}} \lim_{\hat{\mathscr{E}} \to 0} \left\{ \frac{V_{\mathscr{E}}((1 + \hat{\mathscr{E}})\vec{r}) - V_0(\vec{r})}{\mathscr{E}_{ij}} \right\}, \quad (26)$$

between the unperturbed lattice potential $V_0(\vec{r})$ and the lattice potential in the deformed crystal $V_{\mathscr{E}}(\vec{r})$ (Bir and Pikus, 1974, p. 301). In Eq. (26) the potential of the strained material is evaluated in a dilated coordinate system in order to compensate for the shifted atomic positions. The deformation potential constants are found at the Γ point to equal

$$D^{ij}_{n'\alpha'n\alpha} = \left\langle n'\alpha'k = 0 \left| \frac{\hbar^2}{m_0} \frac{\partial^2}{\partial x_i \, \partial x_j} + V_{ij}(\vec{r}) \right| n\alpha k = 0 \right\rangle_B, \quad (27)$$

and the matrix elements of the corresponding electron–phonon interaction are in terms of the Fourier transform, $\mathscr{E}_{ij}(\vec{q})$, of the strain tensor,

$$\langle n'\alpha'\vec{k}' | H_c^{\text{def. pot.}} | n\alpha\vec{k} \rangle = \sum_{ij} \mathscr{E}_{ij}(\vec{k}' - \vec{k}) D^{ij}_{n'\alpha',n,\alpha}. \quad (28)$$

For longitudinal acoustic modes, the deformation potential is the strongest coupling mechanism. Further,

$$\mathscr{E}_{ij}(\vec{q}) = \left(\frac{\hbar}{2\Omega\varrho\omega_{\vec{q}}^{\text{LA}}} \right)^{1/2} i \frac{q_i q_j}{q} (b_{\vec{q}} - b^{\dagger}_{-\vec{q}}), \quad (29)$$

so that

$$\langle n'\alpha'\vec{k}' | H_c^{\text{def. pot.}} | n\alpha\vec{k} \rangle$$

$$= \sum_{ij} \sum_{\vec{q}} \delta_{\vec{k}',\vec{k}+\vec{q}} \left(\frac{\hbar}{2\Omega\varrho\omega_{\vec{q}}^{\text{LA}}} \right)^{1/2} i \frac{q_i q_j}{q} (b_{\vec{q}} - b^{\dagger}_{-\vec{q}}) D^{ij}_{n'\alpha',n,\alpha}. \quad (30)$$

In a spatially nondegenerate band with cubic symmetry such as the conduction band in III–V semiconductors neglecting p-type admixtures from the valence band, only a single deformation potential $D^{ij}_{n\alpha',n\alpha} = E_1 \delta_{i,j} \delta_{\alpha',\alpha}$ is present, and therefore (Zook, 1964)

$$\langle c\vec{k}' | H_c^{\text{def. pot.}} | c\vec{k} \rangle = \sum_{\vec{q}} (M^{\text{acoustic}}_{\vec{q}} b_{\vec{q}} + \{M^{\text{acoustic}}_{-\vec{q}}\}^* b^{\dagger}_{-\vec{q}}) \delta_{\vec{k},\vec{k}+\vec{q}}, \quad (31)$$

where

$$M_{\vec{q}}^{\text{acoustic}} = i\,|\vec{q}|\,\sqrt{\frac{\hbar}{2\Omega\varrho\omega_A(q)}}\,E_1. \tag{32}$$

Hence, the change in electron energy resulting from the local deformation of the lattice is characterized by a single deformation potential constant E_1.

On the other hand, the deformation potential is far more complicated in the case of degenerate or quasi-degenerate bands such as the light and heavy hole system in the vicinity of the Γ point. In this case, an analysis of the lattice symmetry demonstrates that there are three independent deformation potential matrix elements, namely

$$l = \left\langle X \left| -\frac{\hat{p}_x \hat{p}_x}{m_0} + V_{xx}(\vec{r}) \right| X \right\rangle_B, \tag{33}$$

$$m = \left\langle Y \left| -\frac{\hat{p}_x \hat{p}_x}{m_0} + V_{xx}(\vec{r}) \right| Y \right\rangle_B, \tag{34}$$

$$n = \left\langle X \left| -\frac{\hat{p}_x \hat{p}_y}{m_0} + V_{xy}(\vec{r}) \right| Y \right\rangle_B. \tag{35}$$

Defining new constants,

$$D_d = \frac{1}{3}(2m + l),$$

$$D_u = \frac{1}{2}(m - l), \tag{36}$$

$$D'_u = -\frac{n}{2},$$

and

$$P = D_d\,Tr\,\mathscr{E},$$

$$Q = -\frac{1}{3}D_u(\mathscr{E}_{xx} + \mathscr{E}_{yy} - 2\mathscr{E}_{zz}),$$

$$R = \frac{1}{\sqrt{3}}D_u(\mathscr{E}_{xx} - \mathscr{E}_{yy}) - i\frac{2}{\sqrt{3}}D'_u\mathscr{E}_{xy}, \tag{37}$$

$$S = -\frac{2}{\sqrt{3}}D'_u(\mathscr{E}_{zx} - i\mathscr{E}_{zy}),$$

we produce the following 4×4 matrix representation of the interaction with respect to the $j = 3/2$ heavy and light-hole basis set:

$$H_v^{\text{def. pot.}} = \begin{pmatrix} P+Q & -S & R & 0 \\ -S^\dagger & P-Q & 0 & R \\ R^\dagger & 0 & P-Q & S \\ 0 & R^\dagger & S^\dagger & P+Q \end{pmatrix} \quad (38)$$

The second-quantized form of Eq. (37) is obtained from inserting Eq. (29) for the strain tensor. Generally l, m, and n are approximated by E_1. If further $q \ll k$, we find after some manipulations that the matrix element for the interaction of phonons with heavy holes becomes identical in form to that for conduction band electrons, i.e.,

$$\langle h\vec{k}' | H_v^{\text{def. pot.}} | h\vec{k} \rangle = \sum_{\vec{q}} (M_{\vec{q}}^{\text{acoustic}} b_{\vec{q}} + \{M_{-\vec{q}}^{\text{acoustic}}\}^* b_{-\vec{q}}^\dagger) \delta_{\vec{k},\vec{k}+\vec{q}}. \quad (39)$$

3. Electron–Photon Interaction

We now examine the vector photon field in analogy with our treatment of phonons. Introducing the polarization vector $\vec{\xi}(\vec{k}, \lambda)$, the photon frequency ω_k, and creation and annihilation operators $c_{\vec{k},\lambda}^\dagger$ and $c_{\vec{k},\lambda}$, the second quantized vector potential $\vec{A}$ representing the photon field in the Coulomb gauge ($\vec{\nabla} \cdot \vec{A} = 0$) may be written as

$$\vec{A}(\vec{r}, t) = \sum_{\vec{k},\lambda} \left(\frac{2\pi\hbar c^2}{\Omega \omega_k}\right)^{1/2} \vec{\xi}(\vec{k}, \lambda) e^{i\vec{k}\cdot\vec{r}} (c_{-\vec{k},\lambda}^\dagger + c_{\vec{k},\lambda}). \quad (40)$$

The interaction between the photon and the electrons arises to linear order from the term

$$H^{\text{el.-phot.}} = \frac{e}{2m_0 c} \int \Phi^\dagger (\vec{A} \cdot \vec{p} + \vec{p} \cdot \vec{A}) \Phi \, d^3r \quad (41)$$

and is consequently given by — noting that for optical frequencies the transverse electric field varies negligibly over distances on the order of the unit cell dimensions —

$$H^{\text{el.-phot.}} = \sum_{nn'} \sum_{\vec{k},\vec{q}} \sum_{\lambda} M_{\vec{q}}^{\text{phot}} (c_{-\vec{q},\lambda}^\dagger + c_{\vec{q},\lambda}) a_{n',\vec{k}+\vec{q}}^\dagger a_{n,\vec{k}} \langle n', \vec{k}+\vec{q} | \vec{\xi} \cdot \vec{p} | n\vec{k} \rangle_B,$$
$$(42)$$

6. INTRODUCTION TO NONEQUILIBRIUM MANY-BODY ANALYSES

in which we have introduced the electron–photon coupling constant

$$M_{\vec{q}}^{\text{phot}} = \left(\frac{2\pi\hbar c^2}{\Omega \omega_k}\right)^{1/2}. \tag{43}$$

We will employ the $k \cdot p$ model expressions for the overlap matrix elements in the various interaction Hamiltonians introduced above, namely, Eq. (21), and the matrix element,

$$\vec{P}_{\alpha\beta}^{cv} = \frac{1}{\Omega_0} \int_{\Omega_0} d^3\vec{r}\, u_{c,\alpha,\vec{k}}^*(\vec{r})(-i\hbar\vec{\nabla})u_{v,\beta,\vec{k}}(\vec{r}) = \langle c, \alpha, \vec{k}|\,\vec{p}\,|v, \beta, \vec{k}\rangle_B, \tag{44}$$

of the momentum operator in Eq. (42) for vertical (i.e., $q = 0$) conduction-to-valence band transitions of a particle of momentum k. Such matrix elements are evaluated for various pairs of states α and β in Appendix A.

IV. Ensemble Properties

4. Wigner Distribution Function

In optoelectronic devices, a large number of excess carriers are either optically excited or electronically injected into the conduction and valence bands. These carriers relax through phonon interactions to the extrema of their respective bands with a time constant far smaller than that required for interband transitions. As a consequence, each subsystem may be characterized by a one-particle distribution function with a specified effective temperature. To determine either the quasi-equilibrium distribution function or the change in this function in response to additional external perturbations, we must generalize the Boltzmann equation formalism beyond the standard single-particle (diffusion) limit in which carrier–carrier interactions are neglected. This theory can then be easily applied to highly excited electron and hole systems in optoelectronic components. The results of such studies yield bounds on the validity of the quasi-particle model in nonequilibrium contexts and allow the estimation of the nonequilibrium quasi-particle energies and lifetimes. At the same time, we can establish the limitations of concepts typified by the subsystem temperature or distribution function as applied to nonequilibrium systems.

To model the effects of various internal and external fields on the evolution of the particle distribution function in a nonequilibrium system, we now introduce the time-ordered Green's function. The resulting expressions for various physical quantities will then possess a formal similarity to the equilibrium theory, but with a redefinition of operations as, for example, ensemble averaging. Accordingly, we specify some initial equilibrium state at t_0 described by a density matrix ρ_0 and assume that at some finite time in the distant past an interaction is applied that produces the desired nonequilibrium state at a later time t_1. The system relaxes to local equilibrium characterized by a definite quasi-Fermi level within each band at the time of observation; we will study the evolution of this quasi-equilibrium state towards the equilibrium state, as well as the behavior of the quasi-equilibrium system in response to small external perturbations. The time-evolution of the field operator in the Heisenberg representation, $\Phi_H(\vec{r}, t)$, from time t_0 is governed by the formula

$$\Phi_H(\vec{r}, t) = U(t_0, t)\Phi(\vec{r})U(t, t_0), \tag{45}$$

where $U(t, t')$ is the Schrödinger equation evolution operator.

The classical distribution function, $f_{cl}(\vec{p}, \vec{r}, t)$, which gives the particle density at point $\vec{r}$ with momentum $\vec{p}$, does not have meaning in quantum mechanics as a result of the uncertainty principle. A quantity with similar properties that may be used to calculate averages of one-particle operators can, however, be generated from the density function (Danielewicz, 1984)

$$n(\vec{r}, t) = \sum_{\lambda, \lambda'} \rho_{0,\lambda',\lambda} \langle \varphi_\lambda | \Phi_H^\dagger(\vec{r}, t) \Phi_H(\vec{r}, t) | \varphi_{\lambda'} \rangle. \tag{46}$$

Here the statistical operator matrix $\rho_{0,\lambda,\lambda'}$ is given, for example, in the canonical ensemble by

$$\rho_0 = e^{-\beta H}/\text{Tr}\{e^{-\beta H}\}, \tag{47}$$

in which H represents the full many-particle Hamiltonian and $\beta = 1/kT$. Statistical averaging is performed by taking the trace of a given operator with the preceding density matrix and will be denoted by angular brackets.

We now consider for simplicity a homogeneous system for which the $|\varphi_\lambda\rangle$ in Eq. (46) are chosen to be eigenstates of the Hamiltonian and momentum operators with energy E_λ and momentum $\vec{p}_\lambda$. The operator that yields a quantity analogous to the density of particles with energy $\hbar\omega$

6. INTRODUCTION TO NONEQUILIBRIUM MANY-BODY ANALYSES   331

and momentum p can be determined by inserting the projection operator

$$F_\lambda(\omega, \vec{p}) = \sum_\xi (2\pi)^4 \delta(\vec{p} - (\vec{p}_\lambda - \vec{p}_\xi))\delta(\omega - (E_\lambda - E_\xi))|\varphi_\xi\rangle\langle\varphi_\xi| \quad (48)$$

into Eq. (46) to yield

$$f(\vec{R}, \vec{p}, T, \omega) = \sum_\lambda \rho_{0,\lambda\lambda}\langle\varphi_\lambda|\Phi_H^\dagger(\vec{R}, T)F_\lambda(\omega, \vec{p})\Phi_H(\vec{R}, T)|\varphi_\lambda\rangle. \quad (49)$$

Employing the Fourier representation of the delta functions, we recast Eq. (49) into the form

$$f(\vec{R}, \vec{p}, T, \omega) = \sum_\lambda \rho_{0,\lambda\lambda} \sum_\xi \int d^3r \int_{-\infty}^\infty dt\, e^{-i(\vec{p}-(\vec{p}_\lambda-\vec{p}_\xi))\vec{r}} e^{i(\omega-(E_\lambda-E_\xi))t}$$
$$\times \langle\varphi_\lambda|\Phi_H^\dagger(\vec{R}, T)|\varphi_\xi\rangle\langle\varphi_\xi|\Phi_H(\vec{R}, T)|\varphi_\lambda\rangle. \quad (50)$$

Since for a homogeneous system

$$\Phi_H\left(\vec{R} + \frac{\vec{r}}{2}, T + \frac{t}{2}\right) = e^{-(i/2)(\vec{r}\hat{P}-t\hat{H})}\Phi_H(\vec{R}, T)e^{(i/2)(\vec{r}\hat{P}-t\hat{H})}, \quad (51)$$

where $\hat{P}$ and $\hat{H}$ are the momentum and Hamiltonian operators for the system, we finally arrive at the "generalized Wigner function"

$$f(\vec{R}, \vec{p}, T, \omega) = \int d^3r \int_{-\infty}^\infty dt\, e^{-i\vec{p}\vec{r}} e^{i\omega t}$$
$$\left\langle \Phi_H^\dagger\left(\vec{R} - \frac{\vec{r}}{2}, T - \frac{t}{2}\right)\Phi_H\left(\vec{R} + \frac{\vec{r}}{2}, T + \frac{t}{2}\right)\right\rangle, \quad (52)$$

from which the Wigner function is obtained by integrating over energies

$$f(\vec{R}, \vec{p}, T) = \int d^3\vec{r}\, e^{-i\vec{p}\vec{r}}\left\langle\Phi_H^\dagger\left(\vec{R} - \frac{\vec{r}}{2}, T\right)\Phi_H\left(\vec{R} + \frac{\vec{r}}{2}, T\right)\right\rangle. \quad (53)$$

The generalized function is distinguished notationally from the standard Wigner function by the inclusion of the energy variable ω.

While the Wigner function may be negative in certain regions of space, precluding a literal interpretation in terms of a phase space distribution function, it may be employed to calculate classical averages such as the particle density, cf. Eq. (46).

$$n(\vec{r}, t) = \frac{1}{\Omega} \sum_{\vec{p}} f(\vec{r}, \vec{p}, t). \tag{54}$$

The electron current density, which in second-quantized form is $-e/2m_0[(\vec{p} - \hat{\vec{p}}' + e/cA(\vec{r}) + e/cA(\vec{r}')]\Phi^\dagger(\vec{r}')\Phi(\vec{r})|_{\vec{r}'=\vec{r}}$, can additionally be expressed as

$$\vec{j}(\vec{r}, t) = -\frac{e}{\Omega m_0} \sum_{\vec{p}} \left(\vec{p} + \frac{e}{c}\vec{A}(\vec{r}, t)\right) f(\vec{r}, \vec{p}, t). \tag{55}$$

To clarify the relationship between the Wigner function and the classical distribution function, note that the integrand in Eq. (53) may be interpreted in terms of the one-particle Green's function

$$-iG^<(\vec{r}_1, t_1, \vec{r}_2, t_2) \equiv \langle \Phi_H^\dagger(\vec{r}_2, t_2)\Phi_H(\vec{r}_1, t_1)\rangle, \tag{56}$$

which is effectively the correlation function for observing a "hole" at $\vec{r}_2$, t_2 if an electron is removed from the system at $\vec{r}_1$, t_1. In the same manner, the electron Green's function is

$$iG^>(\vec{r}_1, t_1, \vec{r}_2, t_2) \equiv \langle \Phi_H(\vec{r}_1, t_1)\Phi_H^\dagger(\vec{r}_2, t_2)\rangle. \tag{57}$$

If these Green's functions are written in the relative variables $\vec{r} = \vec{r}_1 - \vec{r}_2$, $t = t_1 - t_2$, and average variables $\vec{R} = (\vec{r}_1 + \vec{r}_2)/2$, $T = (t_1 + t_2)/2$, the quantum oscillations of the wavefunctions, and therefore of $G^<$ and $G^>$, are apparent over $\vec{r}$ and t inversely proportional to the electron wavenumber and frequency. Averaging over the oscillations yields the macroscopic behavior of the correlation functions in terms of $\vec{R}$ and T as expressed by the Wigner distribution function. In fact,

$$f(\vec{R}, \vec{p}, T) = -i \int \frac{d\omega}{2\pi} G^<(\vec{p}, \omega, \vec{R}, T), \tag{58}$$

where the transformation from $\vec{r}_1$, t_1, $\vec{r}_2$, t_2 to $\vec{p}$, ω, $\vec{R}$, T is

$$G^\gtrless(\vec{p}, \omega, \vec{R}, T) = \int_{-\infty}^{\infty} dt \int d^3r\, e^{-i\vec{p}\cdot\vec{r}+i\omega t} G^\gtrless\left(\vec{R} + \frac{\vec{r}}{2}, T + \frac{t}{2}, \vec{R} - \frac{\vec{r}}{2}, T - \frac{t}{2}\right). \tag{59}$$

6. INTRODUCTION TO NONEQUILIBRIUM MANY-BODY ANALYSES   333

5. NONEQUILIBRIUM GREEN'S FUNCTIONS

In order to obtain a perturbative analysis of the equation of motion of the Green's function for a system described by a noninteracting one-particle Hamiltonian, H_0, in the presence of a given interaction, H', we introduce the interaction representation in which the field operators $\Phi_I(\vec{r}, t)$ are defined by

$$\Phi_H(\vec{r}, t) = U_I(t_0, t)\Phi_I(\vec{r}, t)U_I(t, t_0), \quad (60)$$

with

$$U_I(t, t_0) = T_C e^{\{-(i/\hbar)\int_{t_0}^{t} H'_I(\tau)\, d\tau\}}. \quad (61)$$

The symbol T_C represents time ordering along the path from t_0 to t for $U_I(t, t_0)$ and the inverse time ordering from t to t_0 for $U_I(t_0, t)$ while $H'_I = U_{H_0}(t_0, t)H'U_{H_0}(t, t_0)$. The statistical operator can be rewritten by employing a complex time evolution operator as

$$\frac{e^{-\beta H}}{\text{Tr}\{e^{-\beta H}\}} = \frac{e^{-\beta H_0}U_I(t_0 - i\beta, t_0)}{\text{Tr}\{e^{-\beta H_0}U_I(t_0 - i\beta, t_0)\}}. \quad (62)$$

Combining Eq. (61) with Eq. (62), we arrive at the following condensed form for the path-ordered Green's functions:

$$iG(t, t') = \frac{\text{Tr}\{e^{-\beta H_0}T_C[U_I(t_0 - i\beta, t_0)\Phi_I(t)\Phi_I^\dagger(t')]\}}{\text{Tr}\{e^{-\beta H_0}U_I(t_0 - i\beta, t_0)\}} \quad (63)$$

The contour C now runs along the real axis from time t_0 to the observation times t' and t, back to t_0 and finally to $t_0 - i\beta$ in the complex time plane, as illustrated in Fig. 1. After expanding the U_I in the preceding formula in powers of the interaction Hamiltonian H'_I, we employ the Wick decomposition to express the expectation values of products of field operators in terms of noninteracting Green's functions.

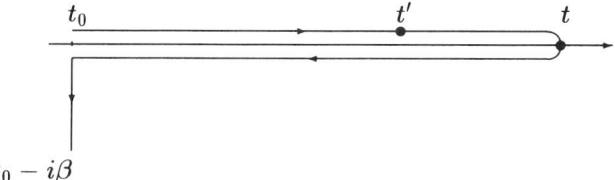

FIG. 1. The contour C for the path-ordered Green's function of Eq. (63).

In order to simplify the formalism, we replace t_0 with $-\infty$, neglect the complex time section from t_0 to $t_0 - i\beta$ (which has a minimal effect on the evolution at far later times; (Danielewicz, 1984) and extend the right-hand limit of the contour to $+\infty$, yielding the so-called Keldysh contour consisting of a positive branch from $-\infty$ to $+\infty$ and a second, negative branch back from $+\infty$ to $-\infty$. While this method formally extends the Feynman diagram technique to nonequilibrium systems, the number of diagrams for a given physical process is far larger than in the equilibrium case, since a different diagram is generated for each possible contour ordering of the internal vertices.

Having discussed the basic concepts of the Keldysh technique, we turn to the derivation of the quantum analog of the Boltzmann equation. The contour Green's function has the structure

$$G(\vec{r}_1, t_1, \vec{r}_2, t_2) = \theta(t_1, t_2) G^>(\vec{r}_1, t_1, \vec{r}_2, t_2) + \theta(t_2, t_1) G^<(\vec{r}_1, t_1, \vec{r}_2, t_2). \tag{64}$$

The correlation functions $G^>$ and $G^<$ are often associated with electron and hole propagation, respectively, while writing $t >_C t'$ if t is further on the contour C than t',

$$\theta(t_1, t_2) = \begin{cases} 1 & t_1 >_C t_2 \\ 1/2 & t_1 = t_2 \\ 0 & t_2 >_C t_1 \end{cases}. \tag{65}$$

Note that when $t_2 = t_1$, the Green's function remains finite. Next, we introduce the self-energy operator Σ through the Dyson's equations

$$G = G_0 \bullet (1 + \Sigma \bullet G) = (1 + G \bullet \Sigma) \bullet G_0 \tag{66}$$

for the one-particle Green's functions. Here "1" represents a contour delta function of position and time, while $G \bullet \Sigma$ is an abbreviation for the integrated product (Langreth, 1976)

$$[G \bullet \Sigma](\vec{r}_1, t_1, \vec{r}_2, t_2) = \int d^3 r \int_C dt \, G(\vec{r}_1, t_1, \vec{r}, t) \Sigma(\vec{r}, t, \vec{r}_2, t_2). \tag{67}$$

Further, G_0 in Eq. (66) denotes the Green's function for the noninteracting system that fulfills

$$\left[i \frac{\partial}{\partial t_1} + \frac{\nabla_1^2}{2m} - U(\vec{r}_1, t_1) \right] G_0(1, 1') = \delta(t_1, t_{1'}) \delta(\vec{r}_1 - \vec{r}_{1'}) \tag{68}$$

6. INTRODUCTION TO NONEQUILIBRIUM MANY-BODY ANALYSES

and

$$\left[-i\frac{\partial}{\partial t_{1'}}+\frac{\nabla^2_{1'}}{2m}-U(\vec{r}_{1'},t_{1'})\right]G_0(1,1') = \delta(t_1,t_{1'})\delta(\vec{r}_1-\vec{r}_{1'}), \quad (69)$$

in which $G_0(1,1') \equiv G_0(\vec{r}_1, t_1, \vec{r}'_1, t'_1)$, $U(\vec{r}, t)$ is the Hartree potential associated with the external sources together with the induced charge density, and $\delta(t_1, t_{1'})$ equals $\delta(t_1 - t_{1'})$ on the chronological and $-\delta(t_1 - t_{1'})$ on the antichronological segments of the contour.

We next transform the Dyson equation for the quantum-mechanical Green's function into an equation of motion for the electron or hole distribution functions $G^>$ and $G^<$. The latter functions satisfy the following equality, cf. Eq. (66):

$$G_0^{-1} \bullet G^{\gtrless} = (\Sigma \bullet G)^{\gtrless} \quad (70)$$

The self-energy Σ may be considered as a functional of G and evaluated by diagrammatic or variational derivative procedures. To simplify the integrals $(G \bullet \Sigma)^>$ or $(G \bullet \Sigma)^<$, we observe that Σ is analytic on the contour C except at $t_1 = t_2$, i.e.,

$$\Sigma(\vec{r}_1, t_1, \vec{r}_2, t_2) = \Sigma^S(\vec{r}_1, t_1, \vec{r}_2, t_2)$$
$$+ \theta(t_1, t_2)\Sigma^>(\vec{r}_1, t_1, \vec{r}_2, t_2) + \theta(t_2, t_1)\Sigma^<(\vec{r}_1, t_1, \vec{r}_2, t_2). \quad (71)$$

The term Σ^S can be represented by a superposition of the time delta function and its derivatives (Danielewicz, 1984) as

$$\Sigma^S(\vec{r}_1, t_1, \vec{r}_2, t_2) = \sum_{n=0}^{\infty} F_i(\vec{r}_1, \vec{r}_2) \frac{d^n}{dt_1^n} \delta(t_1, t_2). \quad (72)$$

Singular contributions to Σ at $t_2 = t_1$ arise, for example, from exchange terms in the Hartree–Fock approximation. Analogous singularities are absent in G_0, but not in G_0^{-1}, for which $(G_0^{-1})^{\gtrless} = 0$ and

$$(G_0^{-1})^S(t_1, t_2) = i\frac{\partial}{\partial t_1}\delta(t_1, t_2) + \left(\frac{\nabla_1^2}{2m_0} - U(\vec{r}_1, t_1)\right)\delta(\vec{r}_1 - \vec{r}_2)\delta(t_1, t_2). \quad (73)$$

The operator $(G_0)^{-1}$ is therefore instantaneous in time. It is possible to demonstrate by inserting Eq. (64) and Eq. (71) into Eq. (67) and transforming integrals over the contour to standard integrals along the real time axis that

$$(G \bullet \Sigma)^>(t_1, t_2) = \int_{-\infty}^{\infty} (G^>(t_1, t')\Sigma^A(t', t_2) + G^R(t_1, t')\Sigma^>(t', t_2))\, dt'. \quad (74)$$

The advanced self-energy and retarded Green's function present in the preceding equality are for real, physical times

$$\Sigma^A(t_1, t_2) = \Sigma^S(t_1, t_2) + \Theta(t_2 - t_1)[\Sigma^<(t_1, t_2) - \Sigma^>(t_1, t_2)] \quad (75)$$

and

$$G^R(t_1, t_2) = \Theta(t_1 - t_2)[G^>(t_1, t_2) - G^<(t_1, t_2)]. \quad (76)$$

The symbol Θ represents the standard Heaviside step function. For an operator, $\mathcal{O}$, defined on the contour it is also useful for diagrammatic analyses to introduce the chronological and antichronological operators $\mathcal{O}^C$ and $\mathcal{O}^{\tilde{C}}$ by

$$\mathcal{O}^A = \mathcal{O}^C - \mathcal{O}^> = -\mathcal{O}^{\tilde{C}} + \mathcal{O}^< \quad (77)$$

and

$$\mathcal{O}^R = \mathcal{O}^C - \mathcal{O}^< = -\mathcal{O}^{\tilde{C}} + \mathcal{O}^>. \quad (78)$$

To derive Eq. (74), we write the integral over t' along C in $(G \bullet \Sigma)^>(t_1, t_2)$ as an integral over a contour extending from $-\infty$ to t_2 and then back to $-\infty$, followed by a second closed path from $-\infty$ to t_1. Along the first section $t' < t_2$, implying $\Sigma = \Sigma^A$, while at the same time $t_1 >_C t'$ so that we may simultaneously replace G by $G^>$. Subsequently, we extend the upper limit of the integral to ∞ since $\Sigma^A(t_1, t_2) = 0$ for $t_1 < t_2$. Analyzing the second path in the same manner, we directly obtain Eq. (76), as discussed in Langreth (1976). Similarly, we can establish that

$$(G \bullet \Sigma)^<(t_1, t_2) = \int_{-\infty}^{\infty} (G^<(t_1, t')\Sigma^A(t', t_2)$$
$$+ G^R(t_1, t')\Sigma^<(t', t_2))\, dt' \equiv G^< \circ \Sigma^A + G^R \circ \Sigma^<. \quad (79)$$

The symbol $\circ$ is employed to denote a standard operator product integrated over physical times. Omitting arguments, we have, since

6. INTRODUCTION TO NONEQUILIBRIUM MANY-BODY ANALYSES

$(G_0^{-1})^{\gtrless} = 0$,

$$G_0^{-1} \circ G^{\gtrless} = \Sigma^{\gtrless} \circ G^A + \Sigma^R \circ G^{\gtrless}. \tag{80}$$

and

$$G^{\gtrless} \circ G_0^{-1} = G^R \circ \Sigma^{\gtrless} + G^{\gtrless} \circ \Sigma^A. \tag{81}$$

Subtracting these two equations produces the most general equation of motion for $G^>$ or $G^<$, and therefore for G^R and G^A. If we write

$$g = \frac{1}{2}(G^R + G^A) \tag{82}$$

and

$$\sigma = \frac{1}{2}(\Sigma^R + \Sigma^A), \tag{83}$$

then we find in view of the identity $G^> - G^< = G^R - G^A$,

$$[G_0^{-1} - \sigma, G^{\gtrless}]_- - [\Sigma^{\gtrless}, g]_- = \frac{1}{2}([\Sigma^>, G^<]_+ - [\Sigma^<, G^>]_+). \tag{84}$$

The preceding commutators and anticommutators are taken with respect to the integrated product, $\circ$, of the two operators.

While our two coupled equation for $G^>$ and $G^<$ are completely general and may be applied to any nonequilibrium system, to obtain the generalized Boltzmann equation we assume further that the space and time variations of the disturbances from equilibrium are slow compared to the scales of the quantum-mechanical fluctuations. We then apply Weyl's formula, which pertains to the Fourier transform of the Green's functions with respect to the relative variables $\vec{r} = \vec{r}_1 - \vec{r}_2$ and $t = t_1 - t_2$. Specializing for simplicity to one dimension, this formula states that

$$(\widetilde{A \circ B})(p, X) \equiv \int_{-\infty}^{\infty} dx e^{-ipx} \int_{-\infty}^{\infty} dx_1 A\left(X + \frac{x}{2}, x_1\right) B\left(x_1, X - \frac{x}{2}\right)$$
$$= e^{(i/2)(\partial_X^A \partial_p^B - \partial_X^B \partial_p^A)} \tilde{A}(p, X) \tilde{B}(p, X), \tag{85}$$

where functions of the variables x and X denoted by tildes are defined by $\tilde{A}(x, X) = A(X + x/2, X - x/2)$, and p is the Fourier transform variable associated with x. The superscripts A and B indicate the operator to which the associated partial derivative is applied. The proof of the preceding expression is given in Appendix B. Short-scale corrections to the long-wavelength limit of the anticommutator are recovered by retaining higher-order terms in the Taylor expansion of the exponential.

We now apply Weyl's formula for the variables T and $\vec{R}$ to Eq. (84). In the quasi-classical limit, we replace the anticommutators on the right-hand side of Eq. (84) by the lowest-order term $\Sigma^{>}G^{<} - \Sigma^{<}G^{>}$. The left-hand side of the equation, however, vanishes in this approximation, and the components of the commutators must instead by expanded as

$$(\widetilde{A \circ B})(\vec{p}, \omega, \vec{R}, T)$$
$$= e^{(i/2)(\vec{\nabla}_R^A \cdot \vec{\nabla}_p^B - \partial_T^A \partial_\omega^B - \vec{\nabla}_R^B \cdot \vec{\nabla}_p^A + \partial_T^B \partial_\omega^A)} \tilde{A}(\vec{p}, \omega, \vec{R}, T) \tilde{B}(\vec{p}, \omega, \vec{R}, T)$$
$$\approx \tilde{A}\tilde{B} + \frac{i}{2}[\vec{\nabla}_R \tilde{A} \cdot \vec{\nabla}_p \tilde{B} - \partial_T \tilde{A} \partial_\omega \tilde{B} - \vec{\nabla}_R(\tilde{B}) \cdot \vec{\nabla}_p \tilde{A} + \partial_T \tilde{B} \partial_\omega \tilde{A}]. \quad (86)$$

Here we assume as in the derivation of the classical Boltzmann equation that the characteristic distance and time scales of the external potentials are much larger than the electron wavelength and oscillation period. We therefore neglect higher-order terms in the gradient expansion. If $E(\vec{p})$ is the one-particle energy and the corresponding velocity $\vec{v} = \vec{\nabla}_p E(\vec{p})$, we obtain with $G_0^{-1} = \omega - E(\vec{p}) - U(\vec{R})$

$$i\left(\frac{\partial \tilde{G}^{<}}{\partial T} + \vec{v} \cdot \vec{\nabla}_R \tilde{G}^{<} - \vec{\nabla}_R \tilde{U} \cdot \vec{\nabla}_p \tilde{G}^{<}\right) - [\widetilde{\Sigma^{<}, g}]_{-} \approx \widetilde{\Sigma^{>}G^{<}} - \widetilde{\Sigma^{<}G^{>}}. \quad (87)$$

Finally, to cast Eq. (87) in the form of the Boltzmann equation, we must assume that the mean free lifetime τ of a quasi-particle is much longer than its quantum-mechanical oscillation period. The value of τ reflects the time required for relaxation from the initial one-particle excitation to a new, orthogonal state of the many-body system. To better understanding the meaning of the variable τ, we introduce the quasi-particle line shape function, $A(\vec{p}, \omega, \vec{R}, T)$, which is also termed the spectral density function and gives the probability that an excitation of momentum $\vec{p}$ has an energy $\hbar\omega$. It is related to the retarded or advanced Green's function by

$$G^{R}_{A}(\vec{p}, \omega, \vec{R}, T) = \int d\omega' \frac{A(\vec{p}, \omega', \vec{R}, T)}{\omega - \omega' \pm i\eta}, \quad (88)$$

where we omit tildes over operators expressed in terms of the

6. INTRODUCTION TO NONEQUILIBRIUM MANY-BODY ANALYSES

$(\vec{p}, \omega, \vec{R}, T)$ coordinate variables. From Eq. (88),

$$A = \frac{i}{2\pi}(G^R - G^A), \tag{89}$$

or, writing variables explicitly,

$$A(\vec{p}, \omega, \vec{R}, T) = \frac{i}{2\pi}(G^>(\vec{p}, \omega, \vec{R}, T) - G^<(\vec{p}, \omega, \vec{R}, T)). \tag{90}$$

It is important to observe that while $G^>$ and $G^<$ uniquely determine the spectral density function, the converse does not hold.

We can rewrite the spectral density function as

$$A = \frac{i}{2\pi} G^R \circ ((G^A)^{-1} - (G^R)^{-1}) \circ G^A$$

$$= \frac{i}{2\pi} G^R \circ (\Sigma^R - \Sigma^A) \circ G^A$$

$$= \frac{i}{2\pi} G^R \circ (\Sigma^> - \Sigma^<) \circ G^A. \tag{91}$$

This formula is often referred to as an optical theorem. If we pass to the semiclassical limit, we can restrict our attention to the first term in the gradient expansion of the operator products. The resulting spectral density function is quasi-Lorentzian, namely,

$$A(\vec{p}, \omega, \vec{R}, T)$$
$$= -\frac{1}{\pi} \frac{\operatorname{Im} \Sigma^R(\vec{p}, \omega, \vec{R}, T)}{(\omega - E(p) - \operatorname{Re} \Sigma^R(\vec{p}, \omega, \vec{R}, T))^2 + (\operatorname{Im} \Sigma^R(\vec{p}, \omega, \vec{R}, T))^2}. \tag{92}$$

We have further

$$\operatorname{Im} \Sigma^R(\vec{p}, \omega, \vec{R}, T) = -\frac{i}{2}(\Sigma^> - \Sigma^<). \tag{93}$$

If $\operatorname{Im} \Sigma^R(\vec{p}, \omega, \vec{R}, T)$ is small compared to the electron energy $\hbar\omega = E(p)$, the quasi-particle energy is determined by solving the self-consistent equation

$$\bar{E}(p) = E(p) + \operatorname{Re} \Sigma^R(\vec{p}, \bar{E}(\vec{p}), \vec{R}, T). \tag{94}$$

The one-particle inverse lifetime is then associated with the linewidth of the spectral density function, that is,

$$\frac{1}{\tau} = -2\,\text{Im}\,\Sigma^R(\vec{p}, \bar{E}(p), \vec{R}, T)$$
$$= i(\Sigma^R(\vec{p}, \bar{E}(p), \vec{R}, T) - \Sigma^A(\vec{p}, \bar{E}(p), \vec{R}, T))$$
$$= i(\Sigma^>(\vec{p}, \bar{E}(p), \vec{R}, T) - \Sigma^<(\vec{p}, \bar{E}(p), \vec{R}, T)). \quad (95)$$

If the interactions are weak, $\tau E(p) \gg 1$, A can be approximated by a delta function. Specifically, representing the renormalized particle energy, which may deviate substantially from the equilibrium bandstructure energy in systems with long-range interactions for which $\text{Re}\,\Sigma$ is large, by $\bar{E}(p)$,

$$A(\vec{p}, \omega, \vec{R}, T) = \delta(\omega - \bar{E}(p) - U(\vec{R}, T)). \quad (96)$$

This expression satisfies the sum rule

$$\int d\omega A(\vec{p}, \omega, \vec{R}, T) = 1, \quad (97)$$

derived from Eqs. (90), (56), (57), together with the commutation relations for field operators. In the same limit, we may neglect the second term on the left-hand side of Eq. (87). Solving Eq. (87) to zeroth order with respect to $\Sigma^{\gtrless}$ and applying Eq. (58), we find

$$iG^<(\vec{p}, \omega, \vec{R}, T) = -2\pi f(\vec{R}, \vec{p}, T)\delta(\omega - \bar{E}(p) - U(\vec{R}, T)). \quad (98)$$

Similarly,

$$iG^>(\vec{p}, \omega, \vec{R}, T) = 2\pi(1 - f(\vec{R}, \vec{p}, T))\delta(\omega - \bar{E}(p) - U(\vec{R}, T)). \quad (99)$$

Corresponding equations for boson operators are, designating the boson distribution function by P and the boson Green's function by D,

$$iD^<(\vec{p}, \omega, \vec{R}, T) = 2\pi P(\vec{R}, \vec{p}, T)\delta(\omega - \bar{E}(p) - U(\vec{R}, T)) \quad (100)$$

and

$$iD^>(\vec{p}, \omega, \vec{R}, T) = 2\pi(1 + P(\vec{R}, \vec{p}, T))\delta(\omega - \bar{E}(p) - U(\vec{R}, T)). \quad (101)$$

6. Introduction to Nonequilibrium Many-Body Analyses   341

If we insert our expressions for $G^>$ and $G^<$ into the approximation to Eq. (87) and integrate over frequencies, we arrive at the Boltzmann equation

$$\left(\frac{\partial}{\partial T} - \vec{\nabla}_{\vec{R}} U \cdot \vec{\nabla}_{\vec{p}} + \vec{v} \cdot \vec{\nabla}_{\vec{R}}\right) f(\vec{R}, \vec{p}, T)$$
$$- i\Sigma^<(\vec{p}, \bar{E}(p), \vec{R}, T)(1 - f(\vec{R}, \vec{p}, T)) -$$
$$i\Sigma^>(\vec{p}, \bar{E}(p), \vec{R}, T) f(\vec{R}, \vec{p}, T). \quad (102)$$

Dynamic effects from the interaction of a particle between collisions with the averaged environment of the other particles are incorporated into the spectral density function, while collision effects enter through the product of the Wigner function with the self-energy functions $\Sigma^{\gtrless}$. Equation (102) allows us to identify $\Sigma^<$ and $\Sigma^>$ as the scattering rate of particles into and out of a state of momentum $\vec{p}$ at position $\vec{R}$ and time T. An analogous equation for bosons is generated by replacing the Fermi distribution functions by the negative of the corresponding boson distribution functions and employing the boson in place of the Fermion self-energy.

We now apply Eq. (102) to minority electron scattering in a p-type III–V semiconductor material in the limit of small excess carrier densities in which the quantum-Boltzmann equation can be considerably simplified. In particular, for electrons in a homogeneous semiconductor,

$$\frac{\partial f(\vec{p})}{\partial T} = -i(1 - f(\vec{p}))\Sigma^<(\vec{p}) - if(\vec{p})\Sigma^>(\vec{p}). \quad (103)$$

For small deviations δf from the equilibrium distribution function,

$$\frac{\partial \delta f}{\partial T} = i[\Sigma^<(\vec{p}) - \Sigma^>(\vec{p})]\delta f - i(1 - f_0(\vec{p}))\delta\Sigma^<(\vec{p}) - if_0(\vec{p})\delta\Sigma^>(\vec{p}). \quad (104)$$

The quantities $\Sigma(\vec{p})$ and f_0 are the time-independent self-energy and distribution functions for local equilibrium, while $\delta f = f(T) - f_0$ and $\delta\Sigma(\vec{p})$ are associated with the nonequilibrium perturbation. For a single additional electron with momentum $\vec{p}$ in an equilibrium system, $\delta f \neq 0$ only at this momentum value. Then, since $\Sigma^{\gtrless}(\vec{p})$ in Eq. (104) scales roughly linearly with carrier concentration, the ratio of $\delta\Sigma^{\gtrless}(\vec{p})$ to $\Sigma^{\gtrless}(\vec{p})$ is on the order of the inverse system volume. We consequently neglect such contributions, obtaining the rate equation

$$\frac{\partial \delta f}{\partial T} = -\frac{1}{\tau_0(\vec{p})} \delta f, \quad (105)$$

in which the one-particle relaxation lifetime $\tau_0(\vec{p})$ is

$$\frac{1}{\tau_0(\vec{p})} = i[\Sigma^>(\vec{p}) - \Sigma^<(\vec{p})] = -2\,\text{Im}\,\Sigma(\vec{p}). \tag{106}$$

Comparing Eqs. (106) and (95) establishes that the particle lifetime determined from the spectral density function is equivalent to the relaxation time of the distribution function for a single electron added to a homogeneous equilibrium system. In the more general situation typified by minority-carrier diffusion or electric current conduction, the deviation from equilibrium is delocalized in momentum space. Therefore, Eqs. (105) and (106) are invalid, and Eq. (84) must instead be solved.

We next consider examples for which the delta-function approximation to the spectral density function, Eq. (96), required to derive the standard Boltzmann equation is not justified. Instead, we employ the quantum Boltzmann equation, Eq. (87), which for a uniform system is

$$i\frac{\partial G^<}{\partial T} = \Sigma^>(\vec{p},\omega)G^<(\vec{p},\omega) - \Sigma^<(\vec{p},\omega)G^>(\vec{p},\omega), \tag{107}$$

to determine the unknown quantities $G^>$ and $G^<$. This formalism is particularly relevant to interband transitions that occur on a time scale much larger than that of intraband relaxation. We accordingly associate the collision terms in the preceding expression solely with interband processes. In other words, intraband scattering is taken into account by approximating the shape of G by that of the stable quasi-equilibrium Green's function defined by the detailed balance relation

$$\Sigma^>_{\text{intra}}G^< - \Sigma^<_{\text{intra}}G^> = 0. \tag{108}$$

The subscript intra refers to the intraband part of the self-energy. From the structure of $\Sigma^>_{\text{intra}}$ and $\Sigma^<_{\text{intra}}$, we derive from Eq. (108) (Danielewicz, 1984)

$$G^>(\vec{p},\omega) = -e^{\beta(\omega - \vec{v}_d\cdot\vec{p} - \mu)}G^<(\vec{p},\omega). \tag{109}$$

Consequently, for fermions using the definition of A,

$$iG^<(\vec{p},\omega) = -2\pi f(\vec{p},\omega)A(\vec{p},\omega) \tag{110}$$

and

$$iG^>(\vec{p},\omega) = 2\pi(1 - f(\vec{p},\omega))A(\vec{p},\omega). \tag{111}$$

6. INTRODUCTION TO NONEQUILIBRIUM MANY-BODY ANALYSES

In these equalities,

$$f(\vec{p}, \omega) = \frac{1}{e^{\beta(\omega - \vec{v}_d \cdot \vec{p} - \mu)} + 1} \qquad (112)$$

is the Fermi–Dirac distribution characterized by local values of the drift velocity $\vec{v}_d$, temperature, and the quasi-Fermi level μ for each band subsystem. For bosons, we have instead

$$iD^<(\vec{q}, \omega) = 2\pi P(\omega) A(\vec{p}, \omega) \qquad (113)$$

and

$$iD^>(\vec{q}, \omega) = 2\pi(1 + P(\omega)) A(\vec{p}, \omega), \qquad (114)$$

with

$$P(\vec{p}, \omega) = \frac{1}{e^{\beta(\omega - \vec{v}_d \cdot \vec{p} - \mu)} - 1}. \qquad (115)$$

The quasiequilibrium functions $G^\gtrless$ or $D^\gtrless$ may be employed as a basis for a perturbative expansion with respect to interband transitions, in exact analogy to the procedure for deriving the standard Boltzmann equation from the unbroadened $G^\gtrless$ of Eq. (98) and Eq. (99).

Relations analogous to Eqs. (110)–(114) hold for the full equilibrium state characterized by the ensemble statistical operator, as illustrated in Appendix C.

V. Electron Self-Energy

6. IMAGINARY PART OF THE SELF-ENERGY AND INELASTIC LOSSES

In previous sections we have related the imaginary part of the electron self-energy to the one-particle relaxation time, and then determined the change in the electron distribution function in terms of the analytic parts of the self-energy function, which give the scattering rates in and out of a given electron state. We accordingly proceed to develop a simple quantum-mechanical approximation for Σ. An appropriate expression may be generated by replacing the classical equation for the power lost by a single highly energetic particle in a polarizable medium characterized by a dielectric function $\varepsilon(\vec{q}, \omega)$ by the corresponding quantum-mechanical formula. We accordingly consider the charge density distribution of a classical point charge $-e$ given by

$$\rho(\vec{r}, t) = -e\delta(\vec{r} - \vec{v}t) \tag{116}$$

with an associated current density

$$\vec{j}(\vec{q}, t) = -e\vec{v}e^{-i\vec{q}\cdot\vec{v}t}. \tag{117}$$

Equivalently, after Fourier transforming with respect to the variables $\vec{r}$ and t,

$$\rho(\vec{q}, \omega) = -2\pi e\delta(\omega - \vec{q}\cdot\vec{v}), \tag{118}$$

and total potential of the moving charged particle is

$$V_{\text{tot}}(\vec{q}, \omega) = -\frac{4\pi e}{q^2}\frac{1}{\varepsilon(\vec{q}, \omega)}2\pi\delta(\omega - \vec{q}\cdot\vec{v}). \tag{119}$$

The dielectric function $\varepsilon(\vec{q}, \omega)$ describes the polarization of the system by each Fourier component of the field of the test charge. After subtracting the bare Coulomb potential from V_{tot}, we generate the change in the potential associated with the response of the surrounding free and bound charges, namely,

$$V_{\text{ind}}(\vec{q}, \omega) = -\frac{4\pi e}{q^2}(\varepsilon^{-1}(\vec{q}, \omega) - 1)\{2\pi\delta(\omega - \vec{q}\cdot\vec{v})\}. \tag{120}$$

Inserting Eq. (120) into $\vec{E}(\vec{q}, \omega) = -i\vec{q}V(\vec{q}, \omega)$, we have

$$\vec{E}_{\text{ind}}(\vec{q}, t) = i\vec{q}\frac{4\pi e}{q^2}(\varepsilon^{-1}(\vec{q}, \vec{q}\cdot\vec{v}) - 1)e^{-i\vec{q}\cdot\vec{v}t}. \tag{121}$$

From Eqs. (121) and (117), the total power absorbed by the system from a high-energy conduction electron at time t equals

$$-\int d^3r \vec{j}(\vec{r}, t)\vec{E}_{\text{ind}}(\vec{r}, t) = \int \frac{d^3q}{(2\pi)^3}(\vec{v}\cdot\vec{q})\frac{4\pi e^2}{q^2}\text{Im}\left\{\frac{-1}{\varepsilon(\vec{q}, \vec{q}\cdot\vec{v})}\right\}$$

$$\approx 2\int \frac{d^3q}{(2\pi)^3}\int_0^\infty \frac{d\omega}{2\pi}\omega\frac{4\pi e^2}{q^2}$$

$$\times \text{Im}\left\{\frac{-1}{\varepsilon(\vec{q}, \omega)}\right\}2\pi\delta(\omega - E_{\vec{k}} + E_{\vec{k}-\vec{q}}). \tag{122}$$

6. INTRODUCTION TO NONEQUILIBRIUM MANY-BODY ANALYSES

In this equation, we have approximated $\vec{v} \cdot \vec{q}$ by $E_{\vec{k}} - E_{\vec{k}-\vec{q}}$ with $\vec{k} = m_c \vec{v}$, with m_c the effective conduction band mass.

By dividing the energy loss component at each frequency ω by the energy, $\omega = E_{\vec{k}} - E_{\vec{k}-\vec{q}}$, lost by the electron in a single scattering event, we determine the inelastic scattering rate for electron–plasma collisions,

$$\frac{1}{\tau_{\text{el-pl}}} = 2 \int \frac{d^3q}{(2\pi)^3} \int_0^\infty \frac{d\omega}{2\pi} \frac{4\pi e^2}{q^2} \text{Im}\left\{\frac{-1}{\varepsilon(\vec{q}, \omega)}\right\} 2\pi \delta(\omega - E_{\vec{k}} + E_{\vec{k}-\vec{q}}). \tag{123}$$

The corresponding quantum-mechanical description of the electron lifetime is provided by the lowest-order term in an expansion with respect to the screened interaction $4\pi e^2 / q^2 \varepsilon(\vec{q}, \omega)$. In this so-called GW approximation (Hedin, 1965),

$$\Sigma(\vec{r}, t, \vec{r}', t') = iG(\vec{r}, t, \vec{r}', t'_+)W(\vec{r}', t', \vec{r}, t). \tag{124}$$

The symbol t'_+ denotes a time infinitesimally later than t', while the screened electron interaction in configuration space is related to the dielectric function by

$$W(\vec{r}', t', \vec{r}, t) = \int \varepsilon^{-1}(\vec{r}', t', \vec{r}'', t) \frac{e^2}{|\vec{r}'' - \vec{r}|} d^3r''. \tag{125}$$

To show that Eq. (124) reduces to Eq. (123) in the classical limit, we first implement the Bloch-state equivalent of Fourier transforming both Σ and G. For example, the Bloch representation of the diagonal matrix elements of an operator $\mathcal{O}(\vec{r} - \vec{r}', t - t')$ is

$$\mathcal{O}_\nu(\vec{k}, t - t') = \frac{1}{\Omega} \int d^3r \int d^3r' u^*_{\nu\vec{k}}(\vec{r}) e^{-i\vec{k}\cdot\vec{r}} u_{\nu\vec{k}}(\vec{r}') e^{i\vec{k}\cdot\vec{r}'} \mathcal{O}(\vec{r} - \vec{r}', t - t'). \tag{126}$$

If we denote the Fourier transform of the screened interaction with respect to the variable $\vec{r} - \vec{r}'$ by $W(\vec{q}, t' - t)$, we find under the assumption that the interband matrix elements of $G^>$ are small,

$$\Sigma^>_c(\vec{k}, t - t') = i\frac{1}{\Omega} \sum_{\nu, \vec{k}'} \int d^3r \int d^3r' u^*_{c\vec{k}}(\vec{r}) u_{\nu\vec{k}'}(\vec{r}) u_{c\vec{k}}(\vec{r}') u^*_{\nu\vec{k}'}(\vec{r}')$$

$$\times \frac{1}{\Omega} \sum_{\vec{q}} e^{i\vec{k}\cdot(\vec{r}' - \vec{r})} e^{-i\vec{k}'\cdot(\vec{r}' - \vec{r})} e^{i\vec{q}\cdot(\vec{r}' - \vec{r})}$$

$$\times G^>_\nu(\vec{k}', t - t') W^<(\vec{q}, t' - t). \tag{127}$$

As the initial and final state electron momenta are close to the Γ point, we ignore Umklapp processes. The double integral then factorizes into the product of an integral over the unit cell of the rapidly varying periodic parts of the Bloch functions and an integral over the slowly varying exponentials. The GW approximation then leads to the following conduction band self-energy:

$$\Sigma_c^{\gtrless}(\vec{k}, \omega) = \frac{i}{\Omega} \sum_{v,\vec{q}} \int \frac{d\omega'}{2\pi} I_{vc}(\vec{k}+\vec{q}, \vec{k}) W^{\gtrless}(\vec{q}, \omega') G_v^{\gtrless}(\vec{k}+\vec{q}, \omega+\omega'). \quad (128)$$

In Eq. (128), the summation over v is performed over all bands, while the overlap matrix element averaged over spins is given by

$$I_{n',n}(\vec{k}+\vec{q}, \vec{k}) = \frac{1}{2} \sum_{\alpha,\alpha'} |\langle n', \alpha', \vec{k}+\vec{q} | n, \alpha, \vec{k} \rangle_B|^2. \quad (129)$$

If we consider the conduction band for which the squared overlap matrix element is unity and apply Eq. (95), we generate the following expression for the contribution to the inverse electron lifetime from plasmon collisions:

$$\frac{1}{\tau_{el-pl}} = 2 \int \frac{d^3q}{(2\pi)^3} \int_{-\infty}^{\infty} \frac{d\omega}{2\pi} \frac{4\pi e^2}{q^2} \mathrm{Im}\left\{\frac{-1}{\varepsilon(q,\omega)}\right\}$$
$$\times (1 - f(E_{\vec{k}-\vec{q}}) + P(\omega)) 2\pi \delta(\omega - E_{\vec{k}} + E_{\vec{k}-\vec{q}}). \quad (130)$$

This equation immediately reduces to Eq. (123) for large $E(\vec{k})$ and $|\beta\omega|$, for which $f(E(\vec{k}-\vec{q})) \to 0$ over most of the integration region while $1 + P(\omega) \to \Theta(\omega)$.

Having demonstrated that the GW approximation coincides with the classical result for the electron lifetime in the limit of large electron energies and may therefore be employed to construct the generalized collision integral appearing in the quantum-mechanical version of the Boltzmann equation, we proceed to discuss the contributions to the self-energy arising from other interactions. Here we first observe from Eq. (122) that the probability of the electron losing energy to a collective excitation of momentum $\vec{q}$ and energy ω is proportional to the dissipative part of the inverse dielectric function, $-\mathrm{Im}\{\varepsilon^{-1}(\vec{q}, \omega)\}$. Consequently, this quantity acts as a spectral density function for plasma excitations induced by charge density fluctuations. Electron–impurity and electron–phonon interactions may be similarly incorporated in the self-energy. In particular, for each electron–phonon interaction characterized by a coupling constant M, cf. Eq. (20), after introducing the phonon field operators $\phi_{\vec{q}} = M_{\vec{q}} b_{\vec{q}} + M^*_{-\vec{q}} b^\dagger_{-\vec{q}}$ and the phonon Green's function

$$iD_{\vec{q}}(t', t) = \langle \phi_{\vec{q}}(t')\phi_{-\vec{q}}(t) \rangle, \qquad (131)$$

we obtain

$$\Sigma_c^{\gtrless}(\vec{k}, \omega) = i \sum_{v,\vec{q}} \int \frac{d\omega'}{2\pi} G_v^{\gtrless}(\vec{k}+\vec{q}, \omega+\omega') D_{\vec{q}}^{\lessgtr}(\omega') I_{cv}(\vec{k}, \vec{k}+\vec{q}). \qquad (132)$$

The overlap integral I_{cv} is only present for long-range interactions, as remarked previously.

The structure of the phonon propagators

$$iD_{\vec{q}}^{>}(\omega) = 2\pi(1 + P(\omega))[\delta(\omega - \omega_{\vec{q}}) - \delta(\omega + \omega_{\vec{q}})]\,|M_{\vec{q}}|^2 \qquad (133)$$

and

$$iD_{\vec{q}}^{<}(\omega) = 2\pi P(\omega)[\delta(\omega - \omega_{\vec{q}}) - \delta(\omega + \omega_{\vec{q}})]\,|M_{\vec{q}}|^2 \qquad (134)$$

resembles the plasmon propagator appearing in Eq. (130). The result of Eq. (132) is equivalent to the replacement

$$\frac{4\pi e^2}{q^2} \operatorname{Im}\left\{-\frac{1}{\pi}\varepsilon^{-1}(\vec{q}, \omega)\right\} \to [\delta(\omega - \omega_{\vec{q}}) - \delta(\omega + \omega_{\vec{q}})]\,|M_{\vec{q}}|^2 \qquad (135)$$

in Eq. (130), as will be discussed in the following section. To model the scattering of electrons from randomly distributed point impurities, we substitute

$$2\pi N_{\text{imp}}\,|v_{\vec{q}}^{\text{imp}}|^2\,\delta(\omega) \qquad (136)$$

for both Eqs. (133) and (134). In Eq. (136), N_{imp} is the impurity concentration while $v_{\vec{q}}^{\text{imp}}$ is given either by the Fourier component of an individual impurity potential or, more accurately, by the squared matrix element of the off-shell scattering t-matrix, e.g., $|v_{\vec{q}}^{\text{imp}}|^2 \to |\langle \vec{k}+\vec{q}|\,t(\omega)\,|\vec{k}\rangle|^2$. Finally, the squared overlap matrix element I in Eq. (132) is associated with the long-range component of the impurity potential, in analogy with our discussion of phonon matrix elements.

7. REAL PART OF THE SELF-ENERGY AND BANDGAP RENORMALIZATION

In this section, we turn from the determination of particle lifetimes in excited semiconductors to quasi-particle energy renormalization (Inkson, 1976; Sernelius, 1986; Abram et al., 1984; Berggren and Sernelius, 1981). Unfortunately, while the one-particle energy level shifts for any electron momenta $\vec{q}$ are in principle obtained by solving Eq. (94), the GW expression for the Green's function is a functional of self-energy from Eq. (66) and Eq. (124). The computation of Σ therefore requires implementation of a complicated self-consistent procedure, and several additional simplifying assumptions are generally employed. For example, since the renormalization is often only required in a small region around a given momentum $\vec{p}$, a single value is adopted for all electronic states in a given band. Further, the method is simplified by replacing the interacting Green's function entering the GW formula by the shifted noninteracting Green's function $G_0(\omega - \Delta, p)$, in which Δ is the momentum-independent energy renormalization (Hedin, 1965). Self-consistency then implies that the renormalized Green's function

$$G(\omega, p) = \frac{1}{\omega - E_p - \Sigma_0(\omega - \Delta, p)}, \qquad (137)$$

where $\Sigma_0 = WG_0$, should have a pole at the same position as the zeroth-order approximation $G_0(\omega - \Delta, p)$, i.e., at $\omega = E_p + \Delta$. Therefore, we require that Δ be given in terms of the unrenormalized electron energy E_p by

$$E_p + \Delta - E_p - \Sigma_0(\omega - \Delta, p) = 0, \qquad (138)$$

or, eliminating ω,

$$\Delta = \Sigma_0(E_p, \vec{p}). \qquad (139)$$

The advantage of this procedure is that $\Sigma_0(\omega, \vec{p})$ is far easier to compute than the GW expression, Eq. (124). Further, Eq. (138) is nonperturbative in nature and may be valid even if Δ is large. While the contributions to the bandgap renormalization from various interactions, including electron–phonon, electron–electron, and electron–impurity scattering, cannot be measured separately, the difference in the state energies between the doped and the intrinsic semiconductor provides a relatively simply interpreted experimental quantity. Applying the G_0W expression for the self energy for p-type material at finite temperature with a negligible minority electron concentration the displacement of the valence band with respect to its $T = 0$ position is (Bardyszewski and Yevick, 1989d, 1987)

$$\Delta E_h = \int \frac{1}{\pi} \left\{ \int_0^\infty dq \int_{-\infty}^\infty \frac{d\omega}{\pi} \text{Im}[\varepsilon_{\text{ext}}^{-1}(q, \omega)(1 + P(\omega)) - \varepsilon_{\text{int}}^{-1}(q, \omega)\Theta(\omega)] \right.$$

$$\times \left(\frac{1}{-\omega + E_h(q) - E_h(0)} + \frac{1}{-\omega + E_l(q) - E_l(0)} \right)$$

$$+ \text{Re}\left[\int_0^\infty dq \{(1 - f_l(q))\varepsilon_{\text{ext}}^{-1}(q, E_l(q) - E_l(0)) \right.$$

$$\left. \left. + (1 - f_h(q))\varepsilon_{\text{ext}}^{-1}(q, E_h(q) - E_h(0))\} \right] \right\}. \tag{140}$$

The quantities $E_h(q)$, $E_l(q)$, $f_h(q)$, and $f_l(q)$ denote the unrenormalized heavy- and light-hole valence band energies and the corresponding Fermi functions for momentum q, respectively, while ε_{ext} and ε_{int} are the longitudinal finite-temperature dielectric function in the extrinsic and the $T = 0$ longitudinal dielectric function in the intrinsic semiconductor, respectively. The displacement of the conduction band is

$$\Delta E_c = \frac{2}{\pi} \int_0^\infty dq \int_{-\infty}^\infty \frac{d\omega}{\pi} \frac{\text{Im}\{\varepsilon_{\text{ext}}^{-1}(q, \omega)(1 + P(\omega)) - \varepsilon_{\text{int}}^{-1}(q, \omega)\theta(\omega)\}}{\omega + E_c(q) - E_c(0)}. \tag{141}$$

All common models for the dielectric function satisfy certain sum rules, such as the f-sum rule

$$\int_0^\infty \omega \, \text{Im}[\varepsilon^{-1}(q, \omega)] \, d\omega = -\frac{\pi}{2} \omega_p^2, \tag{142}$$

where $\omega_p^2 = 4\pi e^2 n/m^*\varepsilon_\infty$ represents the squared plasmon frequency. Calculations of the bandgap shifts are therefore rather insensitive to the model selected. In terms of the dielectric functions presented in the following sections, reasonable estimates of the zero-temperature bandgap renormalization are obtained with a simple plasmon pole description for $\varepsilon_{\text{ext}}^{-1}$, while for finite temperatures a full random phase approximation expression for the dielectric function incorporating both the symmetry of

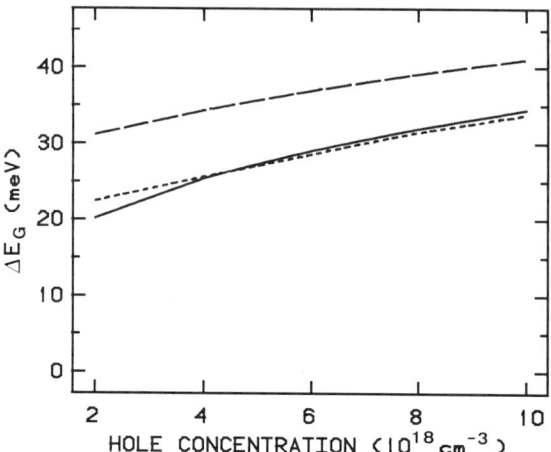

FIG. 2. The $T = 300$ K bandgap shift in milli-electron volts calculated with a simple empirical formula (solid line), the damped plasmon-pole approximation (long-dashed line) and the RPA (short-dashed line) as a function of hole concentration. (From Fig. 2 of Bardyszewski and Yevick, 1989d.)

the valence band wavefunctions and the effects of heavy-to-light hole transitions should be employed (Overhauser, 1971). Results for room-temperature bandgap shifts are presented in Fig. 2.

VI. Dielectric Function Models

We return to the GW expression for the self-energy,

$$\Sigma^{>}(\vec{r}, t, \vec{r}', t') = i \int G^{>}(\vec{r}, t, \vec{r}', t') \{\varepsilon^{-1}(\vec{r}', t', \vec{r}'', t)\}^{<} \frac{4\pi e^2}{|\vec{r}'' - \vec{r}|} d^3 r'', \quad (143)$$

which forms the basis for our calculations of electron–electron interaction effects. In order to develop approximations for the inverse dielectric function, we will first demonstrate that Eq. (143) can be interpreted as the scattering of the electron from charge density fluctuations. We will show that $\varepsilon^{-1}(q, \omega)$ is related to the density–density correlation function, which justifies associating $\text{Im}\{\varepsilon^{-1}(\vec{q}, \omega)\}$ with the plasmon propagator. The fluctuation of the ensemble-averaged one-particle density $\langle \delta \hat{n}(\vec{r}, t) \rangle$ is most simply calculated in the interaction picture in which the electron density operator is, cf. Eq. (60) and (61),

$$\hat{n}_\mathrm{I}(\vec{r}, t) = \hat{\Phi}_\mathrm{I}^{\dagger}(\vec{r}, t) \hat{\Phi}_\mathrm{I}(\vec{r}, t). \quad (144)$$

The statistical operator is evaluated in terms of its value at $t = -\infty$

6. INTRODUCTION TO NONEQUILIBRIUM MANY-BODY ANALYSES

through the equation

$$i\frac{\partial \hat{\rho}_1(t)}{\partial t} = [\hat{H}'_1, \hat{\rho}_1(t)]_-. \qquad (145)$$

In this formula, the perturbation in the Hamiltonian induced by the external potential is

$$\hat{H}'_1 = \int \hat{n}_1(\vec{r}, t) V_{\text{ext}}(\vec{r}, t) d^3r. \qquad (146)$$

Therefore, denoting $\rho(t = -\infty)$ by $\hat{\rho}_0$, to first order in $\hat{H}'_1(t)$,

$$\delta\hat{\rho}_1(t) = \hat{\rho}_1(t) - \hat{\rho}_0 = -i \int_{-\infty}^{t} [\hat{H}'_1(t'), \hat{\rho}_0]_- dt'. \qquad (147)$$

Multiplying both sides of this equation with $\hat{n}_1(\vec{r}, t)$ and taking a trace yields immediately the induced charge distribution,

$$n_{\text{ind}}(\vec{r}, t) = \delta\langle \hat{n}_1(\vec{r}, t) \rangle = -i \int_{-\infty}^{t} dt' \langle [\hat{n}_1(\vec{r}, t), \hat{H}'_1(t')]_- \rangle_0. \qquad (148)$$

To interpret the inverse of the longitudinal dielectric constant in terms of the charge density fluctuations, we recall that ε^{-1} relates the potential present in the system to the external potential through the relationship

$$V_{\text{tot}}(\vec{r}, t) = \int d^3r' \int_{-\infty}^{\infty} dt' \varepsilon^{-1}(\vec{r}, \vec{r}', t, t') V_{\text{ext}}(\vec{r}', t'), \qquad (149)$$

or, equivalently,

$$V_{\text{ext}}(\vec{r}, t) = \int d^3r' \int_{-\infty}^{\infty} dt' \varepsilon(\vec{r}, \vec{r}', t, t') V_{\text{tot}}(\vec{r}', t'). \qquad (150)$$

At the same time, the internal potential can be calculated directly from the induced charge, since

$$V_{\text{tot}}(\vec{r}, t) = V_{\text{ext}}(\vec{r}, t) + \int d^3r' \frac{n_{\text{ind}}(\vec{r}', t)}{|\vec{r} - \vec{r}'|}. \qquad (151)$$

Consequently, in terms of the so-called retarded response function, omitting the subscript I,

$$\chi(\vec{r}'', t, \vec{r}', t') = -i\Theta(t - t')\langle[\hat{n}(\vec{r}'', t), \hat{n}(\vec{r}', t')]_-\rangle_0, \qquad (152)$$

we have

$$V_{\text{tot}}(\vec{r}, t) = V_{\text{ext}}(\vec{r}, t) + \int d^3r'' \int d^3r' \int_{-\infty}^{\infty} dt' \frac{\chi(\vec{r}'', t, \vec{r}', t')V_{\text{ext}}(\vec{r}', t')}{|\vec{r} - \vec{r}''|} \qquad (153)$$

Therefore, comparing Eqs. (149) and (153),

$$\varepsilon^{-1}(\vec{r}, t, \vec{r}', t') = \delta(\vec{r} - \vec{r}')\delta(t - t') + \int d^3r'' \frac{1}{|\vec{r} - \vec{r}''|} \chi(\vec{r}'', t, \vec{r}', t'). \qquad (154)$$

Since the correlation function between electrons that are tightly bound to lattice positions is very short-range, it is generally approximated as a delta function of position and time for the purposes of calculations involving conduction and valence band electrons. This yields finally

$$\varepsilon^{-1}(\vec{r}, t, \vec{r}', t') = \frac{1}{\varepsilon_\infty}\delta(\vec{r} - \vec{r}')\delta(t - t') + \int d^3r'' \frac{1}{|\vec{r} - \vec{r}''|} \chi_v(\vec{r}'', t, \vec{r}', t'). \qquad (155)$$

where ε_∞ incorporates the core polarizability while χ_v, includes only the influence of the valence- and conduction-band electrons.

As the electron density that enters $\varepsilon^{-1}(\vec{r}, t, \vec{r}', t')$ is a product of two fermion field operators, the inverse dielectric function itself behaves as a boson propagator. Therefore, in quasi-equilibrium, formulas analogous to Eqs. (113) and (114) are valid, but with the spectral density function $A(\vec{q}, \omega)$ replaced by $\text{Im}(-(1/\pi)\varepsilon^{-1}(\vec{q}, \omega))$. Consequently, for a homogeneous medium,

$$i[\varepsilon^{-1}(\vec{q}, \omega)]^> = 2\pi(1 + P(\omega))\,\text{Im}\left(-\frac{1}{\pi}\varepsilon^{-1}(\vec{q}, \omega)\right) \qquad (156)$$

and

$$i[\varepsilon^{-1}(\vec{q}, \omega)]^< = 2\pi P(\omega)\,\text{Im}\left(-\frac{1}{\pi}\varepsilon^{-1}(\vec{q}, \omega)\right), \qquad (157)$$

supporting our previous analogy between phonon and plasmon propagators as expressed in Eqs. (133)–(136). That $\mathrm{Im}(-(1/\pi)\varepsilon^{-1}(\vec{q}, \omega))$ describes the magnitude of charge density fluctuations in the system is apparent from

$$[i\varepsilon^{-1}(\vec{q}, t)]^{>} = \frac{4\pi e^2}{q^2} \langle \hat{n}_{\vec{q}}(t)\hat{n}_{-\vec{q}}(0)\rangle. \tag{158}$$

Together with Eq. (156) this formula constitutes the basis of the fluctuation-dissipation theorem in the quasi-equilibrium case.

While we have demonstrated that the inverse dielectric function can be simply related to the correlation between the charge density oscillations in the semiconductor, these fluctuations have a variety of origins. The resulting local field effects that are induced by the charge configuration at small distances from the probe charge are generally difficult to evaluate. Instead, we approximate the interactions among the individual electrons by those between an electron and the self-consistent potential of the surrounding plasma. Mathematically, we substitute V_{tot} in place of V_{ext} and replace χ by the polarizability of the noninteracting gas, χ^0, in Eq. (153). Comparing the resulting expression with Eq. (150) yields the random phase (self-consistent field) approximation (Lindhard, 1954)

$$\varepsilon_{\mathrm{RPA}}(\vec{r}, t, \vec{r}', t') = \varepsilon_\infty \delta(\vec{r} - \vec{r}')\delta(t - t') - \int d^3r'' \frac{1}{|\vec{r} - \vec{r}''|} \chi^0_v(\vec{r}'', t, \vec{r}', t'). \tag{159}$$

Specializing to noninteracting electrons in a semiconductor with statistical operator $\hat{\rho}^0$, we have for the retarded function

$$\chi^{0,R}(\vec{r}', t', \vec{r}, t) = -i\,\mathrm{Tr}\{\hat{\rho}^0[\hat{n}(\vec{r}', t'), \hat{n}(\vec{r}, t)]_-\}\Theta(t' - t). \tag{160}$$

In analogy with the definition of G^R in Eq. (76), we may consider $\chi^{0,R}$ as a combination of the functions $\chi^{\gtrless}$ defined by

$$i\chi^{0>}(\vec{r}', t', \vec{r}, t) = \mathrm{Tr}\{\hat{\rho}^0 \Phi^\dagger(\vec{r}', t')\Phi(\vec{r}', t')\Phi^\dagger(\vec{r}, t)\Phi(\vec{r}, t)\} \tag{161}$$

and

$$i\chi^{0<}(\vec{r}', t', \vec{r}, t) = \mathrm{Tr}\{\hat{\rho}^0 \Phi^\dagger(\vec{r}, t)\Phi(\vec{r}, t)\Phi^\dagger(\vec{r}', t')\Phi(\vec{r}', t')\}. \tag{162}$$

This may be reexpressed in terms of noninteracting Green's functions as

$$i\chi^{0\gtrless}(\vec{r}', t', \vec{r}, t) = G^{\gtrless}(\vec{r}', t', \vec{r}, t)G^{\lessgtr}(\vec{r}, t, \vec{r}', t'). \tag{163}$$

Eq. (163) is easily derived by expressing the field operators in Eqs. (161) and (162) in terms of the eigenstates of the one-particle Hamiltonian and averaging with ρ^0.

In the Bloch wave representation, if $E_n(\vec{k})$ is the energy of band n for momentum $\vec{k}$ and η denotes a positive infinitesimal,

$$\hat{n}(\vec{r}, t) = \frac{1}{\Omega^2} \sum_{n,\vec{k},n',\vec{k}'} u^*_{n,\vec{k}}(\vec{r}) u_{n',\vec{k}'}(\vec{r}) a^\dagger_{n,\vec{k}} a_{n',\vec{k}'} e^{-i(E_{n'}(\vec{k}') - E_n(\vec{k}) - i\eta)t} e^{i(\vec{k}' - \vec{k})\cdot\vec{r}}. \tag{164}$$

We accordingly insert Eq. (164) into Eqs. (161) and (162), evaluate the resulting commutator, and apply the identity $[A, BC]_- = [A, B]_- C + B[A, C]_-$ twice. Further if f denotes the Fermi function, the equality

$$\text{Tr}\{\hat{\rho}^0 a^\dagger_{n_1,\vec{k}_1} a_{n_2,\vec{k}_2}\} = \delta_{n_1,n_2} \delta_{\vec{k}_1,\vec{k}_2} f(E_{n_1}(\vec{k}_1)) \tag{165}$$

implies

$$\text{Tr}\{\hat{\rho}^0 [a^\dagger_{n_1,\vec{k}_1} a_{n_2,\vec{k}_2}, a^\dagger_{n_3,\vec{k}_3} a_{n_4,\vec{k}_4}]_-\}$$
$$= \delta_{n_2,n_3} \delta_{\vec{k}_2,\vec{k}_3} \delta_{n_1,n_4} \delta_{\vec{k}_1,\vec{k}_4} (f(E_{n_1}(\vec{k}_1)) - f(E_{n_2}(\vec{k}_2))). \tag{166}$$

Therefore,

$$\chi_v^{0,R}(\vec{r}', t', \vec{r}, t) = -\frac{i}{\Omega^2} \sum_{n_1,\vec{k}_1} \sum_{n_2,\vec{k}_2} u^*_{n_1,\vec{k}_1}(\vec{r}') u_{n_2,\vec{k}_2}(\vec{r}')$$
$$\times u^*_{n_2,\vec{k}_2}(\vec{r}) u_{n_1,\vec{k}_1}(\vec{r}) e^{i(\vec{k}_2 - \vec{k}_1)\cdot(\vec{r}' - \vec{r})}$$
$$\times e^{-i(E_{n_2}(\vec{k}_2) - E_{n_1}(\vec{k}_1) - i\eta)(t' - t)} \Theta(t' - t)$$
$$\times (f(E_{n_1}(\vec{k}_1)) - f(E_{n_2}(\vec{k}_2))). \tag{167}$$

Introducing the average and relative variables $\vec{R}$ and $\vec{r}_a$ and Fourier transforming with respect to $\vec{r}_a$, we obtain after the transformation $t' - t \to t$

$$\chi_v^{0,R}(\vec{R}, \vec{q}, t) = -\frac{i}{\Omega^2} \sum_{n_1,\vec{k}_1} \sum_{n_2,\vec{k}_2} \Theta(t) e^{-i(E_{n_2}(\vec{k}_2) - E_{n_1}(\vec{k}_1) - i\eta)t}$$
$$\times (f(E_{n_1}(\vec{k}_1)) - f(E_{n_2}(\vec{k}_2))) \times \int d^3 r_a e^{i(\vec{k}_2 - \vec{k}_1 - \vec{q})\cdot\vec{r}_a}$$
$$\times u^*_{n_1,\vec{k}_1}\left(\vec{R} + \frac{\vec{r}_a}{2}\right) u_{n_2,\vec{k}_2}\left(\vec{R} + \frac{\vec{r}_a}{2}\right) u^*_{n_2,\vec{k}_2}\left(\vec{R} - \frac{\vec{r}_a}{2}\right) u_{n_1,\vec{k}_1}\left(\vec{R} - \frac{\vec{r}_a}{2}\right). \tag{168}$$

6. INTRODUCTION TO NONEQUILIBRIUM MANY-BODY ANALYSES 355

If the momentum transfer $\vec{k}_2 - \vec{k}_1 = \vec{q}$ is small compared to the reciprocal lattice vector, the exponential factor varies slowly over the unit cell dimensions so that the integral over positions is approximately

$$\int d^3 r_a e^{i(\vec{k}_2-\vec{k}_1-\vec{q})\cdot \vec{r}_a} u^*_{n_1,\vec{k}_1}\left(\vec{R}+\frac{\vec{r}_a}{2}\right) u_{n_2,\vec{k}_2}\left(\vec{R}+\frac{\vec{r}_a}{2}\right) u^*_{n_2,\vec{k}_2}\left(\vec{R}-\frac{\vec{r}_a}{2}\right) u_{n_1,\vec{k}_1}\left(\vec{R}-\frac{\vec{r}_a}{2}\right)$$

$$\approx \delta_{\vec{k}_2-\vec{k}_1,\vec{q}} \frac{\Omega}{\Omega_0} \int_{\Omega_0} d^3 r_a u^*_{n_1,\vec{k}_1}\left(\vec{R}+\frac{\vec{r}_a}{2}\right) u_{n_2,\vec{k}_2}\left(\vec{R}+\frac{\vec{r}_a}{2}\right)$$

$$\times u^*_{n_2,\vec{k}_2}\left(\vec{R}-\frac{\vec{r}_a}{2}\right) u_{n_1,\vec{k}_1}\left(\vec{R}-\frac{\vec{r}_a}{2}\right), \qquad (169)$$

where Ω_0 denotes the unit cell volume.

We next observe that χ_v^0 is a rapidly oscillating function of $\vec{R}$ with a period given by the primitive cell dimensions. We neglect such variations, which are the source of the local field corrections, by averaging $\chi_v^{0,R}$ over the unit cell, i.e., we compute

$$\chi_v^0(\vec{q}, t) = \frac{1}{\Omega_0} \int_{\Omega_0} \chi_v^{0,R}(\vec{R}, \vec{q}, t)\, d^3 R. \qquad (170)$$

Reexpressing the integrals over $\vec{R}$ and $\vec{r}_a$ in terms of $\vec{r}$ and $\vec{r}'$ transforms Eq. (170) into a product of two identical integrals. We then obtain

$$\chi^{0,R}(\vec{q}, t) = -2\frac{i}{\Omega} \sum_{\vec{k}} \sum_{n_1} \sum_{n_2} I_{n_1,n_2}(\vec{k}, \vec{k}+\vec{q})$$

$$\times [f(E_{n_1}(\vec{k})) - f(E_{n_2}(\vec{k}+\vec{q}))]\Theta(t) e^{-i(E_{n_2}(\vec{k}+\vec{q})-E_{n_1}(\vec{k})-i\eta)t}. \qquad (171)$$

The averaged overlap integral is defined by Eq. (129). After Fourier transforming in frequency, Eq. (171) becomes

$$\chi^{0,R}(\vec{q}, \omega) = 2 \sum_{n_1,n_2} \sum_{\vec{k}} I_{n_1,n_2}(\vec{k}, \vec{k}+\vec{q}) \frac{f(E_{n_1}(\vec{k})) - f(E_{n_2}(\vec{k}+\vec{q}))}{\omega - E_{n_2}(\vec{k}+\vec{q}) - E_{n_1}(\vec{k}) + i\eta}. \qquad (172)$$

Inserting Eq. (172) into the general self-consistent dielectric constant expression, Eq. (159), and employing the notation $f_n(\vec{k}) = f(E_n(\vec{k}) - \mu_n)$ for the quasi-Fermi distribution function for a state in band n with quasi-Fermi level μ_n, we obtain the macroscopic RPA retarded dielectric function

$$\varepsilon_{\text{RPA}}^{\text{R}} = \varepsilon_\infty + \frac{8\pi e^2}{q^2} \sum_{n_1,n_2} \sum_{\vec{k}} I_{n_1,n_2}(\vec{k}, \vec{k}+\vec{q}) \frac{f_{n_2}(\vec{k}+\vec{q}) - f_{n_1}(\vec{k})}{\omega - E_{n_2}(\vec{k}+\vec{q}) - E_{n_1}(\vec{k}) + i\eta}. \quad (173)$$

At frequencies corresponding to zeros of the dielectric functions, collective plasma excitations (charge density oscillations) with an inverse lifetime proportional to the imaginary part of $\varepsilon(\vec{q}, \omega)$ are maintained in the semiconductor in the absence of an applied external electric field. The plasmon frequency is only weakly wavevector-dependent in the $q \to 0$ limit, and is therefore often approximated by a q-independent function. On the other hand, the excitation lifetime varies strongly with both temperature and carrier density. We analyze first the damping in a low-density plasma composed of conduction band electrons. Here, at zero temperature, a longitudinal plasma oscillation of energy ΔE cannot lose energy through the excitation of an individual free electron from an initial state $(\vec{k}, E)$ to a higher conduction band state $(\vec{k} + \Delta \vec{k}, E + \Delta E)$ unless the momentum of the plasmon equals $\Delta \vec{k}$. As the temperature increases, more electrons are thermally excited to states of higher energy where the dispersion relation has a larger slope. Hence these electrons can absorb energy from plasmons of smaller momentum enhancing the plasmon (Landau) damping. A considerably different physical situation is, however, associated with the valence band, as a carrier may at $T = 0$ absorb plasmons with $\vec{q} = \Delta \vec{k} \approx 0$ through vertical light-hole–heavy-hole transitions. (Bardyszewski, 1986; Combescot and Nozières, 1972). In contradiction to the single-band case, increasing the temperature leads to a redistribution of holes away from the valence band maximum and therefore suppresses such processes. Although the intraband plasma absorption partially compensates this effect, its contribution to the damping is in general far smaller for temperatures that are low compared to room temperature. To illustrate, we display the spectral density function $-(1/\pi) \text{Im}(\varepsilon^{-1}(\vec{q}, \omega))$ for GaAs at $q = 0.1 |\vec{q}_p|$ as a function of frequency at a hole concentration of $p = 2 \times 10^{18}$ cm^{-3} for both $T = 0$ K, Fig. 3a, and $T = 200$ K, Fig. 3b (Yevick and Bardyszewski, 1989b). The plasmon momentum q_p is given in terms of the effective heavy- and light-hole density of state masses m_h and m_l through the relationship $q_p^2/2m_l = \hbar \bar{\omega}_p$, with

$$(\hbar \bar{\omega}_p)^2 = \frac{4\pi e^2}{\varepsilon_\infty} \frac{\sqrt{m_l} + \sqrt{m_h}}{m_l^{3/2} + m_h^{3/2}}. \quad (174)$$

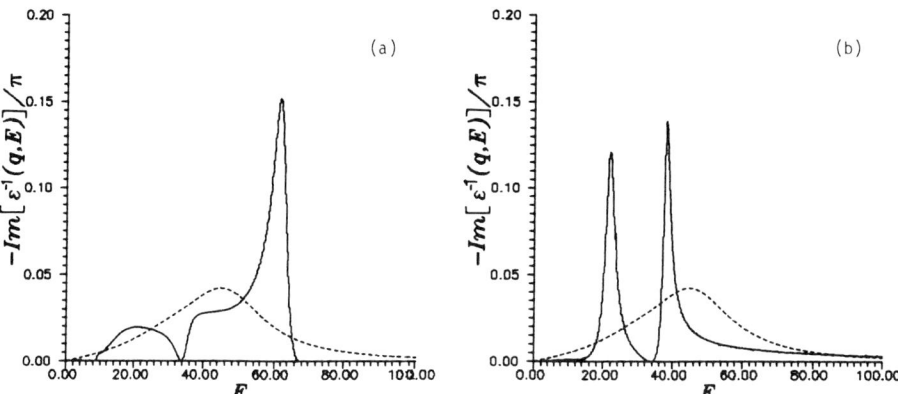

FIG. 3. The room-temperature RPA (solid line) and damped-plasmon pole (dashed line) spectral densities including phonon effects as functions of energy E in milli-electron volts at a hole concentration $p = 2 \times 10^{18}\,\text{cm}^{-3}$ for $q = 0.1q_p$ at (a) $T = 0\,\text{K}$ and (b) $T = 200\,\text{K}$. (From Fig. 2 of Yevick and Bardyszewski, 1989b.)

The two sharp peaks at high temperature in Fig. 3 correspond to nearly undamped plasmon and longitudinal optical phonon modes, which for $q = 0.1\,|\vec{q}_\text{p}|$ are located at 25.9 and 36.5 meV, respectively. At $T = 0\,\text{K}$, these lines are suppressed and a single broad line at a considerably higher energy instead dominates.

To simplify many-body calculations containing multiple integrations over the momenta and energy of the interacting electrons, the dielectric-function spectral density is often approximated by a Lorentzian function of energy centered at the plasmon energy at small momentum transfer and at the electron energy at momenta for which electron-hole pair excitations dominate. In particular, introducing the squared optical phonon frequency, which depends on the hole concentration p through

$$\omega_\text{p}^2 = \frac{2\pi e^2 p}{\varepsilon_\infty}\left(\frac{1}{m_\text{h}} + \frac{1}{m_\text{l}}\right), \qquad (175)$$

we approximate (Overhauser, 1971; Rice, 1974)

$$\varepsilon^{-1}(\vec{q},\omega) = \varepsilon_\infty^{-1}\left[1 + \frac{\omega_\text{p}^2}{2\omega_q}\left(\frac{1}{\omega - \omega_q + i\Gamma_q} - \frac{1}{\omega + \omega_q + i\Gamma_q}\right)\right]. \qquad (176)$$

The plasmon frequency, obtained from

$$\omega_q^2 = \omega_\text{p}^2 - \Gamma_0^2 + (\omega_\text{p}^2/q_\text{TF}^2)q^2 + bq^4, \qquad (177)$$

interpolates between the $q \to 0$ plasma frequency and the free-electron dispersion relation at large q. In Eq. (177), $b = (\hbar/4m_{\rm DV})^2$, with $m_{\rm DV}$ the combined density of states valence band mass, $(m_{\rm h}^{3/2} + m_{\rm l}^{3/2})^{2/3}$, while $\Gamma_q^2 = \Gamma_0^2 + bq^4$ models the broadening of the plasmon resonance associated with the inelastic losses from real light-hole–heavy-hole band transitions. In a p-type III–V semiconductor material at low temperature, the plasmon linewidth broadening factor Γ_q is large. At high T, on the other hand, Γ_q can be instead neglected. Observe, however, that the dashed curve in Fig. 3, which is associated with the standard damped plasmon pole approximation, Eq. (176), is in fact far wider than that obtained from the exact finite temperature calculation.

VII. Intraband Processes and Transport

8. Hot Electron Transport

As noted previously, single-particle lifetimes in a plasma are determined by the electron self-energy. Theoretical results, especially for hot electron lifetimes, are therefore relatively easily derived and can be correlated with recent measurements (Levi and Yafet, 1987). In a p-doped semiconductor, the electron decay rate may, for example, be determined from the dependence of the collector current in a heterojunction bipolar transistor with a large-bandgap emitter and a small-bandgap heavily doped base as a function of the base-collector potential (Hayes et al., 1986; Levi et al., 1987). In this "hot-electron spectroscopy" procedure, a highly monoenergetic beam of electrons is injected into the base region over the slight upward bend in the potential barrier at the emitter side of the emitter–base junction. To monitor the energy distribution of the electrons exiting the base region, the collector-base potential barrier height $\phi_{\rm bc}$ is raised or lowered by changing the collector-base voltage. Since only electrons that have a component, $p_\perp$, of momentum transverse to the interface greater than $\sqrt{2m_e^* \phi_{\rm bc}}$ are registered in the collector, the derivative of the collector current with respect to the base-collector potential is proportional to the electron density distribution $n(p_\perp)$. Measurements for AlGaAs HBTs with a highly doped $N_a = 2 \times 10^{18}\,{\rm cm}^{-3}$ p-GaAs base region indicate that the hot carriers undergo strong scattering leading to complete thermalization in a base length of 900 Å. Quasi-equilibrium carrier injection can also be achieved by compositionally grading the base region to eliminate the emitter–base barrier. In one such measurement, the electron temperature at the collector was found to be 650 K, while a Monte Carlo simulation instead predicted $T_{\text{eff}} = 2,000$ K (Hayes and Harbison, 1988). The origin of the discrepancy was that the theoretical calculation neglected electron–hole and electron–hole plasmon collisions. The carrier heating by the built-in electric field in the base region strongly enhanced such processes, which are modeled by incorporating the valence-band RPA dielectric function into Eq. (130). The resulting formula can subsequently be $\sqrt{}$ewritten as (Bardyszewski and Yevick, 1989b)

$$\frac{1}{\tau(E)} = \frac{2e^2 m_e}{\pi \hbar^2 k} \int_0^\infty \frac{dq}{q} \int_{\omega_{\min}}^{\omega_{\max}} d\omega \, \text{Im}\{-\varepsilon_R^{-1}(\vec{q}, \omega)\}[1 + P(\omega)]. \quad (178)$$

In this equation, the variables $\omega_{\max} = q(2k - q)/2m_e$ and $\omega_{\min} = -q(2k + q)/2m_e$ represent the kinematic limits for the production of an excitation of wavevector q in a collision with an electron of wavevector $\vec{k}$. Electron–optical phonon scattering, which may be parameterized by the longitudinal and transverse optical phonon frequencies ω_{LO} and ω_{TO}, is further included in Eq. (178) by adding the term

$$\Delta \varepsilon^{\text{latt}}(q, \omega) = \varepsilon_\infty (\omega_{\text{LO}}^2 - \omega_{\text{TO}}^2)/(\omega_{\text{TO}}^2 - \omega^2) \quad (179)$$

to the dielectric function. For small $\vec{q}$, energy and momentum conservation restrict the number of available electron–hole scattering channels, and the contribution from electron–plasmon interactions to $\text{Im}(\varepsilon^{-1}(\vec{q}, \omega))$ is dominant.

Since the temperature-dependent dielectric function is relatively complicated to program, theoretical analyses are often performed within the framework of the plasmon pole approximation. This technique greatly simplifies the determination of the electron lifetime in comparison to a direct calculation of Eq. (178) incorporating the full temperature-dependent dielectric function. For p-type GaAs with $p = 2 \times 10^{18}$ cm^{-3}, the inverse electron lifetime varies with electron energy according to Fig. 4 for $T = 0$ K (lower curves), $T = 200$ K (middle curves), and $T = 300$ K (upper curves). The solid and dashed lines are calculated with the full RPA dielectric function and damped plasmon pole approximation, respectively. This graph can be compared to experimental values from HBTs with a built-in field in the 900 Å base that induces an energy gain in the absence of collisions of 180 meV. In this energy range, the electron lifetime is nearly energy-independent, and the room-temperature mean free path of about 280 Å indicates nonballistic transport through the base. The average electron scattering angle can further be estimated by introducing a factor of $\cos \theta$ into the integrand of Eq. (178), where θ is the angle between $\vec{k}$ and $\vec{k} - \vec{q}$. At 300 K, $\bar{\theta} \approx 20°$, which implies an

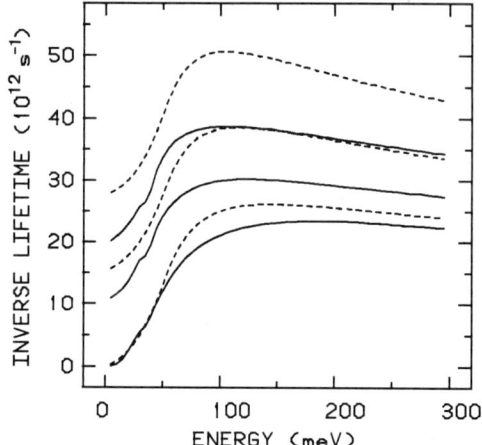

FIG. 4. Inverse inelastic electron lifetimes at $T = 0$ K (lower two curves), $T = 200$ K (middle curves), and $T = 300$ K (upper curves) as a function of electron energy for $p = 2 \times 10^{18}$ cm^{-3}. The RPA and damped plasmon pole results are denoted, respectively, by solid and dashed lines. (From Fig. 1, Bardyszewski and Yevick, 1989b.)

increase in transit time of about 10% relative to that calculated under the assumption of zero scattering angle. The scattering angle increases for lower electron energies and sample temperatures or for higher doping levels.

9. Low-Field Electron Mobility

We next consider the low-field mobility in the p-type base region of standard semiconductor devices such as bipolar transistors. In the diffusion limit, the minority electrons and majority holes are in quasi-equilibrium, and inelastic losses are far smaller than those experienced by a single hot carrier. Further, we assume that the average energy gain arising from the dominant electron scattering mechanisms is far smaller than the average thermal energy. The electron distribution function then approximately retains its equilibrium shape, but is displaced slightly by mv_d in momentum. Experimentally, the electron mobility is generally obtained from the diffusion constant through the Einstein relation. The minority carrier diffusion constant in a p^+n diode can be obtained by etching the base region and plotting the inverse saturation current as a function of the p^+ region thickness (Lundstrom et al., 1990). This technique fails, however, for increased doping once the intrinsic carrier density is affected by bandgap renormalization. For doping levels to 3.6×10^{18} cm^{-3}, the electron diffusion time may instead be extracted from the cutoff frequency of *npn* heterojunction transistors with thick base regions in a common emitter configuration (Nathan *et al.*, 1988). Finally, the electron diffusion constant has recently been determined with the time-of-flight procedure for $p \leq 2.4 \times 10^{19}$ cm^{-3} from the electric signal associated with a short minority electron pulse generated optically near the front surface of a p^+–n diode. If the capacitance of the junction is assumed to be initially constant, the diode current is proportional to the time derivative of the registered voltage. The diffusion constant, minority-carrier lifetime, and surface recombination velocity are then obtained by fitting a numerical model of the diffusion process to the resulting curve (Lovejoy *et al.*, 1990). Similar measurements for finite electric fields were also performed by analyzing the time response of the photocurrent (Furuta *et al.*, 1990; Furuta and Tomizawa, 1990).

These experiments indicate that the electron mobility is far lower for *p*-type than for *n*-type GaAs. For temperatures near or above room temperature, this behavior is predicted by a simple model for the mobility of *p* material in which the holes and host impurity atoms act as $N_A + p \approx 2N_A$ infinite mass scattering centers for electrons. At low temperatures, screening by heavy holes causes a further change in the mobility relative to *n*-type material (Walukiewicz *et al.*, 1979). This model, however, neglects hole recoil and correlation effects, which would be correctly described by a separate coupled transport equation for each subsystem. Therefore, it does not correctly predict experimental results for $N_A > 10^{19}$ cm^{-3}, which display a rise in mobility with doping level. Recent calculations indicate that this enhancement is associated with electron–plasmon scattering, which is less efficient for large plasmon frequencies at the same time that holes below the Fermi energy in the degenerate valence band are effectively removed as scattering centers for electrons by the Pauli principle (Lugli and Ferry, 1985; Lowney and Bennett, 1991; Fischetti, 1991). These calculations are, however, highly sensitive to the value assumed for the plasmon cut-off momentum, defined as the momentum transfer at which electron–hole pair production dominates over plasmon creation.

A further ambiguity is that since electron–plasmon scattering arises from the interaction of an electron with other carriers in its neighborhood, if the system is arbitrarily divided into plasmon and electron subsystems, overall conservation laws may be violated. As we will demonstrate, the plasmons may absorb energy and momentum from an individual electron in this approach without decreasing these quantities for the electron subsystem. These difficulties are, however, absent for effectively independent subsystems of electrons and holes. The energy lost by one subsystem as a result of the interaction is then clearly absorbed by the second, as in minority-carrier scattering by collective majority-carrier plasmon oscillations, or hot electron scattering from an equilibrium plasma.

To formalize the preceding considerations, we return to the Boltzmann equation. The overlap integrals $I(\vec{k}, \vec{k} + \vec{q})$ are approximated by unity in the conduction band, and Eqs. (102) and (124) yield for the Fourier transform of the collision integral

$$\left(\frac{\partial f_{\vec{p}}}{\partial t}\right)_{\text{col}} = -i \sum_{\vec{q}} (1 - f_{\vec{p}+\vec{q}}) f_{\vec{p}} W^<(\vec{q}, E(\vec{p} + \vec{q}) - E(\vec{p}))$$

$$- i \sum_{\vec{q}} f_{\vec{p}+\vec{q}} (1 - f_{\vec{p}}) W^>(\vec{q}, E(\vec{p} + \vec{q}) - E(\vec{p})). \quad (180)$$

Note that W corresponds to a nonequilibrium state for which the simple relationships, Eqs. (156) and (157), between $W^>$ and $W^<$ are replaced by the following "optical theorem," derived from $(\varepsilon^{-1}\varepsilon)^> = 1^> = 0$ and $(\varepsilon^A)^* = \varepsilon^R$:

$$W^{\gtrless}(\vec{q}, \omega) = -\frac{4\pi e^2}{q^2} \frac{\varepsilon^{\gtrless}(\vec{q}, \omega)}{|\varepsilon^R(\vec{q}, \omega)|^2}. \quad (181)$$

Accordingly, from Eqs. (180), (159), and (163), we obtain

$$\left(\frac{\partial f_{\vec{p}}}{\partial t}\right)_{\text{col}} = -2\pi \sum_{\vec{q},\vec{k}} [(1 - f_{\vec{p}+\vec{q}}) f_{\vec{p}} (1 - f_{\vec{k}}) f_{\vec{k}+\vec{q}} - f_{\vec{p}+\vec{q}} (1 - f_{\vec{p}}) f_{\vec{k}} (1 - f_{\vec{k}+\vec{q}})]$$

$$\times \left|\frac{4\pi e^2}{q^2 \varepsilon^R(\vec{q}, E(\vec{p} + \vec{q}) - E(\vec{p}))}\right|^2$$

$$\times \delta(E(\vec{p} + \vec{q}) - E(\vec{p}) - E(\vec{k} + \vec{q}) + E(\vec{k})). \quad (182)$$

This expression, which differs from the usual Born approximation collision integral in that the Coulomb potential is dynamically screened, is subsequently linearized with respect to the deviation

$$f_{\vec{p}} \approx f_{\vec{p}}^0 - \phi_{\vec{p}} \frac{\partial f_{\vec{p}}^0}{\partial E} = f_{\vec{p}}^0 + \frac{1}{kT} \phi_{\vec{p}} f_{\vec{p}}^0 (1 - f_{\vec{p}}^0) \quad (183)$$

of the electron distribution function from its equilibrium position, $f^0(\vec{p})$.

6. Introduction to Nonequilibrium Many-Body Analyses

A few elementary manipulations yield

$$\left(\frac{\partial f_{\vec{p}}}{\partial t}\right)_{\text{col}} = -2\pi\beta \sum_{\vec{q},\vec{k}} f_{\vec{p}}^0(1-f_{\vec{p}+\vec{q}}^0)f_{\vec{k}+\vec{q}}^0(1-f_{\vec{k}}^0)$$
$$\times \{\phi_{\vec{p}} - \phi_{\vec{p}+\vec{q}} + \phi_{\vec{k}+\vec{q}} - \phi_{\vec{k}}\} \left|\frac{4\pi e^2}{q^2 \varepsilon_0^R(\vec{q}, E(\vec{p}+\vec{q}) - E(\vec{p}))}\right|^2$$
$$\times \delta(E(\vec{p}+\vec{q}) - E(\vec{p}) - E(\vec{k}+\vec{q}) + E(\vec{k})). \qquad (184)$$

The label 0 indicates quantities evaluated at quasi-equilibrium.

We may now apply Eq. (181) to separate the collision integral into two parts, namely

$$\left(\frac{\partial f_{\vec{p}}}{\partial t}\right)_{\text{col}} = -\beta \sum_{\vec{q}} f_{\vec{p}}^0(1-f_{\vec{p}+\vec{q}}^0)\{\phi_{\vec{p}} - \phi_{\vec{p}+\vec{q}}\} iW_0^<(\vec{q}, E(\vec{p}+\vec{q}) - E(\vec{p}))$$
$$- 2\pi\beta \sum_{\vec{q},\vec{k}} f_{\vec{p}}^0(1-f_{\vec{p}+\vec{q}}^0)f_{\vec{k}+\vec{q}}^0(1-f_{\vec{k}}^0)\{\phi_{\vec{k}+\vec{q}} - \phi_{\vec{k}}\}$$
$$\times \left|\frac{4\pi e^2}{q^2 \varepsilon_0^R(\vec{q}, E(\vec{p}+\vec{q}) - E(\vec{p}))}\right|^2$$
$$\times \delta(E(\vec{p}+\vec{q}) - E(\vec{p}) - E(\vec{k}+\vec{q}) + E(\vec{k})). \qquad (185)$$

The two expressions on the right-hand side of Eq. (185) reflect, respectively, the change in the distribution functions and in the carrier screening associated with the departure from equilibrium. The first term in the formula can be recast as the electron–plasmon contribution to the Boltzmann equation formalism by applying Eq. (156). We may then derive

$$iW_0^<(\vec{q}, E(\vec{p}+\vec{q}) - E(\vec{p})) = 2\pi\left(\frac{4\pi e^2}{q^2}\right) \int_0^\infty d\omega \, \text{Im}\left[-\frac{1}{\pi}\varepsilon_0^{-1}(\vec{q},\omega)\right]$$
$$\times \{P(\omega)\delta(\omega - E(\vec{p}+\vec{q}) + E(\vec{p}))$$
$$+ (1 + P(\omega))\delta(-\omega - E(\vec{p}+\vec{q}) + E(\vec{p}))\}. \qquad (186)$$

Substituting into the first term of Eq. (185) yields the general formula for the contribution of electron–plasmon processes to the collision integral:

$$-\beta\Sigma_{\vec{q}} f_{\vec{p}}^0(1-f_{\vec{p}+\vec{q}}^0)\{\phi_{\vec{p}} - \phi_{\vec{p}+\vec{q}}\} iW_0^<(\vec{q}, E(\vec{p}+\vec{q}) - E(\vec{p}))$$
$$\to -\beta\Sigma_{\vec{p}'} W_{\text{el-pl}}(\vec{p},\vec{p}') f_{\vec{p}}^0(1-f_{\vec{p}'}^0)(\phi_{\vec{p}} - \phi_{\vec{p}'}), \qquad (187)$$

in which we have identified the probability for electron–plasmon scattering as

$$W_{\text{el-pl}}(\vec{p}, \vec{p}\,') = 2\pi \left(\frac{4\pi e^2}{q^2}\right) \int_0^\infty d\omega \, \text{Im}\left[-\frac{1}{\pi} \varepsilon^{-1}(\vec{q}, \omega)\right]$$
$$\times \{P(\omega)\delta(E(\vec{p}\,') - E(\vec{p}) - \omega)$$
$$+ (1 + P(\omega))\delta(E(\vec{p}\,') - E(\vec{p}) + \omega)\}. \tag{188}$$

We recover the standard expression for electron–phonon scattering through the substitution indicated in Eq. (135), justifying our having separated the electron–plasmon part from the full collision integral of Eq. (184). If scattered carriers participate in collective plasmon excitations, the second term in Eq. (185), associated with the change in the screened Coulomb interaction, must be included in order to insure energy–momentum conservation. In fact, summing Eq. (188) over all momentum $\vec{p}$ yields zero, so that the second term in Eq. (185) may be eliminated without disturbing particle conservation. However, this latter contribution is required to insure that in the absence of external forces $\Sigma_{\vec{p}}\vec{p}(\partial f/\partial t)_{\text{col}} = 0$. Identical arguments apply for energy conservation. In certain physical situations, however, such as that of a single high-energy electron interacting with an equilibrium system, we may safely neglect the departure of the dielectric constant from its equilibrium value and therefore the second contribution in Eq. (185). Similar considerations also apply to a subsystem of minority particles interacting with majority-carrier plasmons.

VII. Interband Processes

10. Auger Processes

After our single-band semiclassical or quantum Boltzmann-equation formulations of intraband decay lifetimes and particle transport processes, we turn in the remainder of this article to interband recombination processes. In contrast to intraband processes, recombination in highly excited solids is inherently a nonequilibrium process. We examine first Auger recombination mechanisms in which recombination occurs through a channel in which the momentum and energy produced by the event is absorbed through a collision process with a second carrier. Such events require the simultaneous presence in a small spatial volume of two electrons and a hole, or two holes and an electron, and they therefore

6. INTRODUCTION TO NONEQUILIBRIUM MANY-BODY ANALYSES

become increasingly probable at high carrier concentration (Beattie and Landsberg, 1959). The calculation of Auger transition rates is a difficult theoretical problem, since the allowable momentum and energy transfer to a given carrier is dependent on the details of the band structure at large quasi-particle momenta. Further, the Auger rate is affected in the degenerate case by occupation factors and by possible many-particle excitations during the collision process that partially remove the momentum and energy restrictions on the recombination process (Takeshima, 1982; Bardyszewski and Yevick, 1985, 1985b, 1985c).

We now develop the kinetic equation describing the conduction band Green's function in a spatially homogeneous semiconductor containing a small number of minority electrons. The derivation of this expression differs from our previous single-band weakly interacting particle case, where we transformed the quantum Boltzmann equation into the classical Boltzmann equation by substituting products of delta functions with corresponding occupation factors for the correlation functions $G^>$ and $G^<$. For interband recombination, each band is separately in quasi-equilibrium, as intraband relaxation is several orders of magnitude faster than interband relaxation. We therefore consider the interband recombination processes as a small perturbation of the quasi-equilibrium state established by the intraband interactions. Here we require that departures from equilibrium are efficiently restored by intraband collisions, which is intrinsically different from the weak interaction limit. We therefore first generate the electronic spectral density functions in the absence of interband processes. We then construct, as in Eqs. (110) and (111), the following quasi-equilibrium Green's functions:

$$iG_v^>(\vec{k}, \omega) = 2\pi(1 - f(\omega - \mu_v))A_v(\vec{k}, \omega), \qquad (189)$$

$$iG_v^<(\vec{k}, \omega) = -2\pi f(\omega - \mu_v)A_v(\vec{k}, \omega). \qquad (190)$$

Labeling the diagonal elements in Σ and G with respect to the band indices with a single subscript, we obtain the following kinetic equation for the Green's function, where G and Σ incorporate the quasi-equilibrium values of the linewidth broadening and quasi-Fermi levels:

$$i\frac{\partial G_c^<(\vec{q}, \omega)}{\partial T} = \Sigma_c^>(\vec{q}, \omega)G_c^<(\vec{q}, \omega) - \Sigma_c^<(\vec{q}, \omega)G_c^>(\vec{q}, \omega). \qquad (191)$$

Equation (191) is generally expanded to lowest order with respect to the interband Coulomb interaction matrix element, while for lightly doped and excited materials the intraband contribution to the spectral density functions is also neglected, leading to a one-particle theory of Auger interactions. In the opposite limit, however, bandgap renormalization and plasmon- and phonon-assisted processes significantly affect the overall Auger transition rate, necessitating an adequate spectral density function model.

To determine the conduction band chemical potential as a function of electron concentration, we integrate the quantum-mechanical electron distribution function, Eq. (58), over all momenta. The resulting Wigner function,

$$f_c^W(\vec{q}) = \int d\omega f(\omega - v_c) A_c(\vec{q}, \omega), \qquad (192)$$

differs from the Fermi–Dirac distribution for nonzero spectral widths. The quantum-mechanical rate equation for conduction band electron concentration n_c is then

$$\frac{dn_c}{dt} = -2 \sum_k \int \frac{d\omega}{2\pi} (\Sigma_c^>(\vec{k}, \omega) G_c^<(\vec{k}, \omega) - \Sigma_c^<(\vec{k}, \omega) G_c^>(\vec{k}, \omega)), \quad (193)$$

with identical formulas for the valence band. While the terms in $\Sigma^>$ and $\Sigma^<$ that give rise to Auger processes are easily separated in the first nontrivial order, in higher order the total number of such contributions in Σ becomes large, motivating a general procedure for extracting the relevant factors (Ziep and Mocker, 1980, 1983). In particular, we observe that $G^<$, $G^>$ and $W^<$, $W^>$ contain products of occupation factors and sharply peaked spectral density functions. These can therefore be associated with the initial or final states of a "real" scattering event that occurs on a much shorter time-scale than the one-particle lifetime (Langer, 1961). In the low-density limit, $W^>$ and $W^<$ approach zero linearly with n as a result of the presence of the Fermi factors in Eq. (186), while W^R and W^A instead approach $4\pi e^2/q^2 \varepsilon_\infty$.

To illustrate the preceding method, we proceed to evaluate the relevant interband contributions to $\Sigma^>$ and $\Sigma^<$ in the GW approximation. In order to make the calculation tractable, we further apply a nonequilibrium version of the RPA that incorporates both the renormalization of the electron and hole states and interband transitions:

$$i\chi_0^\gtrless(\vec{r}', t', \vec{r}, t) = \sum_{n_1, n_2} G_{n_1}^\gtrless(\vec{r}', t', \vec{r}, t) G_{n_2}^\gtrless(\vec{r}, t, \vec{r}', t'), \qquad (194)$$

where n_1, n_2 and $G_n^{\gtrless}$ denote conduction and valence band states and the Green's function for electrons in band n, respectively. Equation (181) consequently implies that

$$iW^<(\vec{q}, \omega) = 4\pi \left|\frac{4\pi e^2}{q^2 \varepsilon^R(\vec{q}, w)}\right|^2 \sum_{\vec{k}, n_1, n_2} \int d\omega_1 \int d\omega_2 I_{n_1, n_2}(\vec{k}, \vec{k} - \vec{q})$$
$$\times A_{n_1}(\vec{k}, \omega_1) A_{n_2}(\vec{k} - \vec{q}, \omega_2)$$
$$\times f_{n_1}(\omega_1)(1 - f_{n_2}(\omega_2))\delta(\omega - \omega_2 - \omega_1). \tag{195}$$

Inserting Eqs. (128) and (195) into the collision integral in Eq. (193) yields

$$\frac{dn_c}{dt} = -8\pi \sum_{vn_1, n_2} \sum_{\vec{k}\vec{q}\vec{k}_1} \int d\omega' \int d\omega_1 \int d\omega_2 \int d\omega I_{cv}(\vec{k}, \vec{k} + \vec{q}) I_{n_1 n_2}(\vec{k}_1, \vec{k}_1 - \vec{q})$$
$$\times \left|\frac{4\pi e^2}{q^2 \varepsilon^R(\vec{q}, \omega' - \omega)}\right|^2$$
$$\times A_c(\vec{k}, \omega) A_v(\vec{k} + \vec{q}, \omega')$$
$$\times A_{n_1}(\vec{k}_1, \omega_1) A_{n_2}(\vec{k}_1 - \vec{q}, \omega_2) \delta(\omega' - \omega - \omega_1 + \omega_2)$$
$$\times [(1 - f_v(\omega'))f_{n_1}(\omega_1)(1 - f_{n_2}(\omega_2))f_c(\omega) - f_v(\omega')$$
$$\times (1 - f_{n_1}(\omega_1))f_{n_2}(\omega_2)(1 - f_c(\omega))]. \tag{196}$$

The first set of Fermi factors inside the integrand function may be identified with a collision event in which a conduction band electron with momentum $\vec{k}$ recombines with a hole with momentum $\vec{k} + \vec{q}$ while simultaneously exciting the second, Auger electron, in band n_1 with momentum $\vec{k}_1$, to band n_2 and momentum $\vec{k}_1 - \vec{q}$. The second group of Fermi factors arises from the inverse process. Auger interactions are classified by the values of the band labels in Eq. (196). For example, if $v = h$ and $n_1 = n_2 = c$ or $v = n_1 = c$ and $n_2 = h$, we calculate the rate for "CHCC" transitions in which the recombining carriers excite the Auger electron to a high conduction-band state. Collisions in which the Auger electron is instead transferred from the split-off to the heavy-hole band ($v = h$, $n_1 = s$, $n_2 = h$) or from the light-hole to the heavy-hole band are termed "CHSH" and "CHLH," respectively. As our derivation of Eq. (196) is based on an extension of the standard RPA formalism, it neglects exchange effects that arise in higher-order approximations for $\Sigma^>$ and $\Sigma^<$. Although the question of whether the Coulomb interaction in the exchange terms is screened by ε^R in analogy with direct interactions or not, as in excitonic theory, has been widely discussed (Haug and Ekardt, 1975), such contributions are generally assumed to be small in direct-gap semiconductors.

The direct theoretical calculation of Auger transition rates is unfortunately a computationally demanding task, since the transition amplitudes between all distinct initial and final states must be summed. At low temperatures, the Fermi factors vary rapidly in the vicinity of the Fermi energy, while the spectral density functions are sharply peaked. In the limit of zero spectral-density function width, the interacting particles are subject to strict constraints derived from kinematic energy and momentum conservation conditions. The absolute value of the final energy and momentum of a CHCC Auger electron or the initial electron energy and momentum in CHSH and CHLH processes must then be greater than certain threshold values for Auger recombination to occur. If the Fermi occupation function is exponentially decreasing at these values, the dominant contribution to Eq. (196) is associated with near-threshold transitions. In terms of the momentum-dependent conduction, heavy-hole and spin split-off band effective masses, $m_\alpha(k) = k^2/2E_\alpha(k)$ with $\alpha = c, h, s$, we have

$$E_c(k_{\text{thresh}}) = \left(1 + \frac{2m_c(0) + m_h}{2m_c(0) + m_h - m_c(k_{\text{thresh}})}\right) E_g \qquad (197)$$

for CHCC interactions and

$$E_s(k_{\text{thresh}}) = \frac{m_c(0) + 2m_h}{m_c(0) + 2m_h - m_s(k_{\text{thresh}})} (\Delta - E_g) - \Delta \qquad (198)$$

for CHSH. All energies are measured with respect to the valence band maximum so that, for example, $E_s(0) = -\Delta$.

Unfortunately, the precise value of the threshold energy and therefore the calculated Auger transition probability depends critically on the band structure far from the Γ point. While attempts have been made to employ sophisticated band-structure models in Eq. (196), most previous work is based on either the tight-binding empirical pseudopotential method or the $k \cdot p$ theory. The $k \cdot p$ overlap integrals between the conduction, heavy-hole, and light-hole bands are

$$I_{cc}(\vec{k}, \vec{k}') = [a'_c a_c + (c'_c c_c + b'_c b_c) \cos \theta]^2$$
$$+ \frac{1}{2}(b'_c b_c - b_c c'_c \sqrt{2} - b'_c c_c \sqrt{2})^2 \sin^2 \theta, \qquad (199)$$

$$I_{\alpha h}(\vec{k}, \vec{k}') = \frac{1}{4}(b_\alpha + c_\alpha \sqrt{2})^2 \sin^2 \theta, \qquad (200)$$

and

$$I_{\text{hh}}(\vec{k}, \vec{k}') = \frac{1}{4}(1 + 3\cos^2\theta). \tag{201}$$

In the preceding equations, α represents either the conduction, split-off, or light-hole band and θ is the angle between $\vec{k}$ and $\vec{k}'$, while a, b, c and a', b', c' refer to the Kane model coefficients corresponding to the vectors $\vec{k}$ and $\vec{k}'$, respectively, cf. Eq. (12).

For significant electronic state broadening, as for example by electron–phonon and electron–plasmon scattering, the electron linewidth in Eq. (196) may be comparable to the plasmon energy. This lowers the effective threshold by $\approx \hbar\omega_p$ and therefore increases the Auger rate. The electron lifetime is determined directly from the GW expression according to the theory of Section 6, Chapter V. If the interacting particle states are assumed to be unbroadened, the Auger process is labeled pure. However, if in particular the heavy-hole levels, which are strongly perturbed by phonons because of the large density of states near the Γ point, are broadened, the collision is instead denoted as phonon-assisted. In both cases the multidimensional phase-space integral in Eq. (196) can be evaluated by first integrating over all frequency delta functions and subsequently restricting the remaining Monte Carlo integrations to a region within a few kT of the threshold energy.

In 1.3 μm $\text{In}_{0.7}\text{Ga}_{0.3}\text{As}_{0.64}\text{P}_{0.36}$, at high excitation levels, CHSH processes arising from the collision of two holes dominate the recombination rate. Theoretical results are therefore conveniently expressed in terms of the Auger coefficient C_p defined by $1/\tau_p = C_p p^2$. The pure collision and phonon-assisted Auger coefficients are given in Fig. 5 for an excess carrier concentration of $p = 2 \times 10^{18}$ e–h pairs/cm^3. In the figure, the dashed line is calculated with the static (Debye) screening of the Coulomb interaction given by $\varepsilon(\vec{q}, \omega) = \varepsilon_0(1 + \lambda_D^2/q^2)$ in Eq. (196), where λ_D and ε_0 are the Debye screening length and low-frequency dielectric constant, respectively. The solid line is instead obtained with dynamical screening modeled by replacing $\varepsilon(\vec{q}, \omega \approx E_g)$ with ε_∞, the dielectric constant at frequencies above the various plasmon and phonon resonances but below ionization frequencies. Since the screening is neither purely static nor purely dynamic, the correct value for the Auger coefficient should interpolate between the two curves in Fig. 5. Experimental values for the Auger coefficient are centered around $C \approx 2.6 \times 10^{-29}$ cm^6/s at 300 K (Sermage et al., 1985). The CHSH calculations only require a proper description of the valence band-structure for which the $k \cdot p$ model is highly reliable. Hence, modeling errors, despite some

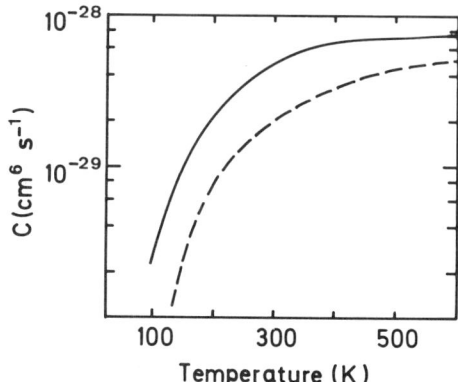

FIG. 5. The full (phonon-assisted and pure collision) CHSH Auger coefficients for $In_{0.7}Ga_{0.3}As_{0.64}P_{0.36}$ vs. temperature. The dashed line corresponds to Debye screening, while the solid line is for an infinite screening length. (From Fig. 2 of Yevick and Bardyszewski, 1987b; © 1987 IEEE.)

uncertainty in certain input parameters, are limited to a factor of two (Yevick and Bardyszewski, 1987b; Bardyszewski and Yevick, 1985a, 1985b). The calculated CHCC recombination rate, which is far more band-structure–dependent, is fortunately smaller by a factor of about three. The CHCC values, unlike the CHSH values, are significantly affected by phonon-assisted processes.

We finally note that a proper treatment of screening is essential in the narrow-gap p-GaSb system for which $\Delta \approx E_g$ (Yevick and Bardyszewski, 1987a). In this material at a certain temperature, an electronic transition from the bottom of the conduction band to an empty state at the valence-band maximum, accompanied by an excitation of a second electron from the spin–orbit split-off band to the valence band extremum, conserves both energy and momentum. For such transitions, $q \approx 0$ and the Debye dielectric function becomes artificially large. Consequently, the calculated Auger rate decreases rapidly with increasing hole concentration for Debye screening while dynamic screening predicts a gradual fall-off, in agreement with experiment (Yevick and Bardyszewski, 1987a).

11. Gain and Stimulated Recombination

We now pass from non-radiative to radiative carrier recombination in III–V materials. Our discussion again incorporates dynamic screening and finite lifetime effects within a quasi-equilibrium framework. In contrast to our Auger calculations, we can here consider the amplitude of the radiation field as a small parameter and expand the Green's function for electron propagation to lowest order in this quantity. In this manner we extend the standard Kubo linear response formalism to nonequilibrium systems (Bardyszewski and Yevick, 1989c). Our technique also corresponds to expanding the generalized collision integrals in Eq. (193) to first-order in the photon propagators (which is equivalent to second-order in the electric field amplitude, since the recombination rate is linear in light intensity). This derivation is, however, more complex and is therefore omitted here.

Accordingly, we consider a monochromatic electric field $\vec{E}(t) = \hat{e}_z E_z e^{-i\omega t}$ applied to a bulk semiconductor with a refractive index n_r. We find the gain $g(\omega)$ from the real part of the conductivity $\text{Re}(\sigma(\omega))$ according to

$$g(\omega) = -\frac{4\pi}{cn_r} \text{Re}(\sigma(\omega)). \quad (202)$$

If the semiconductor crystal possesses cubic symmetry, a z-polarized electric field excitation produces similarly polarized currents, and we may therefore omit the polarization index. The electric current is then, according to Eq. (55),

$$J(t) = i \sum_{n_1,n_2} \sum_{\vec{k}_1,\vec{k}_2} \frac{e}{m_0} \left(p_{n_1\vec{k}_1,n_2\vec{k}_2} + \frac{e}{c} A(t) \delta_{n_1 n_2} \delta_{\vec{k}_1,\vec{k}_2} \right) G^<_{n_2\vec{k}_2,n_1\vec{k}_1}(t, t). \quad (203)$$

In Eq. (203), $p_{n_1\vec{k}_1,n_2\vec{k}_2}$ is the z-component of momentum in the Bloch function representation, and $A(t)$ is the vector potential of the total electric field. Further, the Green's function in Bloch representation is

$$\begin{aligned}-iG^<_{n_2\vec{k}_2,n_1\vec{k}_1}(t, t) &= \langle a^\dagger_{n_1\vec{k}_1}(t) a_{n_2\vec{k}_2}(t) \rangle \\ &= -iG_{n_2\vec{k}_2,n_1\vec{k}_1}(t, t^+).\end{aligned} \quad (204)$$

The time t^+ differs infinitesimally from t but satisfies $t^+ \gtrless_C t$ along the complex time contour. From the variational derivative of G with respect to the electric field, we generate the following lowest-order approximation for the current:

$$J(t) = i\frac{e}{m_0} \sum_{n_1,n_2} \sum_{\vec{k}_1,\vec{k}_2} p_{n_1\vec{k}_1,n_2\vec{k}_2}$$
$$\times \int_C \frac{\delta G_{n_2\vec{k}_2,n_1\vec{k}_1}(t, t^+)}{\delta E(t')} E(t')\, dt' + i\frac{e^2 N}{\omega m_0} E(t), \quad (205)$$

in which the symbol N designates the electron concentration. On the other hand, the coupling Hamiltonian between the electrons and the electromagnetic field is

$$H' = \sum_{n_1,n_2} \sum_{\vec{k}_1,\vec{k}_2} U_{n_1\vec{k}_1,n_2\vec{k}_2}(t) a^{\dagger}_{n_1,\vec{k}_1}(t) a_{n_2,\vec{k}_2}(t), \quad (206)$$

where

$$U_{n_1\vec{k}_1,n_2\vec{k}_2}(t) = \frac{eA(t)}{m_0 c} p_{n_1\vec{k}_1,n_2\vec{k}_2}. \quad (207)$$

The electric field and vector potential, and hence $E(t)$ and $U(t)$, are proportional for a monochromatic excitation. Equation (205) can therefore be rewritten as

$$J(t) = \frac{ie^2}{\omega m_0^2} \sum_{n_1,n_2,n_1',n_2'} \sum_{\vec{k}_1,\vec{k}_2,\vec{k}_1',\vec{k}_2'} p_{n_1\vec{k}_1,n_2\vec{k}_2} p_{n_1'\vec{k}_1',n_2'\vec{k}_2'}$$

$$\times \int_C L_{n_1\vec{k}_1,n_2\vec{k}_2,n_1'\vec{k}_1',n_2'\vec{k}_2'}(t,t') E(t') \, dt' + i\frac{e^2 N}{\omega m_0} E(t). \quad (208)$$

The irreducible linear response function L is defined on the contour C by

$$L_{n_1\vec{k}_1,n_2\vec{k}_2,n_1'\vec{k}_1',n_2'\vec{k}_2'}(t,t') = -i\frac{\delta G_{n_2\vec{k}_2,n_1\vec{k}_1}(t,t^+)}{\delta U_{n_1'\vec{k}_1',n_2'\vec{k}_2'}(t')}$$

$$= -i\langle T_C[a^{\dagger}_{n_1,\vec{k}_1}(t) a_{n_2,\vec{k}_2}(t)$$

$$\times a^{\dagger}_{n_1',\vec{k}_1'}(t') a_{n_2',\vec{k}_2'}(t')]\rangle_{\text{conn, irred}}. \quad (209)$$

The subscript on the angular bracket indicates that the perturbation expansion should be restricted to diagrams that are both connected and irreducible in the sense that removing any single Coulomb interaction line does not divide the diagram into two separate units. Since the variational derivative in Eq. (209) is taken with respect to the total electric field, such graphs can be identified with carrier screening processes.

If we deform the integration contour in Eq. (208) to a loop from $-\infty$ to t and back, we generate two contributions, one for the positive branch of the contour and a second for the negative branch. Taking into account the additional negative sign in the second of these terms associated with

6. Introduction to Nonequilibrium Many-Body Analyses

the integration direction, we find after Fourier transforming and introducing the retarded linear response function L^R,

$$\sigma(\omega) = \frac{ie^2}{\omega m_0^2} \sum_{n_1,n_2,n_1',n_2'} \sum_{\vec{k}_1,\vec{k}_2,\vec{k}_1',\vec{k}_2'} p_{n_1\vec{k}_1,n_2\vec{k}_2} p_{n_1'\vec{k}_1',n_2'\vec{k}_2'}$$
$$\times L^R_{n_1\vec{k}_1,n_2\vec{k}_2,n_1'\vec{k}_1',n_2'\vec{k}_2'}(\omega) + i\frac{e^2 N}{\omega m_0}. \tag{210}$$

As the nonequilibrium linear response function is a two-particle correlation function, it can be evaluated using the Keldysh diagram technique. To determine the gain from Eq. (210), we require the Fourier transform with respect to time of the retarded response function

$$L^R(t, t') = \Theta(t - t')[L^>(t, t') - L^<(t, t')]. \tag{211}$$

Focusing for illustration on $L^>$, we apply the following Keldysh rules:

(1) For $L^>_{n_1\vec{k}_1,n_2\vec{k}_2,n_1'\vec{k}_1',n_2'\vec{k}_2'}(\omega)$, the primed and unprimed labels indicate the positive and negative branches of the contour C, respectively.
(2) Creation/annihilation operators in $L^>$ are connected by outgoing/incoming propagator lines.
(3) These lines are attached to vertices in all possible ways to form irreducible diagrams of the desired order. Vertices are then assigned to either the initial, $(+)$, branch of the integration contour from $-\infty$ to ∞ or the return, $(-)$, path. Each distinguishable ordering of $(-)$ and $(+)$ signs yields a separate Keldysh diagram.
(4) The diagrams are translated into integral expressions by associating the momentum- and frequency-dependent chronological, antichronological, $>$ and $<$ Green's functions iG^c, $iG^{\tilde{c}}$, $iG^<$ and $iG^>$ with lines directed from $(+)$ to $(+)$, $(-)$ to $(-)$, $(-)$ to $(+)$, and $(+)$ to $(-)$ vertices and incorporating an overlap matrix element between the incoming and outgoing particles at each vertex.
(5) The momentum and energy arguments are set in accordance with conservation conditions at each vertex; every loop in the diagram introduces an additional intermediate momentum and an additional energy variable. The photon energy $\hbar\omega$ and momentum $\vec{q}$ enters the diagram at the initial vertex and exits at the final vertex.
(6) Integrations $(1/\Omega)\Sigma_{\vec{k}} \to \int[(d^3k)/(2\pi)^3]$ and $\int(d\omega/2\pi)$ are performed over all intermediate energies and momenta.

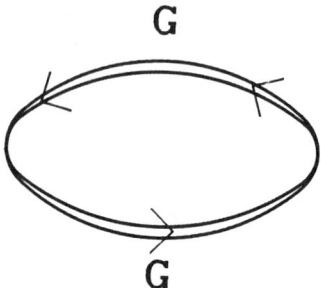

FIG. 6. The zero-order diagram for the gain or spontaneous emission based on the GW self-energy. (From Fig. 1 of Bardyszewski and Yevick, 1989c.)

(7) The resulting integral expression is multiplied by $(-1)^{F+N}$ where F is the number of closed fermion loops and N is the number of $(-)$ vertices to account for the reversed orientation of the integration contour on the $(-)$ axis.

The zeroth-order Keldysh diagram for L with respect to the screened interaction W is presented in Fig. 6. It yields

$$iL^{(1)>}_{c\vec{k},v\vec{k},v\vec{k},c\vec{k}}(\omega) = -\int \frac{d\omega'}{2\pi} iG^{>}_v(\vec{k}, \omega + \omega') iG^{<}_c(\vec{k}, \omega'). \tag{212}$$

Similarly, the four first-order graphs of Fig. 7 generate

$$i \sum_{n_1,n_2,n'_1,n'_2} \sum_{\vec{k}_1,\vec{k}_2,\vec{k}'_1,\vec{k}'_2} p_{n_1\vec{k}_1,n_2\vec{k}_2} p_{n'_1\vec{k}'_1,n'_2\vec{k}'_2} \{L^{(2)}_{n_1\vec{k}_1,n_2\vec{k}_2,n'_1\vec{k}'_1,n'_2\vec{k}'_2}(\omega)\}^{>}$$

$$= -2\hbar^2 \, \text{Im}\left(\sum_{\vec{k}_1,\vec{k}_2} \int_{-\infty}^{\infty} \frac{d\omega_1}{2\pi} \int_{-\infty}^{\infty} \frac{d\omega_2}{2\pi} M_{\text{ch}}(\vec{k}_1, \vec{k}_2) \right.$$

$$\times [G^{c}_{h}(\vec{k}_2, \omega_1 + \omega_2) G^{c}_{c}(\vec{k}_2, \omega_1 + \omega_2 + \omega)$$

$$\times G^{<}_{h}(\vec{k}_1, \omega_1) G^{>}_{c}(\vec{k}_1, \omega_1 + \omega) W^{c}(\vec{q}, \omega_2)$$

$$- G^{<}_{h}(\vec{k}_2, \omega_1 + \omega_2) G^{c}_{c}(\vec{k}_2, \omega_1 + \omega_2 + \omega)$$

$$\left. \times G^{\bar{c}}_{h}(\vec{k}_1, \omega_1) G^{>}_{c}(\vec{k}_1, \omega_1 + \omega) W^{>}(\vec{q}, \omega_2)] \right). \tag{213}$$

Here $\vec{q} = \vec{k}_2 - \vec{k}_1$, and M_{ch} is given by Eq. (240). The frequency integrals are performed after introducing the spectral representations for the

6. INTRODUCTION TO NONEQUILIBRIUM MANY-BODY ANALYSES   375

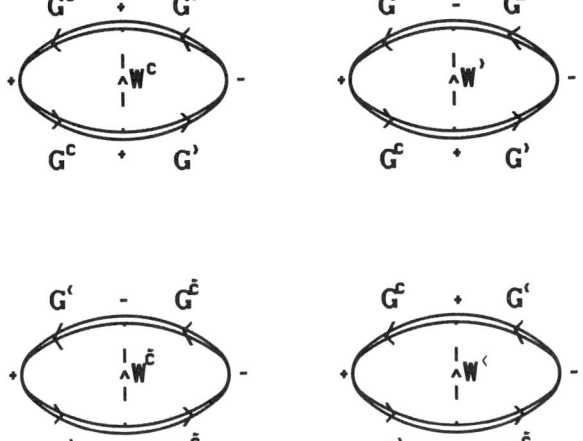

FIG. 7. Keldysh diagrams employed in the evaluation of the spontaneous emission. (From Fig. 3 of Bardyszewski and Yevick, 1989c.)

chronological and antichronological functions W^c and $W^{\bar{c}}$:

$$W^c(\vec{q},\omega) = \{W^{\bar{c}}(\vec{q},\omega)\}^*$$
$$= \frac{4\pi e^2}{\varepsilon_\infty q^2} + i \int_{-\infty}^{\infty} \frac{d\omega'}{2\pi} \left[\frac{W^>(\vec{q},\omega')}{\omega - \omega' + i\eta} - \frac{W^<(\vec{q},\omega')}{\omega - \omega' - i\eta} \right], \quad (214)$$

and similarly for G^c and $G^{\bar{c}}$.

Note that as discussed in the preceding section, the $W^{\gtrless}$ functions correspond to real plasmon creation and emission while W^c and $W^{\bar{c}}$ in the limit of small excess carrier concentration approach the unscreened Coulomb interaction. Consequently, the graphs containing W^c and $W^{\bar{c}}$ in Fig. 7 yield vertex correction terms leading to the renormalization of the electron–photon matrix element, while the remaining, interference, diagrams describe the emission or absorption of real plasmons. Such a division is precluded in the Matsubara formalism which, however, can in principle still be employed to model a quasi-equilibrium state with equal-temperature electron, hole, and phonon subsystems. In this method, the contour is mapped onto an imaginary time interval between $t = -i\beta$ and $t = 0$. Since only one branch is then present, a far smaller number of diagrams are generated. Although the analytical difficulty is therefore decreased, the Matsubara integrations contain multiple complex poles and must therefore be carefully evaluated; further, physical approximations are more easily incorporated in the Keldysh procedure. While we have analyzed both vertex and interference terms to the same order in the screened Coulomb potential, this potential is itself constructed by summing a subclass of diagrams to all orders in the unrenormalized interaction. Consequently, our approximation is most justified in the high-density, weak-interaction limit. For densities such that the average intercarrier distance is much larger than the exciton radius, Coulomb forces are nonperturbative, and exciton formation, which is described by summing ladder diagrams for the vertex corrections to all orders, instead dominates.

To substantiate the preceding discussion, we present analytical results for the gain. After averaging over all directions of the incoming light polarization a combination of the zeroth-order diagram combined with the vertex correction, Fig. 7, yields

$$g^{(1)}(\omega) = \frac{8\pi^2 e^2 P^2}{3cn_r \hbar \omega} \sum_{\vec{k}} \int_{-\infty}^{\infty} d\omega' A_h(\vec{k}, \omega')$$
$$\times A_c(\vec{k}, \omega' + \omega)[f(\omega + \omega' - \mu_c)$$
$$- f(\omega' - \mu_v)][1 + \mathcal{F}(\vec{k}, \omega, \omega')]. \qquad (215)$$

In Eq. (215), A_c and A_h are the electron spectral densities in the conduction and heavy-hole band, P is the Kane model parameter, and the function $\mathcal{F}(\vec{k}, \omega, \omega')$ represents the vertex correction resulting from the modification of the electron–photon matrix element due to plasmon and optical phonon processes, while the conduction and valence quasi-Fermi energies are μ_c and μ_v (Zimmermann, 1978; Schmitt-Rink et al., 1980; Haug and Tran Thoai, 1980; Mahan, 1967). We find

$$\mathcal{F}(\vec{k}, \omega, \omega_1) = \frac{1}{2} \sum_{\vec{q}} (1 + 3\cos^2 \theta) \left(\int_{-\infty}^{\infty} d\omega' \int_{-\infty}^{\infty} d\omega'' \frac{A_h(k, \omega')A_c(k, \omega'')}{\omega - \omega'' + \omega'} \right.$$
$$\times \text{Re}[S(q, \omega + \omega_1 - \mu_c, \omega'' - \mu_c) - S(q, \omega_1 - \mu_v, \omega' - \mu_v)]$$
$$- \frac{\pi^2}{2} \int_{-\infty}^{\infty} d\omega_2 \frac{4\pi e^2}{q^2} A_h(k, \omega_2) A_c(k, \omega + \omega_2)$$
$$\times [1 - 2f(\omega + \omega_2 - \mu_c)][1 - 2f(\omega_2 - \mu_v)]$$
$$\left. \times [1 - 2P(\omega_2 - \omega_1)] \text{Im}\left\{ -\frac{1}{\pi} \varepsilon^{-1}(q, \omega_2 - \omega_1) \right\} \right). \qquad (216)$$

6. INTRODUCTION TO NONEQUILIBRIUM MANY-BODY ANALYSES

The function $S(q, E, E')$ is

$$S(q, E, E') = \frac{4\pi e^2}{q^2} \bigg(f(E') \varepsilon^{-1}(q, E - E')$$

$$- \int_{-\infty}^{\infty} d\omega' \frac{1}{\pi} \text{Im}\bigg[-\varepsilon^{-1}(q, \omega') \bigg] \frac{P(\omega') + 1}{E - E' - \omega' + i\eta} \bigg). \quad (217)$$

Clearly, if we substitute zero in place of $\mathscr{F}$ in Eq. (216), we recover the standard lowest-order formula for the gain as a convolution of spectral densities of conduction- and valence-band electrons (Zee, 1978; Selloni et al., 1984; Aleksanian et al., 1974; Sugimura et al., 1986). The factor $\mathscr{F}$, however, increases as expected the available phase space for the transition.

The two right-hand diagrams in Fig. 7, which vanish in the limit of small electron density, are associated with dynamic screening effects. These yield

$$g^{(3)}(\omega) = \frac{4\pi^2 e^2 P^2}{3cn_r \hbar \omega} \sum_{\vec{k}_1, \vec{k}_2} \frac{4\pi e^2}{q^2} (1 + 3\cos^2 \theta) \int_{-\infty}^{\infty} d\omega_1 \int_{-\infty}^{\infty} d\omega_2 \bigg\{ \int_{-\infty}^{\infty} d\omega' \int_{-\infty}^{\infty} d\omega''$$

$$\times \frac{A_h(k_2, \omega_2) A_c(k_2, \omega') A_h(k_1, \omega'') A_c(k_1, \omega_1)}{\omega - \omega' + \omega_2} \frac{}{\omega - \omega_1 + \omega''}$$

$$- \pi^2 A_h(k_2, \omega_2) A_c(k_2, \omega + \omega_2) A_h(k_1, \omega_1 - \omega) A_c(k_1, \omega_1)$$

$$\times [1 - 2f(\omega_1 - \omega - \mu_v)][1 - 2f(\omega_2 + \omega - \mu_c)]\}$$

$$\times \{f(\omega_2 - \mu_v)[1 - f(\omega_1 - \mu_c)][1 + P(\omega - \omega_1 + \omega_2)]$$

$$- f(\omega_1 - \mu_c)[1 - f(\omega_2 - \mu_v)] P(\omega - \omega_1 + \omega_2)\}$$

$$\times \text{Im}\bigg[-\frac{1}{\pi} \varepsilon^{-1}(q, \omega - \omega_1 + \omega_2) \bigg]. \quad (218)$$

This interference term is large and negative for energies below the absorption edge. If the electron and hole spectral linewidths are narrow, after neglecting the angular dependence of the overlap matrices we arrive at the following simplified formula for the total gain (Aldrich and Silver,

1980; Zimmermann, 1987):

$$g(\omega) = -\frac{4\pi^2 e^2 P^2}{3cn_r \hbar \omega} \sum_{\vec{k}_1,\vec{k}_2} \frac{4\pi e^2}{q^2} \{f_h(k_2)(1 - f_c(k_1))$$

$$\times (1 + P(\omega - E_c(k_1) + E_h(k_2)))$$

$$- f_c(k_1)(1 - f_h(k_2))P(\omega - E_c(k_1) + E_h(k_2)))\}$$

$$\times \text{Im}\left\{-\frac{1}{\pi}\varepsilon^{-1}(q, \omega - E_c(k_1) + E_h(k_2))\right\}$$

$$\times \left| \frac{1}{\omega - E_c^0(k_2) + E_h^0(k_2) - \Sigma_c(k_2, E_h(k_2) + \omega) + \Sigma_h^*(k_2, E_c(k_2) - \omega)} \right.$$

$$\left. - \frac{1}{\omega - E_c^0(k_1) + E_h^0(k_1) - \Sigma_c(k_1, E_h(k_1) + \omega) + \Sigma_h^*(k_1, E_c(k_1) - \omega)} \right|^2.$$

(219)

The interference contribution is given by the cross-term in the preceding squared absolute-value expression, consistent with processes in which a plasmon is emitted by one of the interacting particles and subsequently absorbed by the second particle. The emission of real plasmons is then suppressed (Brinkman and Lee, 1973). In particular, when the plasmon momentum, q, is zero, the electron and hole are confined to a small spatial volume for a long period of time prior to recombination. Consequently, the corresponding amplitude is large, as evidenced by the cancellation of the energy denominators in Eq. (219) (Schaefer et al., 1988).

In view of the preceding discussion, we conclude that interference leads to gain suppression in the plasmon satellite region $E_g - \hbar\omega_p < E < E_g$, as confirmed by experimental observations for highly excited GaAs which also display a shift of the gain maximum toward higher energies from the value calculated in the absence of interference (Göbel, 1974; Henry et al., 1981; Casey and Stern, 1976). These features are generally well described by the so-called no k-selection rule in which $g(\omega)$ is determined by convolving the valence and conduction band density of states as in totally disordered media. Theoretical values for the optical gain are therefore compared with such one-particle results for excitation levels of 3.0×10^{18} cm^{-3} at $T = 300$ K in Fig. 8. The one-particle model calculations are performed for a carrier concentration of 3.3×10^{18} cm^{-3}

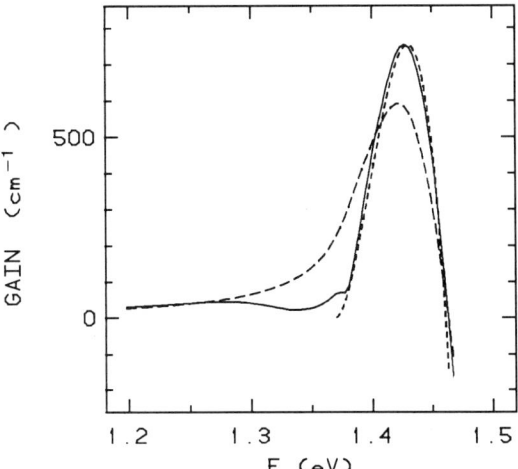

FIG. 8. The results for gain in GaAs as a function of photon energy (solid line), together with the one-particle result multiplied by 0.915 and translated by 12 meV (dashed line) in the shifted G_0W approximation. The long-dashed line is our result in the absence of vertex correction terms. (From Fig. 7 of Bardyszewski and Yevick, 1989c.)

and a properly renormalized energy gap. Unadjusted results therefore possess a certain systematic error.

The preceding formalism is directly extended to the spontaneous emission excess carrier lifetime, which is the integral over all frequencies of the emission rate, $R(\omega)$, defined as the number of spontaneously emitted photons per unit time, volume, and photon energy summed over polarizations and photon propagation directions. The relationship between $R(\omega)$ and the linear response function is

$$R(\omega) = \frac{2e^2 n_r \omega}{c^3 m^2 \hbar \pi} \sum_{n_1,n_2,n_1',n_2'} \sum_{\vec{k}_1,\vec{k}_2,\vec{k}_1',\vec{k}_2'}$$
$$\times p_{n_1 \vec{k}_1, n_2 \vec{k}_2} p_{n_1' \vec{k}_1', n_2' \vec{k}_2'} iL^<_{n_1 \vec{k}_1, n_2 \vec{k}_2, n_1' \vec{k}_1', n_2' \vec{k}_2'}(\omega). \quad (220)$$

In analogy to Eq. (245), we may derive

$$L^>_{CHC'H'} = e^{\beta(\omega - \mu_c + \mu_v)} L^<_{CHC'H'}. \quad (221)$$

Consequently,

$$iL^<(\omega) = i(e^{\beta(\omega - \mu_c + \mu_v)} - 1)^{-1}(L^>(\omega) - L^<(\omega))$$
$$= -2(e^{\beta(\omega - \mu_c + \mu_v)} - 1)^{-1} \operatorname{Im}[L^R(\omega)], \quad (222)$$

which relates $R(\omega)$ to $g(\omega)$ through Eq. (210). One-particle models of the radiative recombination lifetime are generally quite accurate, as many-body effects are largely canceled by the frequency integration. The same remark also applies to the spontaneous emission α factor (Bardyszewski and Yevick, 1989a; Yevick and Bardyszewski, 1989a).

IX. Conclusions

We have examined a variety of processes involving minority and majority carriers in direct-bandgap III–V semiconductors. Our discussion demonstrates that the nonequilibrium Green's function technique gives a unified description of a wide variety of physically interesting electronic and optical phenomena such as hot carrier lifetimes, low-field transport, bandgap renormalization, nonradiative carrier lifetimes, and optical gain spectra. The formalism further allows the interacting electrons and photons to be simply described by screened field–particle interactions and broadened single-particle states. Further, the $k \cdot p$ electronic band structure and self-consistent Fermi levels can be easily incorporated.

While many physical quantities such as the bandgap renormalization and radiative recombination and hot carrier lifetimes are determined by integrals of many-body expressions over energies and wavevectors and therefore can be reasonably well predicted by simplified one-particle models, curves for the gain spectrum and Auger recombination rates are highly sensitive to the details of the many-body interactions and to the electronic band structure. Therefore, for the analysis of these latter processes, especially in the high-density limit where the screened interaction becomes small, an accurate perturbative approach is highly desirable. The Green's function methods are sufficiently straightforward to allow finite-temperature calculations and therefore agree well with experiment. In many cases, however, higher-order perturbative effects, excitonic corrections or complexities in the band structure have not yet been studied. Further work in this area, as well as on the simplification of analytic formulas in certain physically important limits, is expected to be of considerable future interest.

Appendix A

In this appendix, we discuss the matrix elements of Eqs. (21), (44), and (213). We employ the symbols α and β to denote different degenerate states within bands c and v, respectively – for example, $S\uparrow$ and $S\downarrow$ in

6. INTRODUCTION TO NONEQUILIBRIUM MANY-BODY ANALYSES

the case of the conduction band at $k = 0$. While Eqs. (21) and (44) can in principle be calculated from the succeeding two equations for arbitrary orientations of $\vec{k}$ and $\vec{k}' = \vec{k} + \vec{q}$, considerable simplification results if we orient the vector $\vec{k}$ along the z-coordinate axis and average over photon polarizations.

Since $\vec{k}'$ is directed at a direction given by the polar angles (θ, ϕ) with respect to the z axis, the spin and coordinate wavefunctions, Eqs. (10) and (11), must be rotated according to the following prescription:

$$\begin{pmatrix} \uparrow' \\ \downarrow' \end{pmatrix} = \begin{pmatrix} e^{-i(\phi/2)} \cos\frac{\theta}{2} & e^{i(\phi/2)} \sin\frac{\theta}{2} \\ -e^{-i(\phi/2)} \sin\frac{\theta}{2} & e^{i(\phi/2)} \cos\frac{\theta}{2} \end{pmatrix} \begin{pmatrix} \uparrow \\ \downarrow \end{pmatrix} \quad (223)$$

and

$$\begin{pmatrix} R'_+ \\ R'_- \\ Z' \end{pmatrix} = \frac{1}{2} \begin{pmatrix} 1 & i & 0 \\ 1 & -i & 0 \\ 0 & 0 & \sqrt{2} \end{pmatrix} \begin{pmatrix} \cos\theta\cos\phi & \cos\theta\sin\phi & -\sin\theta \\ -\sin\theta & \cos\phi & 0 \\ \sin\theta\cos\phi & \sin\theta\sin\phi & \cos\theta \end{pmatrix}$$

$$\times \begin{pmatrix} 1 & 1 & 0 \\ -i & i & 0 \\ 0 & 0 & \sqrt{2} \end{pmatrix} \begin{pmatrix} R_+ \\ R_- \\ Z \end{pmatrix}$$

$$= \frac{1}{2} \begin{pmatrix} e^{-i\phi}(\cos\theta + 1) & e^{i\phi}(\cos\theta - 1) & -\sqrt{2}\sin\theta \\ e^{-i\phi}(\cos\theta - 1) & e^{i\phi}(\cos\theta + 1) & -\sqrt{2}\sin\theta \\ \sqrt{2}e^{-i\phi}\sin\theta & \sqrt{2}e^{i\phi}\sin\theta & 2\cos\theta \end{pmatrix} \begin{pmatrix} R_+ \\ R_- \\ Z \end{pmatrix}. \quad (224)$$

Unfortunately, this mapping of $u_{n,\alpha}$, $u_{n,\beta}$ to $u'_{n,\alpha}$, $u'_{n,\beta}$, also alters the direction of the quantization axis to coincide with the $\vec{k}'$ vector, complicating the overlap integrals between the $\vec{k}'$ and $\vec{k}$ basis vectors. It is therefore advantageous to rotate the basis functions, and therefore the quantization axis for the $\vec{k}'$ vector, back to the z-direction according to the inverse of Eq. (223):

$$\begin{pmatrix} u_{n,1/2} \\ u_{n,-1/2} \end{pmatrix} = \begin{pmatrix} e^{i(\phi/2)} \cos\frac{\theta}{2} & -e^{i(\phi/2)} \sin\frac{\theta}{2} \\ e^{-i(\phi/2)} \sin\frac{\theta}{2} & e^{-i(\phi/2)} \cos\frac{\theta}{2} \end{pmatrix} \begin{pmatrix} u'_{n,\alpha} \\ u'_{n,\beta} \end{pmatrix}. \quad (225)$$

Considering next the heavy-hole band, we again apply Eqs. (223) and (224) to generate $u'_{h,\alpha}$, $u'_{h,\beta}$, which are referred to the $\vec{k}'$ axis. We now simplify the calculation of the overlap integrals, however, by an additional θ- and ϕ-dependent basis function rotation such that

$$\langle u_{\vec{k},\pm 3/2} | u_{\vec{k}\mp 3/2} \rangle = 0. \tag{226}$$

This is accomplished by

$$\begin{pmatrix} u_{h,3/2} \\ u_{h,-3/2} \end{pmatrix} = \frac{1}{\sqrt{1 + 3\cos^2\theta}}$$

$$\times \begin{pmatrix} (\cos\theta + 1)e^{i(\phi/2)}\cos\frac{\theta}{2} & -(\cos\theta - 1)e^{i(\phi/2)}\sin\frac{\theta}{2} \\ (\cos\theta - 1)e^{-i(\phi/2)}\sin\frac{\theta}{2} & (\cos\theta + 1)e^{-i(\phi/2)}\cos\frac{\theta}{2} \end{pmatrix}$$

$$\times \begin{pmatrix} u'_{h,\alpha} \\ u'_{h,\beta} \end{pmatrix}. \tag{227}$$

Defining the variables $k_\pm = k_x \pm ik_y$ and $\delta = [(k^2 - k_z^2)/(k^2 + 3k_z^2)]^{1/2}$, we arrive at the heavy-hole wavefunctions (Symańska and Dietl, 1978)

$$u_{h,\vec{k}',3/2} = \left[\frac{1}{2}\frac{k'_+}{k'}\delta R_- - \frac{1}{2}\frac{k'_-}{k'}\delta^{-1}R_+ + \sqrt{2}\delta\frac{k'_z}{k'}Z\right]\uparrow$$

$$+ \left[\frac{1}{\sqrt{2}}\frac{k'_+}{k'}\delta Z - \frac{k'_z}{k'}\delta R_+\right]\downarrow. \tag{228}$$

and

$$u_{h,\vec{k}',-3/2} = \left[\frac{1}{2}\frac{k'_-}{k'}\delta R_+ - \frac{1}{2}\frac{k'_+}{k'}\delta^{-1}R_- + \sqrt{2}\delta\frac{k'_z}{k'}Z\right]\downarrow$$

$$- \left[\frac{1}{\sqrt{2}}\frac{k'_-}{k'}\delta Z - \frac{k'_z}{k'}\delta R_-\right]\uparrow. \tag{229}$$

Once the wavefunctions are derived, it is often convenient to express the overlap matrix elements in Eqs. (21) and (44) as matrices with respect to the indices α and β (recall that n denotes c, l, or s). Heavy-hole to

6. Introduction to Nonequilibrium Many-Body Analyses

heavy-hole transitions induced by a long-wavelength potential give rise to

$$B^{hh}(\vec{k}', \vec{k}) = \frac{1}{2} \delta^{-1}(\vec{k}') \sin\theta \begin{pmatrix} e^{i\phi} & 0 \\ 0 & e^{-i\phi} \end{pmatrix}, \qquad (230)$$

while the analogous matrix element for transitions between the heavy-hole subband and any other of the three subbands is

$$B^{nh}(\vec{k}', \vec{k}) = \begin{pmatrix} \dfrac{b'_n + c'_n\sqrt{2}}{2} \sin\theta e^{i\phi} & 0 \\ 0 & \dfrac{b'_n + c'_n\sqrt{2}}{2} \sin\theta e^{-i\phi} \end{pmatrix}. \qquad (231)$$

Here θ denotes the angle between $\vec{k}$ and $\vec{k}'$, while the Kane model coefficients a', b', and c' of Eq. (13) are related to the vector $\vec{k}'$. Finally, for intraband transitions, excluding the heavy-hole band,

$$B^{nn}(\vec{k}', \vec{k}) =$$
$$\begin{pmatrix} a'_n a_n + (c'_n c_n + b'_n b_n)\cos\theta & \frac{1}{2}(b_n b'_n - b_n c'_n\sqrt{2} - c_n b'_n\sqrt{2})\sin\theta e^{-i\phi} \\ -\frac{1}{2}(b_n b'_n - b_n c'_n\sqrt{2} - c_n b'_n\sqrt{2})\sin\theta e^{i\phi} & a'_n a_n + (c'_n c_n + b'_n b_n)\cos\theta \end{pmatrix}. \qquad (232)$$

In the case of electron–photon coupling, Eq. (44), the matrix element of the z-component of momentum between the conduction and valence bands is

$$P_z^{ch}(\vec{k}') = \frac{1}{\sqrt{2}} \frac{m_0 P}{\hbar} a(k')\delta(k') \begin{pmatrix} 2\cos\theta & -\sin\theta e^{-i\phi} \\ \sin\theta e^{i\phi} & 2\cos\theta \end{pmatrix}, \qquad (233)$$

while that of the combinations $P_\pm = (1/\sqrt{2})(P_x \pm iP_y)$ is given by

$$P_+^{ch}(\vec{k}') = \frac{m_0 P}{2\hbar} a(k')\delta(k') \begin{pmatrix} e^{i\phi}\sin\theta & 2\cos\theta \\ 0 & -\dfrac{1}{\delta^2(k')} e^{i\phi}\sin\theta \end{pmatrix}, \qquad (234)$$

and

$$P_-^{ch}(\vec{k}') = (P_+^{hc}(\vec{k}'))^\dagger. \qquad (235)$$

Finally, in perturbation theory calculations we encounter traces over products of overlap matrices that are evaluated through multiplication of the preceding expressions. To illustrate, the heavy-hole to heavy-hole contribution to the polarizability and self-energy in the *GW* approximation contains

$$\text{Tr}\{B^{hh}(\vec{k}, \vec{k}')B^{hh}(\vec{k}', \vec{k})\} = \frac{1}{2}(1 + 3\cos^2\theta). \tag{236}$$

Transitions between the light-hole and heavy-hole band lead instead to

$$\text{Tr}\{B^{lh}(\vec{k}, \vec{k}')B^{lh}(\vec{k}', \vec{k})\} = \frac{1}{2}(b_l + c_l\sqrt{2})^2 \sin^2\theta, \tag{237}$$

which reduces for small k to $(3/2)\sin^2\theta$. For intraband transitions not involving heavy holes,

$$\text{Tr}\{B^{nn}(\vec{k}, \vec{k}')B^{nn}(\vec{k}', \vec{k})\} = 2\Big\{[a'_s a_s + (c'_s c_s + b'_s b_s)\cos\theta]^2$$

$$+ \frac{1}{4}(b_s b'_s - b_s c'_s\sqrt{2} - c_s b'_s\sqrt{2})^2 \sin^2\theta\Big\}. \tag{238}$$

In optical absorption and gain calculations, we must evaluate $\text{Tr}\{P_z^{ch}(\vec{k})P_z^{hc}(\vec{k})\}$. The average over polarizations is

$$\frac{1}{3}\sum_{\gamma=x,y,z}\text{Tr}\{P_\gamma^{ch}(\vec{k})P_\gamma^{hc}(\vec{k})\} = \frac{2}{3}\frac{m_0^2}{\hbar^2}P^2. \tag{239}$$

Optical processes accompanied by plasmon or phonon absorption or emission instead give rise to the averaged overlap matrix element

$$M_{ch}(\vec{k}, \vec{k}') = \frac{1}{3}\sum_{\gamma=x,y,z}\text{Tr}\{P_\gamma^{hc}(\vec{k})B^{cc}(\vec{k}, \vec{k}')P_\gamma^{ch}(\vec{k}')B^{hh}(\vec{k}', \vec{k})\}$$

$$= \frac{2}{3}\frac{m_0^2}{\hbar^2}P^2\frac{1}{4}(1 + 3\cos^2\theta), \tag{240}$$

in which θ is the angle between $\vec{k}$ and $\vec{k}'$.

Appendix B

To prove Weyl's theorem, we first observe that

$$(\widetilde{A \circ B})(p, X) = \int_{-\infty}^{\infty} dx e^{-ipx} \int_{-\infty}^{\infty} dx_1 \tilde{A}\left(X + \frac{x}{2} - x_1, X - \frac{1}{2}\left(X - \frac{x}{2} - x_1\right)\right)$$
$$\times \tilde{B}\left(x_1 - X + \frac{x}{2}, X - \frac{1}{2}\left(X + \frac{x}{2} - x_1\right)\right). \quad (241)$$

In terms of the new variables $y = X + (x/2) - x_1$ and $z = -X + (x/2) + x_1$,

$$(\widetilde{A \circ B})(p, X) = \int dy \int dz e^{-ip(y+z)} \tilde{A}\left(y, X + \frac{z}{2}\right)\tilde{B}\left(z, X - \frac{y}{2}\right). \quad (242)$$

Rewriting this equality using displacement operators yields

$$(\widetilde{A \circ B})(p, X) = \int dy \int dz e^{-ip(y+z)} e^{(1/2)(z\partial_X^A - y\partial_X^B)} \tilde{A}(y, X)\tilde{B}(z, X), \quad (243)$$

which together with the identity

$$\int dz e^{-ipz} e^{(1/2)z\partial_X^A} \tilde{B}(z, X) = e^{(i/2)\partial_p \partial_X^A} \int dz e^{-ipz} \tilde{B}(z, X) \quad (244)$$

proves Eq. (85).

Appendix C

We demonstrate in this appendix that the correlation function $G^>$, which is analytic in the region from $t = 0$ to $t = -i\beta$, obeys the relation

$$G^>(\omega) = -e^{\omega\beta} G^<(\omega), \quad (245)$$

where energy is for simplicity measured from the Fermi level. In particular, we demonstrate that

$$G^>(\omega) = \int_{-\infty}^{\infty} e^{i\omega t} G^>(t, 0)\, dt \quad (246)$$

equals

$$-\int_{-\infty}^{\infty} e^{i\omega t} G^{<}(t + i\beta, 0)\, dt = -e^{\omega \beta} G^{<}(\omega). \tag{247}$$

We first observe that since H is time-independent, we may replace t_0 by zero and express the evolution operator $U(t_0 - i\beta, t_0)$ in Eq. (63) in terms of the difference of its time arguments. Accordingly,

$$iG^{>}(t, 0) = \frac{\mathrm{Tr}\{U(-i\beta, t)\Phi U(t, 0)\Phi^{\dagger}\}}{\mathrm{Tr}\{U(-i\beta, 0)\}}. \tag{248}$$

In Eq. (248) $U(-i\beta, 0) = e^{-\beta H}$ is the Schrödinger evolution operator, and the contour explicitly proceeds from $t_0 = 0$ to t and then to $-i\beta$. By virtue of the cyclical property of the trace, we can rewrite the preceding formula as

$$iG^{>}(t, 0) = \frac{\mathrm{Tr}\{U(-i\beta, 0)U(0, -i\beta)\Phi^{\dagger}U(-i\beta, t)\Phi U(t, 0)\}}{\mathrm{Tr}\{U(-i\beta, 0)\}}$$
$$= -iG^{<}(t, -i\beta) = -iG^{<}(t + i\beta, 0). \tag{249}$$

The $<$ ordering associated with the last two Green's functions reflects the structure of the transformed contour, which passes first through t and then $-i\beta$.

References

Abram, R., Childs, G., and Saunderson, P. (1984). *J. Phys. C: Solid State Phys.* **17**, 6105–6125.
Aldrich, C., and Silver, R. (1980). *Phys. Rev. B* **21**, 600–614.
Aleksanian, A., Poluetkov, I., and Popov, Y. M. (1974). *IEEE J. Quantum Electron.* **QE-10**, 297–305.
Bardyszewski, W. (1986). *Solid State Comm.* **57**, 873–876.
Bardyszewski, W., and Yevick, D. (1985a). *IEEE Journal of Quantum Electronics*, **QE-21**, 1131–1134.
Bardyszewski, W., and Yevick, D. (1985b). *J. Appl. Phys.* **58**, 2713.
Bardyszewski, W., and Yevick, D. (1985c). *J. Appl. Phys.* **57**, 4820–4822.
Bardyszewski, W., and Yevick, D. (1987). *Phys. Rev. B* **35**, 619–625.
Bardyszewski, W., and Yevick, D. (1989a). *J. Appl. Phys.* **66**, 988–991.
Bardyszewski, W., and Yevick, D. (1989b). *Appl. Phys. Lett.* **54**, 837–839.
Bardyszewski, W., and Yevick, D. (1989c). *Phys. Rev. B* **39**, 10839–10851.
Bardyszewski, W., and Yevick, D. (1989d). *Phys. Rev. B* **39**, 5861–5864.

6. Introduction to Nonequilibrium Many-Body Analyses

Baym, G., and Kadanoff, L. (1961). *Phys. Rev.* **124**, 287–299.
Beattie, A., and Landsberg, P. (1959). *Proc. R. Soc. London, Ser. A* **249**, 16–29.
Berggren, K.-F. and Sernelius, B. (1981). *Phys. Rev. B* **24**, 1971–1986.
Bir, G., and Pikus, G. (1974). *Symmetry and Strain-Induced Effects in Semiconductors*, Wiley, New York.
Brinkman, W., and Lee, P. (1973). *Phys. Rev. Lett.* **31**, 237–239.
Casey, H., and Stern, F. (1976). *J. Appl. Phys.* **47**, 631–643.
Combescot, M., and Nozières, P. (1972). *Solid State Comm.* **10**, 301–305.
Craig, R. (1968). *J. Math. Phys.*, **9**, 605–611.
Danielewicz, P. (1984). *Annals of Physics* **152**, 239–304.
Fetter, A., and Walecka, J. (1971). *Quantum Theory of Many-Particle Systems*, McGraw-Hill, New York.
Fischetti, M. (1991). *Phys. Rev. B* **44**, 5527–5539.
Furuta, T., and Tomizawa, M. (1990). *Appl. Phys. Lett.* **56**, 824–826.
Furuta, T., Taniyama, H., and Tomizawa, M. (1990). *J. Appl. Phys.* **67**, 293–299.
Göbel, G. (1974). *Appl. Phys. Lett.* **24**, 492–494.
Haug, A., and Ekardt, W. (1975). *Solid State Comm.* **17**, 267–268.
Haug, H., and Tran Thoai, D. (1980). *Phys. Stat. Sol. (b)* **98**, 581–589.
Hayes, J., and Harbison, J. (1988). *Appl. Phys. Lett.* **53**, 490–492.
Hayes, J., Levi, A., Gossard, A., and English, J. (1986). *Appl. Phys. Lett.* **49**, 1481–1483.
Hedin, L. (1965). *Phys. Rev.* **139**, A796–A823.
Henry, C., Logan, R., and Bertness, K. (1981). *J. Appl. Phys.* **52**, 4457–4461.
Inkson, J. (1976). *J. Phys. C: Solid State Phys.* **9**, 1177–1183.
Kadanoff, L., and Baym, G. (1962). *Quantum Statistical Mechanics*. Benjamin, New York.
Kane, E. O. (1957). *J. Phys. Chem. Solids* **1**, 249–261.
Kane, E. O. (1966). "The $k \cdot p$ Method", in *Semiconductors and Semimetals*, Vol. 1 (Williamson, R. and Beer, A., eds.), pp. 75–101. Academic Press, New York.
Keldysh, L. (1965). *Sov. Phys. JETP* **20**, 1018–1026. [*J. Exptl. Teoret. Phys. (USSR)* **47**, 1515–1527 (1964)].
Langer, J. (1961). *Phys. Rev.* **124**, 997–1003.
Langreth, D. C. (1976). "Linear and Nonlinear Response Theory with Applications," in *Linear and Nonlinear Electron Transport in Solids*, Vol. 17 (Devreese, J. and van Doren, V., eds.) pp. 3–32. NATO Advanced Study Institutes, Series B: Physics. Plenum, New York.
Levi, A., and Yafet, Y. (1987). *Appl. Phys. Lett.* **51**, 42.
Levi, A., Hayes, J., Gossard, A., and English, J. (1987). *Appl. Phys. Lett.* **50**, 98–100.
Lindhard, J. (1954). *Dan. Vidensk. Selskab. Mat.-Fys. Medd.* **28**, 8.
Lovejoy, M., Keyes, B., Klausmeier-Brown, M., Melloch, M., Ahrenkiel, R., and Lundstrom, M. (1990). In *Extended Abstracts for the 22nd International Conference of Solid State Devices and Materials, Sendai, Japan*, pp. 613–616.
Lowney, J., and Bennett, H. (1991). *J. Appl. Phys.* **69**, 7102–7110.
Lugli, P., and Ferry, D. (1985). *Appl. Phys. Lett.* **46**, 594–596.
Lundstrom, M., Klausmeier-Brown, M., Melloch, M., Ahrenkiel, R., and Keyes, B. (1990). *Solid-State Electronics* **33**, 693–704.
Mahan, G. (1967). *Phys. Rev.* **153**, 882–889.
Nathan, M., Dumke, W., Wrenner, K., Tiwari, S., Wright, S., and Jenkins, K. (1988). *Appl. Phys. Lett.* **52**, 654–656.
Overhauser, A. (1971). *Phys. Rev. B* **3**, 1888–1897.
Rice, T. (1974). *Il Nuovo Cimento* **23B**, 226–233.
Rode, D. (1975). "Low Field Electron Transport," In *Semiconductors and Semimetals*, Vol. 10 (Williamson, R. and Beer, A., eds.), pp. 1–89. Academic Press, New York.

Schaefer, W., Binder, R., and Schuldt, K. (1988). *Z. Phys. B-Condensed Matter* **70**, 145–157.
Schmitt-Rink, S., Tran Thoai, D., and Haug, H. (1980). *Z. Physik B—Condensed Matter* **39**, 25–31.
Selloni, A., Modesti, S., and Capizzi, M. (1984). *Phys. Rev. B* **30**, 821–831.
Sermage, B., Heritage, J., and Dutta, N. (1985). *J. Appl. Phys.* **57**, 5443–5449.
Sernelius, B. (1986). *Phys. Rev. B* **34**, 5610–5620.
Sugimura, A., Patzak, E., and Meissner, P. (1986). *J. Phys. D: Appl. Phys.* **19**, 7–16.
Szymańska, W., and Dietl, T. (1978). *J. Phys. Chem. Solids* **39**, 1025–1040.
Takeshima, M. (1982). *Phys. Rev. B* **26**, 917–930.
Walukiewicz, W., Lagowski, J., Jastrzebski, L., and Gatos, H. (1979). *J. Appl. Phys.* **50**, 5040–5042.
Yevick, D., and Bardyszewski, W. (1987a). *Appl. Phys. Lett.* **51**, 124–126.
Yevick, D., and Bardyszewski, W. (1987b). *IEEE J. Quant. Elec.* **QE-23**, 168–170.
Yevick, D., and Bardyszewski, W. (1989a). *IEEE J. Quant. Elec.* **QE-25**, 1813–1821.
Yevick, D., and Bardyszewski, W. (1989b). *Phys. Rev. B* **39**, 8605–8608.
Zee, B. (1978). *IEEE J. Quantum Elect.* **QE-14**, 727–736.
Ziep, O., and Mocker, M. (1980). *Phys. Stat. Sol.* (*b*) **98**, 133–142.
Ziep, O., and Mocker, M. (1983). *Phys. Stat. Sol.* (*b*) **119**, 299–305.
Zimmermann, R. (1978). *Phys. Stat. Sol.* (*b*) **86**, K63–K66.
Zimmermann, R. (1987). *Many-Particle Theory of Highly Excited Semiconductors*. BSB Teubner, Leipzig.
Zook, J. (1964). *Phys. Rev.* **136**, A869–A878.

Index

A

Absorption, 5–15
Absorption edge, 302
Affinity, electron, 196
AlGaAs, HBT, 358
Alpha factor, spontaneous emission, 380
Anticommutation relations, 323
Asymmetry factor, effective, 200
Auger
  coefficient, 369
  electron, 367
  processes, 364
  recombination rates, 56, 319

B

Balistic transport (non), 359
Band filling, 302
Band stretching, 275
Band structure, 3
  perturbed, 201–202
Band tail, 291, 300, 303, 311
Bandgap, renormalization, 319, 348, 366
Bandgap narrowing
  bulk GaAs, 270–279, 301–306, 309–312
  capacitance measurements, 311
  doped bulk GaAs, 272–276, 301–305, 309–312
  effective, 310
  electrical measurements, 309–312
  electron–hole plasma in GaAs, 276–279, 305–306
  electrostatic models, 279
  finite temperatures, 278–279
  GaAs quantum well, 282–288, 307–308
  nonspectroscopic measurement, 308
  optical measurements, 301–308
  theory, 264, 270–279, 282–288
Bands
  conduction, 320
  light-hold, 320
  heavy-hole, 320
  nondegenerate, 326
  spin–orbit split-off, 320
Baym, 318
Bipolar transistor
  base delay, 225, 226, 248–254
  collector current measurements, 309–311
  GaAs devices, 309–311
  gain–bandwidth product, 222, 225, 226
  graded-base, 234–238
  heavily doped base, 309–311
  heterojunction devices, 311
    AlGaAs/GaAs, 229–231
    InP/InGaAs, 247–254
  homojunction, 215
  pseudo-heterojunction, 231–234
Bloch, 318, 320
Bloch functions, periodic parts, 325
Boltzmann equation, 197–198, 319, 334, 362
  quantum, 341, 342
  standard, 342
Born approximation, 362

Boson
    operators, 340
    propagator, 352
Brooks–Herring model, 179
Built-in field, 359
Burstein–Moss shift, 203, 215

C

Carrier injection, 358
Cell, primitive unit, 355
Charge density fluctuations, 350
CHCC, CHSH, CHLH processes, 368, 370
Collective excitations, 318
Collective plasma mode, 178
Collision integral, 363
Collision terms, 342
Conductivity, 371
Confinement structures
    analysis of, 72–76
    isotype double heterostructure, 74
    PL lifetime, 82-85
Conservation laws, 361, 364, 368
Conwell–Weisskopf model, 179
Core polarizability, 352
Correlation, 264
    functions, 334, 385
    term, 271
Coulomb gauge, 328
Coulomb hole term, 270
    in X valley of $Al_xGa_{1-x}As$, 306
Coulomb potential model, 179
Coupled flow, carriers and photons, 205–213
Coupled plasmon–LO phonon mode, 178
Cutoff wave number, 178

D

Debye screening, 369, 370
Deformation potentials, 325
    constants, 326
Degeneracy, 158, 180, 185
Density fluctuations, 346
Density matrix, 330
Density-of-states
    effect, 199
    effective mass, 356, 358
Density–density correlation function, 350
Detailed balance, 342

Dielectric function, 264, 266–270, 292, 343, 346
    damped plasmon pole approximation, 270, 275
    electron gas, 267–270, 281–282
    intrinsic semiconductor, 267
    inverse, 268
    lattice contribution, 267
    Lindhard formula, 268, 273
    longitudinal, 346
    plasma pole approximation, 268–270, 273–275, 277, 279, 282, 285, 349, 357, 359
    random phase approximation (RPA), 349
    retarded RPA, 355
    two-dimensional electron gas, 281–282
    valence band temperature dependent, 359
Diffusion transit time, 100
Direct bandgap, 319
Distribution function, Boson, 340
Doping
    compensation, 289
    modulation, 299, 307–308
    spatial distribution, 288–289, 291, 312
    substitutional impurity, 289
    useful measure of density ($N^{1/3}a$), 289
Drift velocity of minority electrons
    in p-GaAs, 156–159
    in quantum-well structure, 159–162
Drift–diffusion equation, 198–200
Drifted Maxwellian distribution, 180, 185
Dynamical screening, 369, 370
Dynamically screened potential, 362, 370, 377
Dyson's equation, 334

E

ε_0 approximation, 267
Effective mass, 237
    position dependent: 197
Electron density, 323
Electron gas,
    dielectric function, 267–270, 281–282
    high-density regime, 263
    response function, 267
Electron temperature,
    versus electric field, 165
    time-evolution of, 175
Electron, field, 323

Electron–electron interaction
  direct, 264, 293, 294
  dynamically screened, 264, 266, 282
  exchange, 264
  impurity band, 289
  poles in, 270
  quantum well, 281
  screening, 269–272
Electron–hole
  droplets, 261, 263, 273, 276
  interaction, 152–153
  plasma
    bulk GaAs, 276–278, 305–307
    quantum well, 285–286, 307–308
    semiconductor laser, 305, 307–308
Electron-impurity interaction, 290–292, 346
  factorized form, 297
  linear screening, 290, 299
  matrix element, 292
  screening in bulk, 290, 295
  screening in quantum well, 299
Electron–phonon coupling, 324, 329, 346
Electron–photon
  interaction, 324, 325, 328
  scattering, 364
Electron–plasma collisions, 345
Energy gap
  apparent, 203
  effective, 200, 214, 217
  electrical, 203
  shrinkage, 213–217
Energy level broadening, 265, 302
Energy loss rate,
  due to electron–hole interaction, 167
  due to LO phonon cascade process, 167
Energy relaxation process, 170–177
Energy relaxation time
  as a function of minority-electron energy, 169
Evolution operator, 386
Exchange terms, 335
  screening of, 367
Exchange-correlation energy, 277
  universal formula for, 277, 305
Exciton formation, 376

F

f-sum rule, 349
Fermi–Dirac distribution function, 163, 343

Fermion, self energy, 341
Feynman, diagram technique, 334
Field
  effective, 237
  quasi-electric, 197
Fluctuation–dissipation theorem, 353

G

GaAs, 12–17, 359
  lasers, 33
  lifetime, see lifetime, GaAs
Gain measurement, 305, 308
Gain, 14, 371, 384
  spectra, 319, 370
  suppression, 378
GaSb, p-type, 370
Gradient expansion, 319
Green's function, 319, 330
  antichronological, 373
  Bloch representation, 371
  bosons, 340
  chronological, 373
  nonequilibrium, 333
  one-particle, 332
  quasi-equilibrium, 342, 365
  renormalized, 348
GW, approximation, 345, 346

H

Hartree potential, 286–288, 300
Hartree–Fock
  approximation, 335
  exchange term, 271
HBT (heterojunction bipolar transistor), 151, 158, 189
Heavy doping effects,
  on effective energy gap, 214–217
  on heterojunction bipolar transistors, 229–231, 234–238
  on minority-carrier injection, 213, 214
  on minority-carrier mobility, 226–228
  on solar cells, 241–244
Heisenberg representation, 330
Heterostructure, semiconductor, 195–197
High injection, definition, 45
  radiative lifetime effects, 108–112
Hole temperature, 163

Hot electron
  scattering, 362
  spectroscopy, 358
  transport, 358
Hydrodynamic approximation, 168

I

Impurity band
  distinct band, 289, 296
  electron–electron interactions in, 289
  merged with parent band, 289, 296
  metal–insulator transition, 289
  tight-binding model, 290
Impurity effects on band structure
  bulk GaAs, 288–299
  GaAs quantum well, 299–301
  semiclassical models, 290–291, 300
  variational models, 291
Impurity scattering, 291–297, 347
  high density limit, 296
  multiple scattering theory, 291–297
  self-consistent calculations, 294–297
  single and double scattering, 292, 294, 296
Impurity state
  binding energy, 289
  effective Bohr radius, 289
  shallow level, 289
In(0.7)Ga(0.3)As(0.64)P(0.36), 369
Inelastic losses, 343
Inelastic scattering, 345
InGaAs, 56
InGaAsP, 18–22, 56
  lasers, 34
Injection level, definition, 47
Input power per electron, 167
Interband processes, 342
Interference diagrams, 375, 377
Intraband plasma absorption, 356
Intraband relaxation, 342
Intrinsic carrier concentration, effective, 203

J

Junction devices, PL lifetime, 97–102

K

k-selection rule, 378
$k \cdot p$ method
  band structure, 319
  Hamiltonian, 319
  overlap integrals, 368
Kadanoff, 318
Kane model, 7
  parameter, 321, 376
Keldysh, contour, 334, 386
Keldysh diagram technique, 373, 374
Kramers conjugates, 321
Kubo, 371

L

Ladder diagrams, 376
Landau damping, 356
Lifetime
  addition of, 51, 59, 73, 74
  Auger, 56
  diffusion, 101
  GaAs,
    electrons in p-type, 129
    epitaxial, 129–134
    figure of merit, 122, 127
    holes in n-type, 132
    surface, 122
    temperature dependence, 128–129
    undoped DH, 121–129
    wafer, 41, 119–121
  hot electron, 358
  instantaneous, 46, 62–64, 80, 107
  low injection, minority-carrier, 46, 47, 72, 73
  mean free, 338
  measurement technique, 65–69
    inductively coupled probe, 69
    time-correlated single photon counting, 66–68
    time-resolved photoluminescence (TRPL), 61, 65–68
  minority carrier transport, 319
  nonradiative, 62, 74, 83, 209
  one-particle, 341, 342
  photoluminescence (PL), 65, 71, 84, 87, 96
    analysis for DH devices, 74–84
    analysis for junction device, 97–102

analysis for wafers, 102–108
bulk, minority-carrier, 84, 101, 102
radiative, 22, 59, 60, 85, 208
  high injection, 62
  low injection, 61
  temperature dependence, 64, 65
recombination, 44
recycling, see Photon recycling
Shockley–Read–Hall (SRH), 50, 51
  AlGaAs, 117–119
  effective, 113, 115
  intensity dependence of, 112–119
  minority-carrier, 114, 118
  saturation effects, 114
  spontaneous emission, 379
  surface or interface, 81, 85, 96
  temperature dependence: 123
Line shape function, 338
Linear response formalism, 371
Linewidth broadening, 365
Local field
  corrections, 355
  effects, 353
Long-range interaction, 324, 347
Lorentzian, 339, 357
Low injection, definition, 45
Low-field drift mobilities,
  as a function of hole concentration, 186

M

Many-body theory
  bulk GaAs, 261–279
  GaAs quantum well, 279–288
Matrix element, 5, 380
  electron–phonon, 375
  overlaps, 346, 382
  traces over products of overlaps, 384
Matsubara formalism, 318, 375
Mobility
  edge, 202
  low-field electron, 360
  minority electron in GaAs, 226, 227
  minority hole in GaAs, 227, 228
  minority-carrier, 217–228
Momentum relaxation time
  as a function of minority-electron energy, 170
Monte Carlo integration, 369

Monte Carlo simulation, 358
  calculated results, 180
  drift velocity, 181–187
  electron temperature, 187–189
  integration, 369
  scattering rate, 181
Moss–Burstein shift, 302, 310

N

Negative differential resistance, 157–158, 160, 167

O

Occupation factors, 365
One-electron Green function, 264, 265, 270–271, 291–292
  ensemble average, 292
  perturbation expansion, 292
  poles in, 270
One-particle properties, 262
Optical absorption, 5–6, 59–60, 302
Optical theorem, 339

P

Pauli's exclusion principle, 159, 180, 185
Peak velocity, 157, 161
Phonon
  absorption and emission, 384
  branch, 324
  creation and annihilation operators, 323, 324, 347
  field, 324
  frequency, 324
  Green's function, 347
  longitudinal optical, 324, 325, 357
  optical frequency, 357
  polarization vector, 324
  propagator, 347
Photoluminescence (PL), 57, 302, 304–305, 307–308
  excitation spectroscopy, 302
  spectroscopy, 69
  time resolved, 65–68, 71, 306
Photoluminescence spectra
  as a function of electric field, 164
  time-resolved, 175

Photon
  absorption and emission, 384
  annihiliation and creation operators, 328
  field, 328
  frequency, 328
  polarization, 328
Photon recycling, 41, 84–85, 117
  calculation of factor phi, 90–94
  definition, 85, 194
  effects on lifetime, 84–85, 137–141, 209–210
  effects on photodiode quantum efficiency, 245–247
  effects on transport, 210, 212
  factor, 210
  high injection in DH, 108–112
  junction devices, 102
  in n-GaAs, 132–133
  in p-GaAs, 131
  theory, 206–212
  in undoped GaAs, 124
Piezoelectric
  constant, 325
  coupling, 325
Plasmon
  absorption and emission, 375, 384
  contribution to Boltzman equation, 363
  cut-off momentum, 361
  damped plasmon pole approximation, 358
  damping, 356
  frequency, 356, 357
  linewidth broadening factor, 358
  majority-carrier excitations, 362
  pole approximation, 268–270, 273–275, 277, 282, 285
    damped, 270, 275
    nondegenerate gas, 279
  propagator, 347, 350
  satellite

Q

Quantum efficiency, photodiode, 245–247
Quantum well, 2
  energy, 339
Quasi-equilibrium, 318, 363
Quasi-Fermi level, 355, 365
Quasiparticle, 262, 263, 323, 329
  energy, 339

R

Random phase approximation, 353
Rate equation, 341
Rate
  spontaneous emission, 379
Reabsorption, photon, see Photon recycling
Recombination
  Auger, 56
  midgap center, 50
  nonradiative, 74
  radiative, 57, 370
  Shockley–Read–Hall (SRH), 47, 48
  SRH rate, 48
  surface and interface, 1
  velocity, 52
Relative permittivity
  optical, 267
  static, 267
Response function
  linear, 372, 373, 379
  retarded, 352, 373
Rigid band shift, 274, 277
RPA
  macroscopic dielectric function, 355
  screening, 319

S

Scattering
  effects on base delay, 247–254
  electron–plasmon, 361, 363
  plasma, 218, 219
Scattering angle, electron, 359
Screened exchange term, 271
  in X valley of $Al_x Ga_{1-x} As$, 306
Screening, of Auger processes, 370
Second-quantization, 322
Self absorption, 23, 41, 59, 62, 84
  data, 86, 90
  DH devices in high injection: 108
  effects in TRPL, 87–88, 90
  one-dimensional model, 85–88
  thick films and bulk crystals, 106
  three-dimensional calculation, 90–93
Self-consistent field, 353
Self-energy
  fermion, 341
  imaginary part, 343

operator, 334
real part, 348
Self-energy (electron–electron)
  Coulomb hole term, 270
  finite temperatures, 278
  formalism, 264–266, 281–284
  imaginary part, 265, 302
  real part, 265, 302
  screened exchange term, 271
Self-energy (electron-impurity), 291–295
  formalism, 291–295
  imaginary part, 293
  real part, 293
Semiconductor
  doped, intrinsic, 348, 349
  laser measurements, 305, 307–308
Silicon lifetime, 40
Single particle transitions, 269, 282
  interband, 269
Solar cell
  back-surface-field, 219–226
  dark current, 240
  efficiency, 241
  fill factor, 241
  GaAs, 238–245
  open-circuit voltage, 242
Spatial overlap of wavefunctions
  for minority electrons and majority holes, 161
Spectral density function, 339
Spin–orbit interaction, 320
Spontaneous emission, 5–15
Statistical averaging, 330
Statistical operator, 330, 350
Stimulated emission, 370
  condition, 7
  gain–current relation, 17, 32
Strain tensor, 326
Symmetrically doped diode, 311

T

t-matrix, off-shell scattering, 347
Thermally nonequilibrium energy states, 165
Thomas–Fermi screening, 269, 295, 299

Threshold
  energy and momentum in Auger processes, 368
  field, 157, 161
Time evolution, 330
Time-dependent diffusion equation
  solution for confinement structures, 72
Time-of-flight measurement, 155–156
  zero-field, 219–226
Time-resolved photoluminescence
  dependence on hole concentration, 172
  dependence on photon energy, 174
  generated analysis of, 71–74
Transistors (*see also* Bipolar transistor)
  bipolar heterojunction, 358, 361
Transitions
  heavy-hole to heavy-hole, 382
  light-hole to heavy-hole, 384
Transport
  ballistic, 249–254
  in quasi-neutral regions, 205, 234–238
  minority-carrier, 204, 205

U

Umklapp processes, 346
Up-conversion technique, 171

V

van Roosbroeck–Shockley relationship, 59–60, 91, 207–208
Vector potential, 328
Velocity overshoot effects, 189
Vertex correction, 375, 376
Vertices, 373

W

Wave functions, 322
  heavy-hole, 362
Wavevector conservation, 301, 305
  lack of, 305, 307
Weyl's formula, 337, 385
Wick decomposition, 333
Wigner distribution function, 329, 366
  generalized, 331

Contents of Volumes in this Series

Volume 1 Physics of III–V Compounds

C. Hilsum, Some Key Features of III–V Compounds
Franco Bassani, Methods of Band Calculations Applicable to III–V Compounds
E. O. Kane, The k-p Method
V. L. Bonch-Bruevich, Effect of Heavy Doping on the Semiconductor Band Structure
Donald Long, Energy Band Structures of Mixed Crystals of III–V Compounds
Laura M. Roth and Petros N. Argyres, Magnetic Quantum Effects
S. M. Puri and T. H. Geballe, Thermomagnetic Effects in the Quantum Region
W. M. Becker, Band Characteristics near Principal Minima from Magnetoresistance
E. H. Putley, Freeze-Out Effects, Hot Electron Effects, and Submillimeter Photoconductivity in InSb
H. Weiss, Magnetoresistance
Betsy Ancker-Johnson, Plasma in Semiconductors and Semimetals

Volume 2 Physics of III–V Compounds

M. G. Holland, Thermal Conductivity
S. I. Novkova, Thermal Expansion
U. Piesbergen, Heat Capacity and Debye Temperatures
G. Giesecke, Lattice Constants
J. R. Drabble, Elastic Properties
A. U. Mac Rae and G. W. Gobeli, Low Energy Electron Diffraction Studies
Robert Lee Mieher, Nuclear Magnetic Resonance
Bernard Goldstein, Electron Paramagnetic Resonance
T. S. Moss, Photoconduction in III–V Compounds
E. Antončik and J. Tauc, Quantum Efficiency of the Internal Photoelectric Effect in InSb
G. W. Gobeli and F. G. Allen, Photoelectric Threshold and Work Function
P. S. Pershan, Nonlinear Optics in III–V Compounds
M. Gershenzon, Radiative Recombination in the III–V Compounds
Frank Stern, Stimulated Emission in Semiconductors

Volume 3 Optical of Properties III–V Compounds

Marvin Hass, Lattice Reflection
William G. Spitzer, Multiphonon Lattice Absorption
D. L. Stierwalt and R. F. Potter, Emittance Studies
H. R. Philipp and H. Ehrenveich, Ultraviolet Optical Properties
Manuel Cardona, Optical Absorption above the Fundamental Edge
Earnest J. Johnson, Absorption near the Fundamental Edge
John O. Dimmock, Introduction to the Theory of Exciton States in Semiconductors
B. Lax and J. G. Mavroides, Interband Magnetooptical Effects

H. Y. Fan, Effects of Free Carries on Optical Properties
Edward D. Palik and George B. Wright, Free-Carrier Magnetooptical Effects
Richard H. Bube, Photoelectronic Analysis
B. O. Seraphin and H. E. Bennett, Optical Constants

Volume 4 Physics of III–V Compounds

N. A. Goryunova, A. S. Borschevskii, and D. N. Tretiakov, Hardness
N. N. Sirota, Heats of Formation and Temperatures and Heats of Fusion of Compounds $A^{III}B^V$
Don L. Kendall, Diffusion
A. G. Chynoweth, Charge Multiplication Phenomena
Robert W. Keyes, The Effects of Hydrostatic Pressure on the Properties of III–V Semiconductors
L. W. Aukerman, Radiation Effects
N. A. Goryunova, F. P. Kesamanly, and D. N. Nasledov, Phenomena in Solid Solutions
R. T. Bate, Electrical Properties of Nonuniform Crystals

Volume 5 Infrared Detectors

Henry Levinstein, Characterization of Infrared Detectors
Paul W. Kruse, Indium Antimonide Photoconductive and Photoelectromagnetic Detectors
M. B. Prince, Narrowband Self-Filtering Detectors
Ivars Melngalis and T. C. Harman, Single-Crystal Lead-Tin Chalcogenides
Donald Long and Joseph L. Schmidt, Mercury-Cadmium Telluride and Closely Related Alloys
E. H. Putley, The Pyroelectric Detector
Norman B. Stevens, Radiation Thermopiles
R. J. Keyes and T. M. Quist, Low Level Coherent and Incoherent Detection in the Infrared
M. C. Teich, Coherent Detection in the Infrared
F. R. Arams, E. W. Sard, B. J. Peyton, and F. P. Pace, Infrared Heterodyne Detection with Gigahertz IF Response
H. S. Sommers, Jr., Macrowave-Based Photoconductive Detector
Robert Sehr and Rainer Zuleeg, Imaging and Display

Volume 6 Injection Phenomena

Murray A. Lampert and Ronald B. Schilling, Current Injection in Solids: The Regional Approximation Method
Richard Willliams, Injection by Internal Photoemission
Allen M. Barnett, Current Filament Formation
R. Baron and J. W. Mayer, Double Injection in Semiconductors
W. Ruppel, The Photoconductor-Metal Contact

Volume 7 Application and Devices
PART A

John A. Copeland and Stephen Knight, Applications Utilizing Bulk Negative Resistance
F. A. Padovani, The Voltage-Current Characteristics of Metal-Semiconductor Contacts
P. L. Hower, W. W. Hooper, B. R. Cairns, R. D. Fairman, and D. A. Tremere, The GaAs Field-Effect Transistor
Marvin H. White, MOS Transistors
G. R. Antell, Gallium Arsenide Transistors
T. L. Tansley, Heterojunction Properties

PART B

T. Misawa, IMPATT Diodes
H. C. Okean, Tunnel Diodes
Robert B. Campbell and Hung-Chi Chang, Silicon Carbide Junction Devices
R. E. Enstrom, H. Kressel, and L. Krassner, High-Temperature Power Rectifiers of $GaAs_{1-x}P_x$

Volume 8 Transport and Optical Phenomena

Richard J. Stirn, Band Structure and Galvanomagnetic Effects in III–V Compounds with Indirect Band Gaps
Roland W. Ure, Jr., Thermoelectric Effects in III–V Compounds
Herbert Piller, Faraday Rotation
H. Barry Bebb and E. W. Williams, Photoluminescence 1: Theory
E. W. Williams and H. Barry Bebb, Photoluminescence II: Gallium Arsenide

Volume 9 Modulation Techniques

B. O. Seraphin, Electroreflectance
R. L. Aggarwal, Modulated Interband Magnetooptics
Daniel F. Blossey and Paul Handler, Electroabsorption
Bruno Batz, Thermal and Wavelength Modulation Spectroscopy
Ivar Balslev, Piezopptical Effects
D. E. Aspnes and N. Bottka, Electric-Field Effects on the Dielectric Function of Semiconductors and Insulators

Volume 10 Transport Phenomena

R. L. Rode, Low-Field Electron Transport
J. D. Wiley, Mobility of Holes in III–V Compounds
C. M. Wolfe and G. E. Stillman, Apparent Mobility Enhancement in Inhomogeneous Crystals
Robert L. Petersen, The Magnetophonon Effect

Volume 11 Solar Cells

Harold J. Hovel, Introduction; Carrier Collection, Spectral Response, and Photocurrent; Solar Cell Electrical Characteristics; Efficiency; Thickness; Other Solar Cell Devices; Radiation Effects; Temperature and Intensity; Solar Cell Technology

Volume 12 Infrared Detectors (II)

W. L. Eiseman, J. D. Merriam, and R. F. Potter, Operational Characteristics of Infrared Photodetectors
Peter R. Bratt, Impurity Germanium and Silicon Infrared Detectors
E. H. Putley, InSb Submillimeter Photoconductive Detectors
G. E. Stillman, C. M. Wolfe, and J. O. Dimmock, Far-Infrared Photoconductivity in High Purity GaAs
G. E. Stillman and C. M. Wolfe, Avalanche Photodiodes
P. L. Richards, The Josephson Junction as a Detector of Microwave and Far-Infrared Radiation
E. H. Putley, The Pyroelectric Detector–An Update

Volume 13 Cadmium Telluride

Kenneth Zanio, Materials Preparation; Physics; Defects; Applications

Volume 14 Lasers, Junctions, Transport

N. Holonyak, Jr. and M. H. Lee, Photopumped III–V Semiconductor Lasers
Henry Kressel and Jerome K. Butler, Heterojunction Laser Diodes
A. Van der Ziel, Space-Charge-Limited Solid-State Diodes
Peter J. Price, Monte Carlo Calculation of Electron Transport in Solids

Volume 15 Contacts, Junctions, Emitters

B. L. Sharma, Ohmic Contacts to III–V Compound Semiconductors
Allen Nussbaum, The Theory of Semiconducting Junctions
John S. Escher, NEA Semiconductor Photoemitters

Volume 16 Defects, (HgCd)Se, (HgCd)Te

Henry Kressel, The Effect of Crystal Defects on Optoelectronic Devices
C. R. Whitsett, J. G. Broerman, and C. J. Summers, Crystal Growth and Properties of $Hg_{1-x}Cd_xSe$ alloys
M. H. Weiler, Magnetooptical Properties of $Hg_{1-x}Cd_xTe$ Alloys
Paul W. Kruse and John G. Ready, Nonlinear Optical Effects in $Hg_{1-x}Cd_xTe$

Volume 17 CW Processing of Silicon and Other Semiconductors

James F. Gibbons, Beam Processing of Silicon
Arto Lietoila, Richard B. Gold, James F. Gibbons, and Lee A. Christel, Temperature Distributions and Solid Phase Reaction Rates Produced by Scanning CW Beams
Arto Leitoila and James F. Gibbons, Applications of CW Beam Processing to Ion Implanted Crystalline Silicon
N. M. Johnson, Electronic Defects in CW Transient Thermal Processed Silicon
K. F. Lee, T. J. Stultz, and James F. Gibbons, Beam Recrystallized Polycrystalline Silicon: Properties, Applications, and Techniques
T. Shibata, A. Wakita, T. W. Sigmon, and James F. Gibbons, Metal-Silicon Reactions and Silicide
Yves I. Nissim and James F. Gibbons, CW Beam Processing of Gallium Arsenide

Volume 18 Mercury Cadmium Telluride

Paul W. Kruse, The Emergence of $(Hg_{1-x}Cd_x)Te$ as a Modern Infrared Sensitive Material
H. E. Hirsch, S. C. Liang, and A. G. White, Preparation of High-Purity Cadmium, Mercury, and Tellurium
W. F. H. Micklethwaite, The Crystal Growth of Cadmium Mercury Telluride
Paul E. Petersen, Auger Recombination in Mercury Cadmium Telluride
R. M. Broudy and V. J. Mazurczyck, (HgCd)Te Photoconductive Detectors
M. B. Reine, A. K. Soad, and T. J. Tredwell, Photovoltaic Infrared Detectors
M. A. Kinch, Metal-Insulator-Semiconductor Infrared Detectors

Volume 19 Deep Levels, GaAs, Alloys, Photochemistry

G. F. Neumark and K. Kosai, Deep Levels in Wide Band-Gap III–V Semiconductors
David C. Look, The Electrical and Photoelectronic Properties of Semi-Insulating GaAs
R. F. Brebrick, Ching–Hua Su, and Pok-Kai Liao, Associated Solution Model for Ga–In–Sb and Hg–Cd–Te
Yu. Ya. Gurevich and Yu. V. Pleskon, Photoelectrochemistry of Semiconductors

Volume 20 Semi-Insulating GaAs

R. N. Thomas, H. M. Hobgood, G. W. Eldridge, D. L. Barrett, T. T. Braggins, L. B. Ta, and S. K. Wang. High-Purity LEC Growth and Direct Implantation of GaAs for Monolithic Microwave Circuits
C. A. Stolte, Ion Implantation and Materials for GaAs Integrated Circuits
C. G. Kirkpatrick, R. T. Chen, D. E. Holmes, P. M. Asbeck, K. R. Elliott, R. D. Fairman, and J. R. Oliver, LEC GaAs for Integrated Circuit Applications
J. S. Blakemore and S. Rahimi, Models for Mid-Gap Centers in Gallium Arsenide

Volume 21 Hydrogenated Amorphous Silicon Part A

Jacques I. Pankove Introduction
Masataka Hirose, Glow Discharge; Chemical Vapor Deposition

Yoshiyuki Uchida, dc Glow Discharge
T. D. Moustakas, Sputtering
Isao Yamada, Ionized-Cluster Beam Deposition
Bruce A. Scott, Homogeneous Chemical Vapor Deposition
Frank J. Kampas, Chemical Reactions in Plasma Deposition
Paul A. Longeway, Plasma Kinetics
Herbert A. Weakliem, Diagnostics of Silane Glow Discharges Using Probes and Mass Spectroscopy
Lester Guttman, Relation between the Atomic and the Electronic Structures
A. Chenevas-Paule, Experiment Determination of Structure
S. Minomura, Pressure Effects on the Local Atomic Structure
David Adler, Defects and Density of Localized States

Part B

Jacques I. Pankove, Introduction
G. D. Cody, The Optical Absorption Edge of a-Si: H
Nabil M. Amer and Warren B. Jackson, Optical Properties of Defect States in a-Si: H
P. J. Zanzucchi, The Vibrational Spectra of a-Si: H
Yoshihiro Hamakawa, Electroreflectance and Electroabsorption
Jeffrey S. Lannin, Raman Scattering of Amorphous Si, Ge, and Their Alloys
R. A. Street, Luminescence in a-Si: H
Richard S. Crandall, Photoconductivity
J. Tauc, Time-Resolved Spectroscopy of Electronic Relaxation Processes
P. E. Vanier, IR-Induced Quenching and Enhancement of Photoconductivity and Photoluminescence
H. Schade, Irradiation-Induced Metastable Effects
L. Ley, Photoelectron Emission Studies

Part C

Jacques I. Pankove, Introduction
J. David Cohen, Density of States from Junction Measurements in Hydrogenated Amorphous Silicon
P. C. Taylor, Magnetic Resonance Measurements in a-Si: H
K. Morigaki, Optically Detected Magnetic Resonance
J. Dresner, Carrier Mobility in a-Si: H
T. Tiedje, information about band-Tail States from Time-of-Flight Experiments
Arnold R. Moore, Diffusion Length in Undoped a-Si: H
W. Beyer and J. Overhof, Doping Effects in a-Si: H
H. Fritzche, Electronic Properties of Surfaces in a-Si: H
C. R. Wronski, The Staebler-Wronski Effect
R. J. Nemanich, Schottky Barriers on a-Si: H
B. Abeles and T. Tiedje, Amorphous Semiconductor Superlattices

Part D

Jacques I. Pankove, Introduction
D. E. Carlson, Solar Cells

G. A. Swartz, Closed-Form Solution of I–V Characteristic for a-Si: H Solar Cells
Isamu Shimizu, Electrophotography
Sachio Ishioka, Image Pickup Tubes
P. G. LeComber and W. E. Spear, The Development of the a-Si: H Field-Effect Transistor and Its Possible Applications
D. G. Ast, a-Si: H FET-Addressed LCD Panel
S. Kaneko, Solid-State Image Sensor
Masakiyo Matsumura, Charge-Coupled Devices
M. A. Bosch, Optical Recording
A. D'Amico and G. Fortunato, Ambient Sensors
Hiroshi Kukimoto, Amorphous Light-Emitting Devices
Robert J. Phelan, Jr., Fast Detectors and Modulators
Jacques I. Pankove, Hybrid Structures
P. G. LeComber, A. E. Owen, W. E. Spear, J. Hajto, and W. K. Choi, Electronic Switching in Amorphous Siliocn Junction Devices

Volume 22   Lightwave Communications Technology
Part A

Kazuo Nakajima, The Liquid-Phase Epitaxial Growth of IngaAsp
W. T. Tsang, Molecular Beam Epitaxy for III–V Compound Semiconductors
G. B. Stringfellow, Organometallic Vapor-Phase Epitaxial Growth of III–V Semiconductors
G. Beuchet, Halide and Chloride Transport Vapor-Phase Deposition of InGaAsP and GaAs
Manijeh Razeghi, Low-Pressure Metallo-Organic Chemical Vapor Deposition of $Ga_xIn_{1-x}AsP_{1-y}$ Alloys
P. M. Petroff, Defects in III–V Compound Semiconductors

Part B

J. P. van der Ziel, Mode Locking of Semiconductor Lasers
Kam Y. Lau and Ammon Yariv, High-Frequency Current Modulation of Semiconductor Injection Lasers
Charles H. Henry, Spectral Properties of Semiconductor Lasers
Yasuharu Suematsu, Katsumi Kishino, Shigehisa Arai, and Fumio Koyama, Dynamic Single-Mode Semiconductor Lasers with a Distributed Reflector
W. T. Tsang, The Cleaved-Coupled-Cavity (C^3) Laser

Part C

R. J. Nelson and N. K. Dutta, Review of InGaAsP InP laser Structures and Comparison of Their Performance
N. Chinone and M. Nakamura, Mode-Stabilized Semiconductor Lasers for 0.7–0.8- and 1.1–1.6-μm Regions
Yoshiji Horikoshi, Semiconductor Lasers with Wavelengths Exceeding 2 μm
B. A. Dean and M. Dixon, The Functional Reliability of Semiconductor Lasers as Optical Transmitters

R. H. Saul, T. P. Lee, and C. A. Burus, Light-Emitting Device Design
C. L. Zipfel, Light-Emitting Diode-Reliability
Tien Pei Lee and Tingye Li, LED-Based Multimode Lightwave Systems
Kinichiro Ogawa, Semiconductor Noise-Mode Partition Noise

Part D

Federico Capasso, The Physics of Avalanche Photodiodes
T. P. Pearsall and M. A. Pollack, Compound Semiconductor Photodiodes
Takao Kaneda, Silicon and Germanium Avalanche Photodiodes
S. R. Forrest, Sensitivity of Avalanche Photodetector Receivers for High-Bit-Rate Long-Wavelength Optical Communication Systems
J. C. Campbell, Phototransistors for Lightwave Communications

Part E

Shyh Wang, Principles and Characteristics of Integratable Active and Passive Optical Devices
Shlomo Margalit and Amnon Yariv, Integrated Electronic and Photonic Devices
Takaoki Mukai, Yoshihisa Yamamoto, and Tatsuya Kimura, Optical Amplification by Semi-conductor Lasers

Volume 23 Pulsed Laser Processing of Semiconductors

R. F. Wood, C. W. White, and R. T. Young, Laser Processing of Semiconductors: An Overview
C. W. White, Segregation, Solute Trapping, and Supersaturated Alloys
G. E. Jellison, Jr., Optical and Electrical Properties of Pulsed Laser-Annealed Silicon
R. F. Wood and G. E. Jellison, Jr., Melting Model of Pulsed Laser Processing
R. F. Wood and F. W. Young, Jr., Nonequilibrium Solidification Following Pulsed Laser Melting
D. H. Lowndes and G. E. Jellison, Jr., Time-Resolved Measurements During Pulsed Laser Irradiation of Silicon
D. M. Zebner, Surface Studies of Pulsed Laser Irradiated Semiconductors
D. H. Lowndes, Pulsed Beam Processing of Gallium Arsenide
R. B. James, Pulsed CO_2 Laser Annealing of Semiconductors
R. T. Young and R. F. Wood, Applications of Pulsed Laser Processing

Volume 24 Applications of Multiquantum Wells, Selective Doping, and Superlattices

C. Weisbuch, Fundamental Properties of III–V Semiconductor Two-Dimensional Quantized Structures: The Basis for Optical and Electronic Device Applications
H. Morkoc and H. Unlu, Factors Affecting the Performance of (Al, Ga)As/GaAs and (Al, Ga)As/InGaAs Modulation-Doped Field-Effect Transistors: Microwave and Digital Applications

N. T. Linh, Two-Dimensional Electron Gas FETs: Microwave Applications
M. Abe et al, Ultra-High-Speed HEMT Integrated Circuits
D. S. Chemla, D. A. B. Miller, and P. W. Smith, Nonlinear Optical Properties of Multiple Quantum Well Structures for Optical Signal Processing
F. Capasso, Graded-Gap and Superlattice Devices by Band-Gap Engineering
W. T. Tsang, Quantum Confinement Heterostructure Semiconductor Lasers
G. C. Osbourn et al., Principles and Applications of Semiconductor Strained-Layer Superlattices

Volume 25 Diluted Magnetic Semiconductors

W. Giriat and J. K. Furdyna, Crystal Structure, Composition, and Materials Preparation of Diluted Magnetic Semiconductors
W. M. Becker, Band Structure and Optical Properties of Wide-Gap $A^{II}_{1-x}Mn_xB^{VI}$ Alloys at Zero Magnetic Field
Saul Oseroff and Pieter H. Keesom, Magnetic Properties: Macroscopic Studies
T. Giebultowicz and T. M. Holden, Neutron Scattering Studies of the Magnetic Structure and Dynamics of Diluted Magnetic Semiconductors
J. Kossut, Band Structure and Quantum Transport Phenomena in Narrow-Gap Diluted Magnetic Semiconductors
C. Riquaux, Magnetooptical Properties of Large-Gap Diluted Magnetic Semiconductors
J. A. Gaj, Magnetooptical Properties of Large-Gap Diluted Magnetic Semiconductors
J. Mycielski, Shallow Acceptors in Diluted Magnetic Semiconductors: Splitting, Boil-off, Giant Negative Magnetoressitance
A. K. Ramdas and R. Rodriquez, Raman Scattering in Diluted Magnetic Semiconductors
P. A. Wolff. Theory of Bound Magnetic Polarons in Semimagnetic Semiconductors

Volume 26 III–V Compound Semiconductors and Semiconductor Properties of Superionic Materials

Zou Yuanxi, III–V Compounds
H. V. Winston, A. T. Hunter, H. Kimura, and R. E. Lee, InAs-Alloyed GaAs Substrates for Direct Implantation
P. K. Bhattachary and S. Dhar, Deep Levels in III–V Compound Semiconductors Grown by MBE
Yu. Yu. Gurevich and A. K. Ivanov-Shits, Semiconductor Properties of Superionic Materials

Volume 27 High Conducting Quasi-One-Dimensional Organic Crystals

E. M. Conwell, Introduction to Highly Conducting Quasi-One-Dimensional Organic Crystals
I. A. Howard, A Reference Guide to the Conducting Quasi-One-Dimensional Organic Molecular Crystals
J. P. Pouquet, Structural Instabilities
E. M. Conwell, Transport Properties

C. S. Jacobsen, Optical Properties
J. C. Scott, Magnetic Properties
L. Zuppiroli, Irradiation Effects: Perfect Crystals and Real Crystals

Volume 28 Measurement of High-Speed Signals in Solid State Devices

J. Frey and D. Ioannou, Materials and Devices for High-Speed and Optoelectronic Applications
H. Schumacher and E. Strid, Electronic Wafer Probing Techniques
D. H. Auston, Picosecond Photoconductivity: High-Speed Measurements of Devices and Materials
J. A. Valdmanis, Electro-Optic Measurement Techniques for Picosecond Materials, Devices, and Integrated Circuits
J. M. Wiesenfeld and R. K. Jain, Direct Optical Probing of Integrated Circuits and High-Speed Devices
G. Plows, Electron-Beam Probing
A. M. Weiner and R. B. Marcus, Photoemissive Probing

Volume 29 Very High Speed Integrated Circuits: Gallium Arsenide LSI

M. Kuzuhara and T. Nazaki, Active Layer Formation by Ion Implantation
H. Hasimoto, Focused Ion Beam Implantation Technology
T. Nozaki and A. Higashisaka, Device Fabrication Process Technology
M. Ino and T. Takada, GaAs LSI Circuit Design
M. Hirayama, M. Ohmori, and K. Yamasaki, GaAs LSI Fabrication and Performance

Volume 30 Very High Speed Integrated Circuits: Heterostructure

H. Watanabe, T. Mizutani, and A. Usui, Fundamentals of Epitaxial Growth and Atomic Layer Epitaxy
S. Hiyamizu, Characteristics of Two-Dimensional Electron Gas in III–V Compound Heterostructures Grown by MBE
T. Nakanisi, Metalorganic Vapor Phase Epitaxy for High-Quality Active Layers
T. Nimura, High Electron Mobility Transistor and LSI Applications
T. Sugeta and T. Ishibashi, Hetero-Bipolar Transistor and Its LSI Application
H. Matsueda, T. Tanaka, and M. Nakamura, Optoelectronic Integrated Circuits

Volume 31 Indium Phosphide: Crystal Growth and Characterization

J. P. Farges, Growth of Discoloration-free InP
M. J. McCollum and G. E. Stillman, High Purity InP Grown by Hydride Vapor Phase Epitaxy
T. Inada and T. Fukuda, Direct Synthesis and Growth of Indium Phosphide by the Liquid Phosphorous Encapsulated Czochralski Method
O. Oda, K. Katagiri, K. Shinohara, S. Katsura, Y. Takahashi, K. Kainosho, K. Kohiro, and R. Hirano, InP Crystal Growth, Substrate Preparation and Evaluation

K. Tada, M. Tatsumi, M. Morioka, T. Araki, and T. Kawase, InP Substrates: Production and Quality Control
M. Razeghi, LP-MOCVD Growth, Characterization, and Application of InP Material
T. A. Kennedy and P. J. Lin-Chung, Stoichiometric Defects in InP

Volume 32 Strained-Layer Superlattices: Physics

T. P. Pearsall, Strained-Layer Superlattices
Fred H. Pollack, Effects of Homogeneous Strain on the Electronic and Vibrational Levels in Semiconductors
J. Y. Marzin, J. M. Gerárd, P. Voisin, and J. A. Brum, Optical Studies of Strained III–V Heterolayers
R. People and S. A. Jackson, Structurally Induced States from Strain and Confinement
M. Jaros, Microscopic Phenomena in Ordered Superlattices

Volume 33 Strained-Layer Superlattices: Materials Science and Technology

R. Hull and J. C. Bean, Principles and Concepts of Strained-Layer Epitaxy
William J. Schaff, Paul J. Tasker, Mark C. Foisy, and Lester F. Eastman, Device Applications of Strained-Layer Epitaxy
S. T. Picraux, B. L. Doyle, and J. Y. Tsao, Structure and Characterization of Strained-Layer Superlattices
E. Kasper and F. Schaffler, Group IV Compounds
Dale L. Martin, Molecular Beam Epitaxy of IV–VI Compound Heterojunction
Robert L. Gunshor, Leslie A. Kolodziejski, Arto V. Nurmikko, and Nobuo Otsuka, Molecular Beam Epitaxy of II–VI Semiconductor Microstructures

Volume 34 Hydrogen in Semiconductors

J. I. Pankove and N. M. Johnson, Introduction to Hydrogen in Semiconductors
C. H. Seager, Hydrogenation Methods
J. I. Pankove, Hydrogenation of Defects in Crystalline Silicon
J. W. Corbett, P. Deák, U. V. Desnica, and S. J. Pearton, Hydrogen Passivation of Damage Centers in Semiconductors
S. J. Pearton, Neutralization of Deep Levels in Silicon
J. I. Pankove, Neutralization of Shallow Acceptors in Silicon
N. M. Johnson, Neutralization of Donor Dopants and Formation of Hydrogen-Induced Defects in n-Type Silicon
M. Stavola and S. J. Pearton, Vibrational Spectroscopy of Hydrogen-Related Defects in Silicon
A. D. Marwick, Hydrogen in Semiconductors: Ion Beam Techniques
C. Herring and N. M. Johnson, Hydrogen Migration and Solubility in Silicon
E. E. Haller, Hydrogen-Related Phenomena in Crystalline Germanium
J. Kakalios, Hydrogen Diffusion in Amorphous Silicon
J. Chevalier, B. Clerjaud, and B. Pajot, Neutralization of Defects and Dopants in III–V Semiconductors

G. G. DeLeo and W. B. Fowler, Computational Studies of Hydrogen-Containing Complexes in Semiconductors
R. F. Kiefl and T. L. Estle, Muonium in Semiconductors
C. G. Van de Walle, Theory of Isolated Interstitial Hydrogen and Muonium in Crystalline Semiconductors

Volume 35 Nanostructured Systems

Mark Reed, Introduction
H. van Houten, C. W. J. Beenakker, and B. J. van Wees, Quantum Point Contacts
G. Timp, When Does a Wire Become an Electron Waveguide?
M. Büttiker, The Quantum Hall Effect in Open Conductors
W. Hansen, J. P. Kotthaus, and U. Merkt, Electrons in Laterally Periodic Nanostructures

Volume 36 The Spectroscopy of Semiconductors

D. Heiman, Laser Spectroscopy of Semiconductors at Low Temperatures and High Magnetic Fields
Arto V. Nurmikko, Transient Spectroscopy by Ultrashort Laser Pulse Techniques
A. K. Ramdas and S. Rodriguez, Piezospectroscopy of Semiconductors
Orest J. Glembocki and Benjamin V. Shanabrook, Photoreflectance Spectroscopy of Microstructures
David G. Seiler, Christopher L. Littler, and Margaret H. Weiler, One- and Two-Photon Magneto-Optical Spectroscopy of InSb and $Hg_{1-x}CD_xTe$

Volume 37 The Mechanical Properties of Semiconductors

A.-B. Chen, Arden Sher and W. T. Yost, Elastic Constants and Related Properties of Semiconductor Compounds and Their Alloys
David R. Clarke, Fracture of Silicon and Other Semiconductors
Hans Siethoff, The Plasticity of Elemental and Compound Semiconductors
Sivaraman Guruswamy, Katherine T. Faber and John P. Hirth, Mechanical Behavior of Compound Semiconductors
Subhanh Mahajan, Deformation Behavior of Compound Semiconductors
John P. Hirth, Injection of Dislocations into Strained Multilayer Structures
Don Kendall, Charles B. Fleddermann, and Kevin J. Malloy, Critical Technologies for the Micromachining of Silicon
Ikuo Matsuba and Kinji Mokuya, Processing and Semiconductor Thermoelastic Behavior

Volume 38 Imperfections in III/V Materials

Udo Scherz and Matthias Scheffler, Density-Functional Theory of sp-Bonded Defects in III/V Semiconductors
Maria Kaminska and Eicke R. Weber, EL2 Defect in GaAs
David C. Look, Defects Relevant for Compensation in Semi-Insulating GaAs

R. C. Newman, Local Vibrational Mode Spectroscopy of Defects in III/V Compounds
Andrzej M. Hennel, Transition Metals in III/V Compounds
Kevin J. Malloy and Ken Khachaturyan, DX and Related Defects in Semiconductors
V. Swaminathan and Andrew S. Jordan, Dislocations in III/V Compounds
Krzysztof W. Nauka, Deep Level Defects in the Epitaxial III/V Materials

Volume 39 Minority Carriers in III–V Semiconductors: Physics and Applications

Niloy K. Dutta, Radiative Transitions in GaAs and Other III–V Compounds
Richard K. Ahrenkiel, Minority-Carrier Lifetime in III–V Semiconductors
Tomofumi Furuta, High Field Minority Electron Transport in p-GaAs
Mark S. Lundstrom, Minority-Carrier Transport in III–V Semiconductors
Richard A. Abram, Effects of Heavy Doping and High Excitation on the Band Structure of GaAs
David Yevick and Witold Bardyszewski, An Introduction to Non-Equilibrium Many-Body Analyses of Optical Processes in III–V Semiconductors

ISBN 0-12-752139-9